CHEMICAL MODIFICATION, PROPERTIES, AND USAGE OF LIGNIN

CHEMICAL MODIFICATION, PROPERTIES, AND USAGE OF LIGNIN

Edited by

Thomas Q. Hu

Pulp and Paper Research Institute of Canada
Vancouver, British Columbia, Canada

Kluwer Academic / Plenum Publishers
New York, Boston, Dordrecht, London, Moscow

Library of Congress Cataloging-in-Publication Data

Chemical modification, properties, and usage of lignin/edited by Thomas Q. Hu.
p. cm.
Includes bibliographical references (p.).
ISBN 0-306-46769-0
1. Lignin. I. Hu, Thomas Q.

QK898.L5 .C54 2002
572′.56682—dc21

2001058348

ISBN 0-306-46769-0

233 Spring Street, New York, N.Y. 10013

http://www.wkap.com

10 9 8 7 6 5 4 3 2 1

A C.I.P. record for this book is available from the Library of Congress

Printed in the United States of America

To Xuan and Nicholas

PREFACE

One of the most significant challenges facing mankind in the twenty-first century is the development of a sustainable global economy. Within the scientific community, this calls for the development of processes and technologies that will allow the sustainable production of materials from renewable natural resources. Plant material, in particular lignin, is one such resource. During the annual production of about 100 million metric tons of chemical wood pulps worldwide, approximately 45 and 2 million metric tons/year of kraft lignin and lignosulfonates, respectively, are also generated. Although lignosulfonates have found many applications outside the pulp and paper industry, the majority of kraft lignin is being used internally as a low-grade fuel for the kraft pulping operation. A surplus of kraft lignin will become available as kraft mills increase their pulp production without expanding the capacity of their recovery boilers that utilize lignin as a fuel. There is a tremendous opportunity and an enormous economic incentive to find better uses of kraft lignin, lignosulfonates and other industrial lignins.

The pulp and paper industry not only produces an enormous amount of lignins as by-products of chemical wood pulps, but it also utilizes about 10 million metric tons of lignin per year as a component of mechanical wood pulps and papers. Mechanical wood pulps, produced in a yield of 90-98% with the retention of lignin, are mainly used to make low-quality, non-permanent papers such as newsprint and telephone directories because of the light-induced photooxidation of lignin and the yellowing of the papers. Effective photostabilization of lignin will enable the wider use of mechanical wood pulps in high-quality paper products.

In Asia and other areas of the world where forest resources are less abundant, non-wood plant materials such as straw, bagasse and bamboo have been used to make paper products for many centuries. The effective isolation and product development of lignin from chemical pulping of non-wood plant materials have become increasingly more important because internal use of the non-wood pulping spent liquors is difficult and the traditional way of discharging them to the environment is no longer acceptable.

This book covers recent developments in chemical modification and utilization of lignin. Chapters 1 to 11 discuss the opportunities and new findings on the use of lignin as a component of polymeric materials such as starch films, conducting polymers, polyurethanes and thermoplastics, and as a key building block to make carbon fibers, soil conditioners, nitrogenous fertilizers, and pulping catalysts. Chapters 12 and 13 describe approaches to inhibit the yellowing of mechanical wood pulps and papers by chemical modification and photostabilization of lignin functional groups, while the last chapter provides a thorough literature review of efforts undertaken to characterize and use lignin from non-wood resources.

It is hoped that this book will provide references on the properties, state-of-art chemical modifications and uses of lignin to both academic and industrial researchers who work in the fields of polymers, wood and paper, material science and chemicals. It is also hoped that this volume will stimulate further efforts in the development of new processes and technologies to make value-added products from lignin, an abundant, readily available and renewable natural resource.

Acknowledgements

I would like to thank my former supervisor, the late Professor Larry Weiler of the University of British Columbia for his encouragement during the planning and organization of a symposium on Chemical Modification, Properties and Usage of Lignin at the 2000 International Chemical Congress of Pacific Basin Societies. I would also like to thank all the contributors of this book for their time, effort and enthusiasm in writing their chapters; without their active participation and support, this book would not have been possible. Thanks are also due to Professor Brian R. James of the University of British Columbia and Professor Simo Sarkanen of the University of Minnesota for their helpful discussions, and Thanh Trung of Paprican for his assistance in formatting the chapters electronically. Finally and most importantly, I would like to thank my wife, Xuan, for her patience and support during the preparation of this book and our little son, Nicholas, for his entertainment and inspiration.

Thomas Q. Hu
Pulp and Paper Research Institute of Canada
Vancouver, British Columbia, Canada V6S 2L9
October 25, 2001

CONTENTS

STARCH-LIGNIN FILMS

Stéphanie Baumberger*

1. INTRODUCTION

The increasing need for environmental protection together with the concerns over the future availability of petrochemical feedstock have led to the design and development of new degradable thermoplastic materials, based on renewable resources that are more friendly to the environment than the conventially used petroleum-based plastics (Chapman, 1994; Fritz *et al.*, 1995; Chiellini and Solaro, 1996; Krochta and De Mulder-Johnston, 1996).

Among the biopolymers available from agricultural resources, the potential of starch to give thermoplastics has been demonstrated (Protzman *et al.*, 1964; Lourdin *et al.*, 1995; Van Soest, 1996). Indeed, the transformation of this raw material either by casting (polymer solubilization followed by evaporation of the solvent) or by thermomechanical treatments in the presence of a plasticizer can lead to the production of films and molded items. The industrial development of such starchy materials as food packaging is hindered by their swelling and partial dissolution in moist environments. Chemical modification of starch has proven to be an effective way to reduce its water affinity but costly and detrimental in biodegradability (Funke and Lindhauer, 1994; Rivard *et al.*, 1995). Another approach consists of blending starch with hydrophobic materials such as synthetic polymers, lipids or natural polymers. In this context, an early generation of starch-synthetic polymer composites, such as starch/polyethylene, starch/poly (ethylene-co-acrylic acid) (Otey *et al.*, 1987; Doane, 1992; Kim and Pometto, 1994) or starch/poly (ethylene-co-vinyl acetate) (Jauregui *et al.*, 1995), has been intensively studied since the 1970's. More recently, efforts have been devoted to replace the petroleum-based polymer of the system by a biopolymer, with the advantage of producing totally biodegradable materials. In this context, biodegradable polyesters such as poly (hydroxyalkanoate), poly (lactic acid) or polycaprolactam (Hänggi, 1995; Chiellini and Solaro, 1996) are very attractive because they are not only biodegradable and biocompatible but also have good mechanical properties. Recent papers reported on the properties of polyhydroxybutyrate-co-valerate-starch (Averous *et al.*, 1998) and poly (lactic acid)-starch composites (Ke and

* Stéphanie Baumberger, UMR de Chimi Biologique INRA / INA PG, Institut National Agronomique Paris-Grignon, 78850 Thiverval-Grignon, France.

Chemical Modification, Properties, and Usage of Lignin
Edited by Thomas Q. Hu, Kluwer Academic/Plenum Publishers, 2002

Sun, 2000; Fang and Hanna, 2000). However, these polymers derived from fermentations are not economically competitive nowadays. Garcia *et al.* (2000) reported on the possibility to increase the barrier properties to water vapor of starch films by adding lipids (sunflower oil) to the system, as previously observed with various matrix (Callegarin *et al.*, 1997). In an attempt to combine the properties of two low cost polymers while preserving the biodegradability of the final materials, starch has also been blended with natural polymers derived from agricultural resources: cereal proteins (Lim and Jane, 1993), pectins (Fishman *et al.*, 1996) and lignins. Among these plant polymers, lignins have the advantages of being capable of playing an important role as a hydrophobic agent and available in large amount at low cost as co-products from the pulp and paper industry. Their hydrophobic properties have already been exploited to confer water resistance to chitosan films grafted with lignin model compounds or kraft lignins (Muzzarelli and Ilari, 1993).

Two main types of polymeric systems including both starch and lignins are described in literature: the first one concerns polyurethanes obtained after liquefaction of wood and starch in the presence of polyhydric alcohols, with wood/starch ratios ranging from 0/100 to 60/40 (Yao *et al.*, 1993; 1994; 1995). These highly cross-linked systems can generate rigid thermosetting foams with mechanical properties similar to those of the conventional polyurethane foams used for isolation. The second type of systems consists of thermoplastics based on a starch matrix filled with isolated lignins. In this chapter we review our studies aiming at incorporating lignins into a starch matrix to produce natural materials both biodegradable and water resistant.

Our chapter addresses the following questions: are starch/lignin systems blends or composites? Does lignin structure affect the properties of the films ? Finally, where does the interest of lignin incorporation lie? In the first part of our chapter, the methods used for laboratory and pilot film preparation are presented, together with the morphology of the resulting materials. This part is illustrated by microscopic observations in the fluorescence mode. The second and third parts of our chapter report on the consequence of lignin incorporation on the thermomechanical properties and the hydrophilic character of the starch matrix, respectively. Highlights on the various interaction mechanisms resulting from the molecular heterogeneity of lignins are given, based on the fractionation experiments. In the final concluding part of our chapter, the potential of various industrial lignins to improve the performances of starchy materials is discussed, in relation with different target applications.

2. SOURCE OF COMPONENTS AND FORMULATION OF THE MATERIALS

2. 1. Lignin Preparations

The different lignins used so far for the elaboration of starch/lignin materials are listed in Table 1. They are either industrial lignins provided by the pulp and paper industry (kraft lignins, lignosulfonates, Alcell lignins and alkali Granit lignins) or lignins isolated and purified at the laboratory or pilot scale (dioxane and ethanol reflux lignins, kraft lignin fractions) (Table 1). As a consequence of the diversity of the botanical origins and isolation process, all these lignins are very likely to differ in molecular weight distribution, interunit linkages, functional groups and amount of associated polysaccharides. The structure of the three lignin fractions F1, F2 and F3 recovered from the industrial kraft lignins has been intensively investigated in order to elucidate the impact of lignin structural variations on the film properties (Baumberger *et al.*, 1998a). The three lignin fractions represent the low molecular fragments dissolved at the initial delignification stage (F1), the condensed carbohydrate-enriched lignins dissolved in the final delignification stage (F3) and the intermediate lignins likely corresponding to the bulk delignification step (F2).

The alkali Granit lignins present major environnemental advantages: they are recovered from sulfur-free black liquors obtained from agricultural resources (wheat and hemp). Hemp lignin fractions H1 and H2 are obtained using two different acidic precipitation pH's (pH 4 and pH 6.6, respectively). Lignosulfonates form a singular lignin family since the presence of sulfonate groups make them water-soluble and surface active, in particular at neutral pH (Fengel and Wegener, 1989).

Table 1. Summary of the different starches/lignin systems investigated

Starch origin	Lignins designation and origin	Lignin content[a]	Glycerol Content[b]	References
unknown	Na Lignosulfonates (Borregard, Sweden)	0-100%	0	Volperts *et al.*, 2001
Wheat[c]	Pine sodium Lignosulfonates (N18) Pine calcium Lignosulfonates (C12) (Avebene, Tartas, France)	0-30%	33%-extrusion 23%-casting	Baumberger *et al.*, 1997
Wheat[c]	Pine Kraft Lignin (KRL) (Westvaco, Charleston SC, USA)	0-30%	0	Baumberger *et al.*, 1998b; Rutledge *et al.*, 1999
Wheat[c]	F1 F2 kraft lignin fractions[f] F3	0-20%	0	Baumberger *et al.*, 1998a
High-amylose maize[d]	Pine Kraft Lignin (KRL) (Westvaco, Charleston SC, USA)	20%		Baumberger, 1999
Potato purified amylose	Pine Kraft Lignin (KRL) (Westvaco, Charleston SC, USA) F2 F3	20%	0	Baumberger, 1999
Wheat[c]	Wheat dioxan lignin (LD)[e] Wheat Alcell® lignin (Alcell)	0-20%	24%	Baumberger *et al.*, 1999
Wheat[c]	Wheat and hemp alkali lignins (H1 and H2) (Granit, Suisse) Wheat reflux ethanol lignins (ERL)[g]	20%	22%	Baumberger *et al.*, 2000

[a] With respect to the total polymer weight.
[b] With respect to dry starch.
[c] Prepared on a laboratory scale using aqueous dioxan according to a published procedure (Monties, 1988).
[d] Type A, 87.5% dry wt., 27% amylose, F33, Amylum.
[e] 88.2% dry wt., 69% amylose, EURYLON 7, Roquette.
[f] Recovered from sequential solvent fractionation according to Mörck *et al.* (1988).
[g] 1h cooking under reflux in a 50/50 (v/v) water/ethanol mixture.

2. 2. Starches

Commercial starches (wheat and high amylose) together with an amylose preparation purified from potato were used in our studies. They differ from one another mainly in their amylose/amylopectin ratio but also in the minor non-polysaccharidic fraction. Indeed, potato starch is nearly deprived of lipid fraction, which is not the case for wheat and maize starch (Zobel and Stephen, 1995).

2. 3. Film Composition

Glycerol was used as a plasticizer, in a glycerol/starch ratio ranging from 78/22 to 76/24. Such a proportion of glycerol is usually employed for starch plasticization (Lourdin *et al.*, 1997). The physico-chemical properties of starch films, particularly mechanical and thermal properties, are known to be highly dependent on the amount of plasticizer and water present in the film (Lourdin *et al.*, 1997a; 1997b). Consequently, the investigation of the impact of lignins on starch films is based on the comparison of films prepared in the presence of the same proportion of glycerol with respect to starch and stored at the same relative humidity. After storage at room temperature over a saturated sodium bromide solution (equilibrium relative humidity RH of 57.7% at 25°C) the materials exhibit an average water content of 10.6 to 13.8% which rises up to 15-16% when the relative humidity is set at 70.8% RH.

3. PREPARATION AND PHASE MORPHOLOGY OF THE FILMS

3.1. Film Preparation

Two types of methods are conventionally used for the production of thermoplastic films based on natural polymers. In the case of starch, the aim is to destructurize starch granules, either by a thermomechanical treatment (extrusion, molding...) or by a thermal treatment in an excess of solvent (gelatinization). The dispersed polymers, amylose and amylopectin, can form an amorphous three-dimensional network, by rapid cooling in the case of a thermomechanical treatment or by solvent evaporation in the case of evaporative-casting. Starch/lignin films can be obtained using both of these methods.

Thermal molding (140 °C, 250 bar, 10 min) is performed after powder mixing and extrusion at 120 °C in the presence of 20/25% water with respect to dry starch.

For mixtures containing water-insoluble lignins, dimethylsulfoxide is employed as casting solvent using a 5 wt% of polymer concentration. The use of such a good solvent of lignins and starch allows the components to mix intimately at a molecular level. However, it requires a long drying step (5 hours at 90 °C and atmospheric pressure followed by a 12-hour drying at 60 °C under vacuum). It could be advantageously replaced by an alkaline aqueous medium in the case of starch/alkali lignins mixtures.

The main differences between these techniques are the thickness of the final films (around 100 μm for cast films and 800 μm for thermally molded films) and the phase morphology of the materials. Macroscopic properties such as water resistance and mechanical behavior can also be affected as will be discussed later. All the films exhibit uniform and homogenous appearance together with a brown to yellow coloration depending on the lignin origin and on the lignin incorporation level (Figure 1).

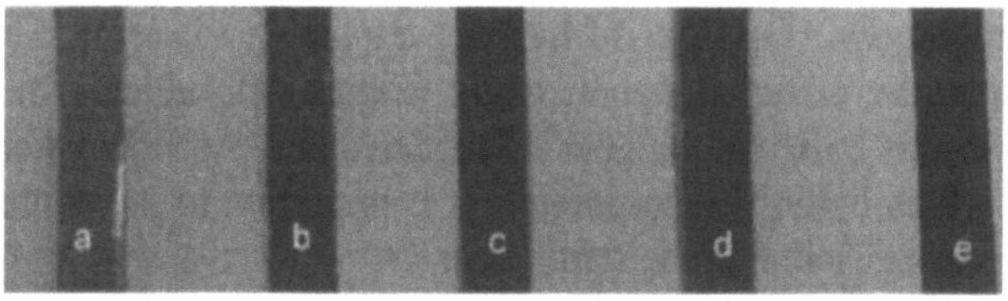

Figure 1. Film samples obtained by solvent casting of 95/5 (a), 90/10 (b), 80/20 (c), 70/30 (d) and 60/40 (e) starch/Na lignosulfonates blends.

3.2. Phase Morphology

Microscopic observations of the systems including different starches (wheat, high amylose starch or purified amylose) and pine kraft lignins (crude or fractionated) reveal that the phase morphology of the system depends on three main parameters: the technique employed for the film formation, the amylose content of the starch and the type of lignins incorporated.

3. 2. 1. Microscopic Techniques

In contrast to most polymeric systems, the phase morphology of starch-lignin films can be investigated directly with light microscopy, without requiring phase contrast or immunochemical techniques. The natural brown coloration of lignin components ensures enough contrast to distinguish lignin rich domains from the starch matrix (Figure 2d). Specific properties of the components can also be used for further in depth investigations: the complexation of the linear polysaccharidic chains of starch by iodine, leading to a blue to purple staining, and the fluorescence of phenolic compounds when excited under UV radiations.

3. 2. 2. Influence of the Mode of Film Preparation

The changes in the morphology of the polymer materials with the procedures used for the film formation are clearly illustrated in the starch/kraft lignin systems (Figure 2). The micrographs of the wheat starch cast films reveal a two-phase morphology, whereas extrusion of the same initial material leads to a single-phase morphology in which it is almost impossible to detect phase separation. High-amylose maize starch-based films also exhibit two phases (a dark lugol stained phase corresponding to amylose and a light phase corresponding to amylopectin). The macrophase separation observed in cast films can be explained by the incompatibility between amylose and amylopectin (Kalichevsky and Ring, 1987; Leloup *et al.*, 1991). Although solution blending provides mixing of the two components at the molecular level in a common solvent, the slow drying step of the casting procedure allows amylose and amylopectin to undergo separation. For kinetic reasons, this separation phenomenon is not observed within the extruded systems: the high viscosity of the melted blends and the short length of the thermomechanical treatment prevent the polymers from rearranging, thus keeping the system far from its thermodynamics equilibrium. Whatever the system, lignin incorporation generates particles with an average diameter of 0.6-1.4 µm, as determined by image analysis of scanned micrographs (IMAQ-vision software, National Instrument). These particles are evenly dispersed in the extruded starch matrix (Figure 2l), which is not the case when casting is used. In this case, the number of particles and their diameter depends on the type of phase in which lignin is incorporated. It is important to notice that some phenolic lignin compounds are miscible in the starch matrix.

Starch/lignosulfonates systems also exhibit important ultrastructural differences depending on whether casting or thermal molding is used for the film preparation. A remarkable characteristic of starch/lignosulfonates cast films is the presence of spindle-shaped structures, of about 35 micrometers in length, evenly distributed in the starch matrix (Figure 3b). These structures are not preserved when the film is immersed in water. Moreover, they are totally absent from the highly homogeneous thermally molded films (Figure 3a). It cannot be ruled out that this structures could be formed during the slow drying step of the casting procedure as a result of a complexation phenomenon involving amylose and phenolics (Whistler, 1965).

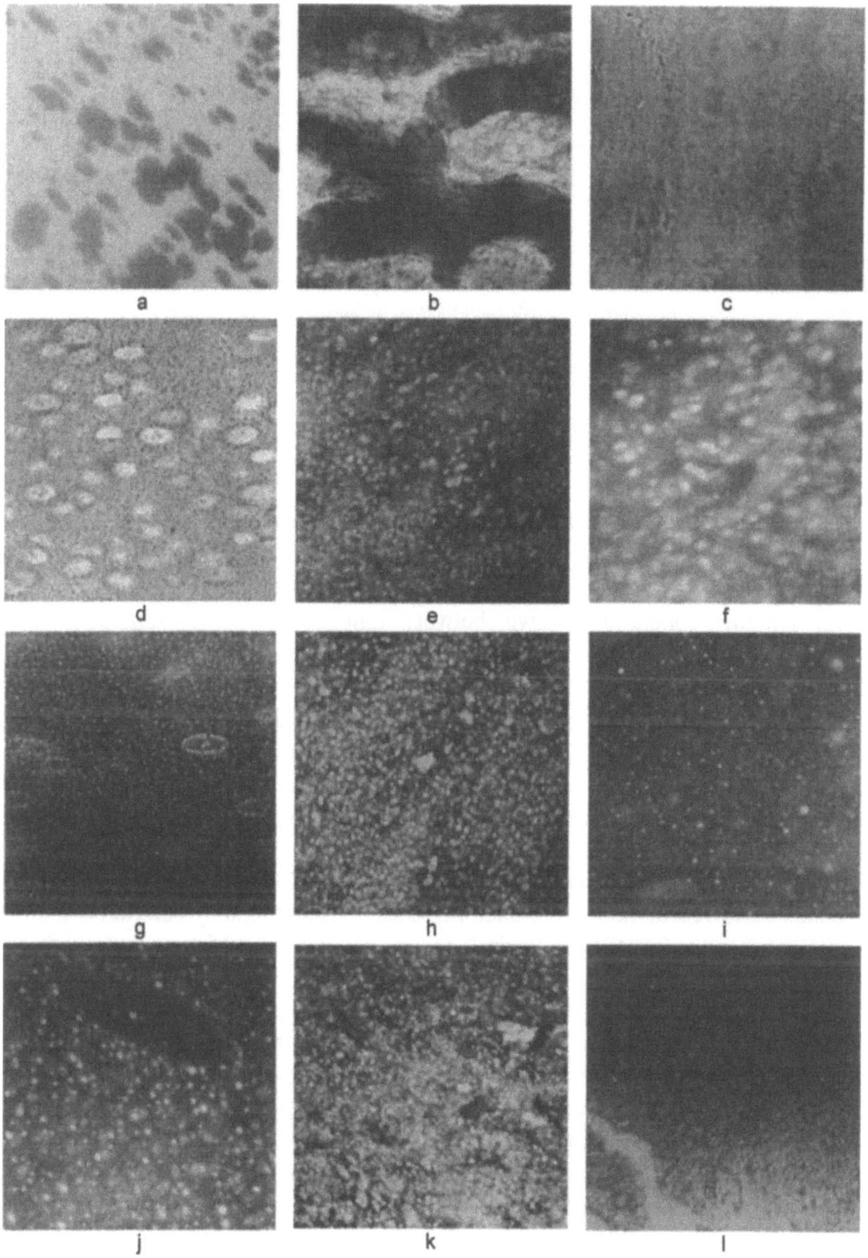

Figure 2. Comparison of morphologies, by light microscopy, of thin cross-sections of pure starch (**a-c**) and starch/lignin cast films (**d-k**) together with a starch/lignin extruded blends (**l**): wheat starch WS (**a**, lugol staining), high amylose starch HAS (**b**, lugol staining), purified amylose AM (**c**, lugol staining), 80/20 WS/KRL (**d**, fluorescence), 80/20 HAS/KRL (**e**, fluorescence), 80/20 AM/KRL (**f**, fluorescence), 80/20 WS/F2 (**g**, fluorescence), 80/20 HAS/F2 (**h**, fluorescence), 90/10 WS/F1 (**i**, fluorescence), 80/20 WS/F3 (**j**, fluorescence), 80/20 HAS/F3 (**k**, fluorescence), 70/30 WS/KRL (**l**, lugol staining). Magnification X 280 (**a-e**, **g**, **h**, **k**), 400x (**l**), 410x (**j**) and 466x (**i**, **f**).

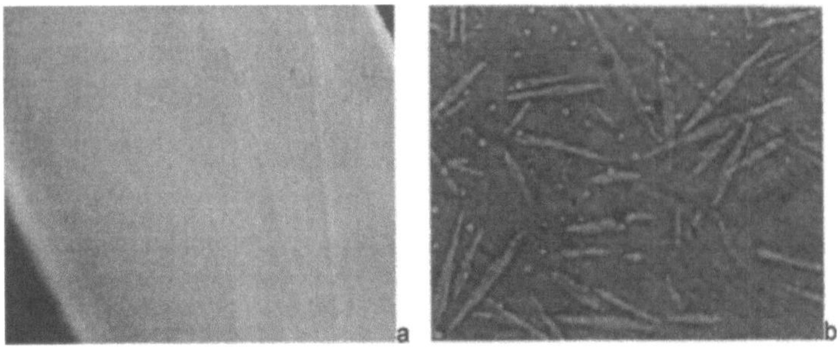

Figure 3. Comparison of the morphologies by light microscopy of thin cross sections of films obtained from 70/30 starch/Na lignosulfonates blends by extrusion-thermal molding (a, magnification X 105) and casting (b, magnification X 420).

3. 2. 3. Impact of the Starch and Lignin Composition

Whatever the composition of the starch constituting the matrix, the morphology of amylose and amylopectin phase is preserved in the presence of lignin. A consequence of an increase in amylose content in starch is the increase of the average diameter of lignin particles (0.9 µm for wheat starch against 1.4 µm for purified amylose). This observation suggests a higher compatibility level between lignins and amylopectin than between lignins and amylose, consistent with the highly branched structures of both lignins and amylopectin.

The size and distribution of lignin particles are also dependant on lignin structural traits, as illustrated in Figure 2 by the micrographs **i**, **g** and **j** that correspond respectively to the 90/10 starch/F1, 80/20 starch/F2 and 80/20 starch/F3 cast films. Two interesting phenomena can be observed: the increase of the average particle diameters from 0.6 µm (starch/F1) to 1.0 µm (starch/F3) and the tendency of fraction F2 to form particles concentrated at the interface between amylose and amylopectin, without any phenolic compounds dissolved in the matrix. Such particles surrounding amylose-rich domains are also present when the initial kraft lignin is incorporated instead of the separated fractions (micrograph **d**, Figure 2).

In summary, the thermal molding of starch/lignosulfonates and starch/kraft lignins mixtures leads in the first case to homogeneous blends and in the second to composites materials where part of the lignins form small domains of a few micrometers. The casting procedure introduces an additional level of heterogeneity, as a result of amylose/amylopectin phase separation. The alteration of the phase morphology as a function of starch and lignin composition suggests the possibility to modulate the compatibility of the systems and to control variation in the macroscopic properties.

4. THERMAL AND TENSILE PROPERTIES AS REFLECTING STARCH AND LIGNIN COMPATIBILITY

4. 1. Mechanical Tests

4. 1. 1. Experimental

The mechanical properties of starch/lignin films obtained either by casting or by extrusion-thermal molding have been investigated by stress-strain tests using an INSTRON 1122 universal testing machine and different stretching speeds (0.2 to 5 mm/min depending on the extensibility of the film). Elongation ε and stress σ at break were determined from the stress-strain curves as follows: the stress at break was calculated on the basis of the original cross-sectional area of the test specimen and the elongation was calculated on the basis of the length of the narrow parallel portion. As frequently observed with starch-based films (Fishman *et al.*, 1996; Lourdin *et al.*, 1997; Chang *et al.*, 2000), a rather high dispersion is generally observed for measurements run on three to ten samples of the same starch/lignin film (average standard deviation of 15% for σ and 25% for ε). This dispersion mainly arises from the difficulty to precisely measure small deformations and from the sample thickness variations responsible for stress non- uniformity. It can be dramatically reduced (average standard deviation of 14% for σ and 5% for ε) when measurements are run on thermally molded, highly extensible films, such as starch/lignosulfonates films.

4. 1. 2. Influence of Lignin Incorporation

The mechanical behavior of starch/lignin films is primarily governed by the presence of glycerol interacting with the starch matrix and by the procedure used for the film formation. It is also affected by the type and amount of lignin incorporated (Figures 4 and 6) and by the origin of the starch matrix (Figure 5). As far as wheat starch-based films are concerned, the incorporation of lignins up to 30% induces an increase in elongation at break of the starch films, with a maximum 2.7 fold increase observed for starch/sodium lignosulfonates (WS/N18) cast films. Such a positive effect on extensibility is diagnostic of partial miscibility between the starch and lignin fractions and can be explained by the presence of hydrophilic groups in lignins.

In the case of kraft lignins (Figure 4), the elongation increase for the films stored at 58% RH is rather low and associated with an increase in stress at break. This suggests that kraft lignins behave as a filler. At higher water contents (71% relative humidity, Figure **4b**), lignin incorporation detrimentally affects the mechanical resistance of the film. This negative effect most probably reflects the incompatibility between the highly hydrophilic moist starch matrix and the hydrophobic lignin fraction. A negative effect of lignin incorporation can also be observed at 58% RH when high-amylose maize starch or purified amylose is used as matrix (Figure 5). This confirms the poor compatibility of amylose-rich/lignins systems suggested by the previous microscopic observations.

Tests run at 58% RH indicate that all the lignin samples obtained from wheat (Alcell, LD, ethanol reflux and alkali) induce a decrease in stress at break together with the elongation increase. These lignins would thus have the potential to plasticize glycerol/starch films. In contrast to wheat lignins, alkali hemp lignins (H1 and H2) present the singularity to induce an important increase of both elongation and stress at break.

Finally, incorporating separately the three kraft lignin fractions F1, F2 and F3 leads to distinct behaviors, which has helped in understanding the properties of the starch/unfractionated kraft lignin films: F1 increases elongation at break while decreasing stress at break, thereby acting as a plasticizer, whereas F2 does not affect the characteristics at break and F3 leads to fragile films with poor mechanical properties (data not shown, Baumberger *et al.*, 1998a).

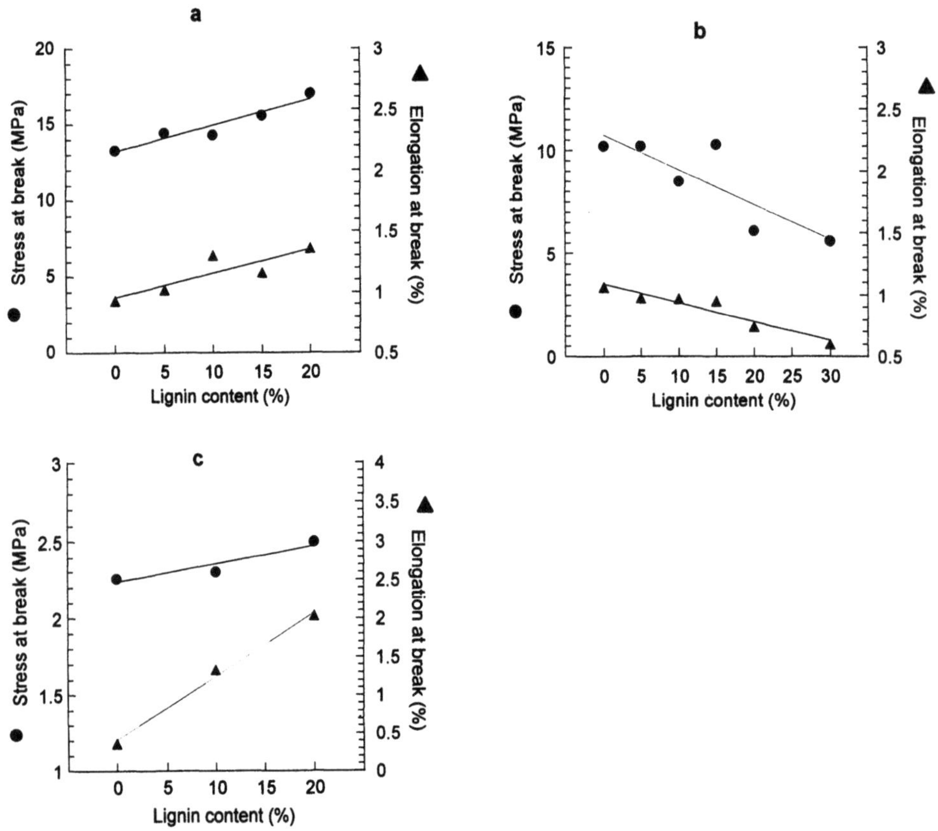

Figure 4. Mechanical properties of thermally molded (a, b) and cast (c) starch/kraft lignin films stored at 58% (a, c) and 71% (b) relative humidity. Film water content: 10.6-12.6% at 58% RH and 15-16% at 71% RH.

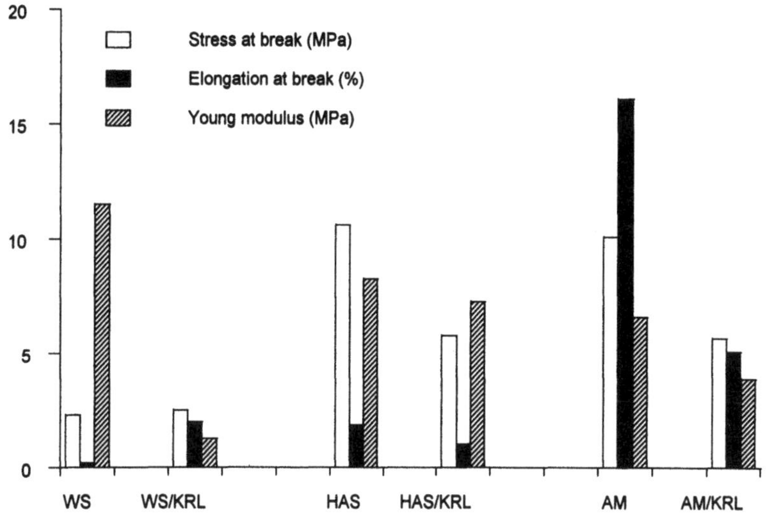

Figure 5. Mechanical properties of wheat starch- (WS), high amylose starch- (HAS) and purified amylose- (AM) based cast films containing 0 or 20% kraft lignin (KRL).

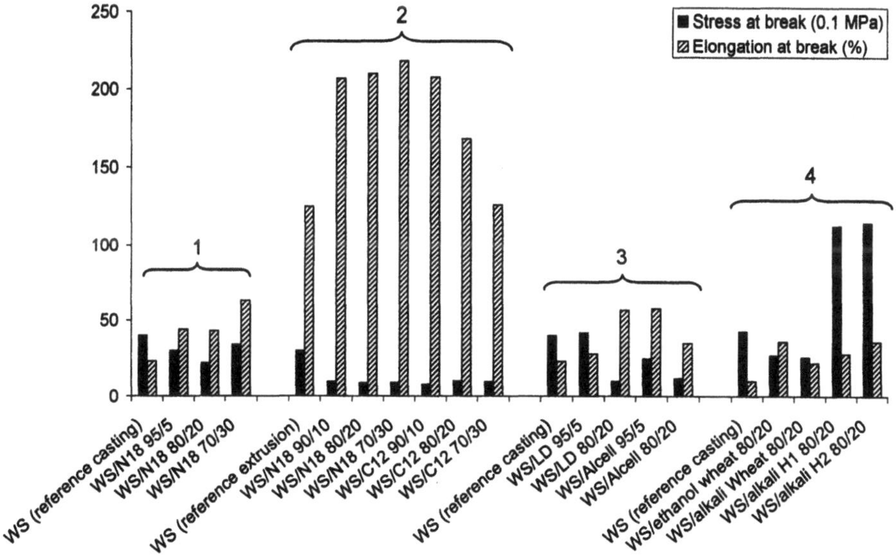

Figure 6. Mechanical properties of various thermally molded (2) and cast (1, 3, 4) films obtained from mixtures of glycerol/wheat starch (WS) with sodium- (N18) and calcium- (C12) lignosulfonates, wheat dioxan- (LD), wheat ethanol Alcell- (Alcell), wheat ethanol reflux- (ethanol) and alkali- (alkali wheat, alkali H1 and alkali H2) lignins.

4. 2. Thermal Properties

4. 2. 1. Experimental

Two kinds of techniques can be used to investigate the thermal transitions of lignin-starch blends, and more generally of polymer blends: dynamic mechanical thermal analysis (DMTA) and differential scanning calorimetry (DSC). DMTA shows the advantage to be directly applicable to film samples and to allow the detection of transition of low intensity, such as the low temperature transition, β relaxation, involving groups or short chain segments (Lourdin *et al.*, 1997a). DMTA measurements were performed on a Dynamic Mechanical Thermal Analyzer (DMTA MKIII, Polymer Lab.) with the vibration frequency in tension mode set at 1Hz, the solicitation amplitude at 0.01%, and the heating rate at 3 °C/min. A static force was applied, with an intensity depending on the mechanical properties of the film. As the peak of the loss factor tan δ is difficult to precisely detect, in particular for samples showing a high T_g, the temperature of α relaxation associated with the glass transition T_g is defined here as the onset of the drop in the storage modulus E' (Figure 7).

4. 2. 2. Influence of Lignin on the Glass Transition Temperature

For all references and starch/lignin materials, a single relaxation is observed in the -20 - 80 °C temperature range. This relaxation is supposed to be associated with the glass transition of the matrix. Except for the alkali hemp lignin fraction H2, which does not affect the thermal properties of starch, lignin incorporation alters the temperature T_g at which this relaxation occurs (Figure 7 and Table 2). This observation is, here again, diagnostic of partial miscibility of the components.

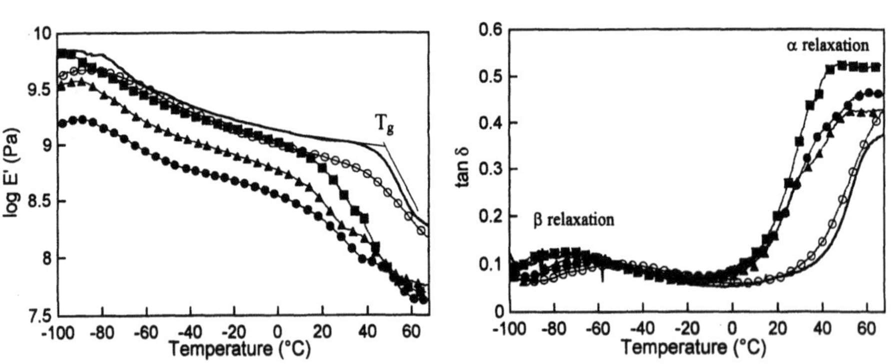

Figure 7. Influence of lignins on elastic modulus (log E') and loss factor (tan δ) as determined by Dynamic Mechanical Thermal Analysis (DMTA) of a control starch/glycerol film (—) and of starch/lignin/glycerol films containing different lignin preparations: wheat ethanol lignins (▲); wheat Granit lignin (■); hemp Granit lignin (●) and one of its fractions (○).

The increase of T_g as a function of lignosulfonates concentration was first revealed by DMTA measurements performed on starch/glycerol/lignosulfonates films (Table 2, Baumberger *et al.*, 1997) then confirmed by a calorimetric T_g determination of glycerol-free lignosulfonates/starch mixtures containing similar amount of residual water (Volperts *et al.*, 2001). This last study further demonstrates that the dependence of T_g as a function of the lignosulfonates concentration follows, over all the polymers ratio range, the Fox equation (Eq. 1)

$$1 / T_g = (w1 / T_g1) + (w2 / T_g2) \quad (1)$$

where w1 and w2 are weight fractions and T_g1 and T_g2 the glass transition temperatures of the blend components. This indicates that starch/lignosulfonates systems behave as ideal homogeneous blends. In the case of kraft lignin/starch mixtures, a deviation from the Fox equation is observed.

DMTA measurements performed on cast films including starch and various non sulfonated lignins reveal a T_g decrease of 27 to 33°C relative to the reference when lignin (20% / total polymers) are incorporated (Figure 7, Baumberger *et al.*, 2000). Such a T_g decrease suggests that lignin incorporation increases the mobility of amylose and amylopectin chains, thus leading to a higher plasticity.

Table 2. Composition and temperature of α relaxation T_g of starch / sodium (N18) and calcium (C12) lignosulfonates blends (RH 58% at 25°C)

Polymers ratio	Residual water %[a]	Glycerol % / starch[b]	T_g
WS/N18 90/10	13.5	32.9	3.4
WS/N18 80/20	13.8	31.4	7.0
WS:N18 70/30	13.5	34.5	10.0
WS/C12 90/10	14.1	43.7	-11
WS/C12 80/20	13.8	43.0	-1.5
WS/C12 70/30	13.8	30.2	4.0

[a] Determined by the Karl Fisher method.
[b] Determined by HPLC according to Lourdin *et al.* (1997b).

5. IMPACT OF LIGNIN INCORPORATION ON THE WATER AFFINITY

There are several approaches for evaluating the impact of lignins on the water sensitivity of either starch or glycerol/starch matrix. They include swelling, sorption and solubility tests as well as contact angle measurements. Contact angle measurements are used to study the film wettability and can give an insight into the surface hydrophobicity. However, in the case of thin and hydrophilic cast films, swelling phenomena hinder these measurements. We report here on sorption, swelling and solubility tests, which provide relevant information on the bulk properties of the materials in relation with their potential applications (packaging, mulching, etc.)

5. 1. Sorption

Sorption isotherms of starch/lignosulfonates and starch/kraft lignins thermally molded films (Figures 8 and 9) indicate a significant diminution of the film water content. In the case of lignosulfonates, this apparent decrease of water sorption most probably results from the diminution of the total amount of glycerol. Indeed, this hydrophilic compound is known to contribute significantly to sorption (Lourdin *et al.*, 1997b).

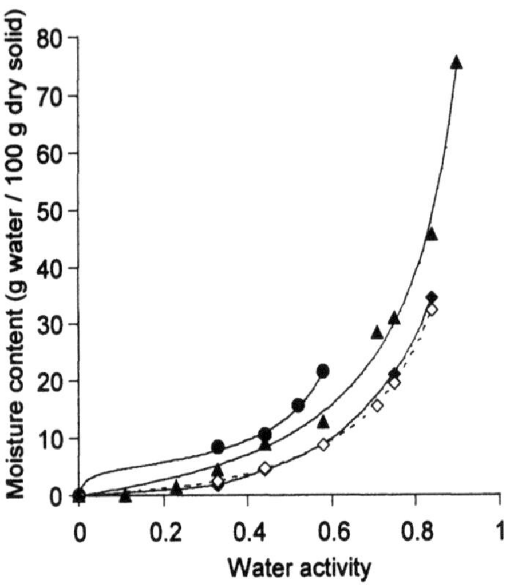

Figure 8. Sorption isotherms (22 °C) of thermally molded wheat starch/lignosulfonates films containing 33%wt. glycerol with respect to dry starch: WS (•), WS/N18 90/10 (▲), WS/N18 70/30 (♦), WS/C12 70/30 (◊).

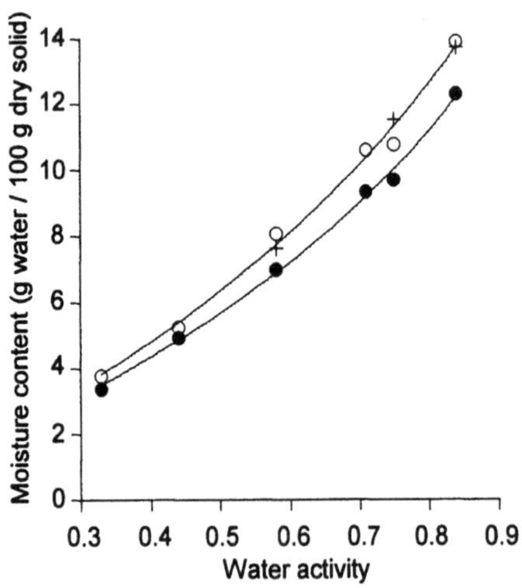

Figure 9. Sorption isotherms (22 °C) of thermally molded wheat starch films containing 0% (o), 15% (+) and 30% (•) KRL

In the starch/kraft lignin glycerol-free systems, the diminution of water content is less pronounced and proportional to the lignin incorporation rate. In this case, the presence of hydrophobic lignins in the matrix can explain the shift of the isotherm. The same moderate effect of lignin incorporation has also been observed with the following systems: WS/F1, HAS/KRL, HAS/F2 and HAS/F3 (Baumberger, 1999).

5. 2. Swelling tests and solubility

A consequence of starch hydrophilic character is the swelling of starch-based films in contact with water. This property is illustrated in Figure 10 where the pure starch film shows a 2.5 fold size increase while the glycerol/starch film exhibits an uncohesive gel-like behavior. The Alkali, organosolv and kraft lignins have a dramatic positive impact on the dimensional stability of cast films (Figure 10). Several phenomena can explain the high dimensional stability imparted by lignins in the case of cast films:

1) the miscibility of phenolic compounds with starch,
2) the possible presence of an hydrophobic barrier at the surface of the film, resulting from the migration of amphiphilic phenolic compounds at the air/liquid interface during casting,
3) the possible existence of cross-linkages formed through phenolic compounds.

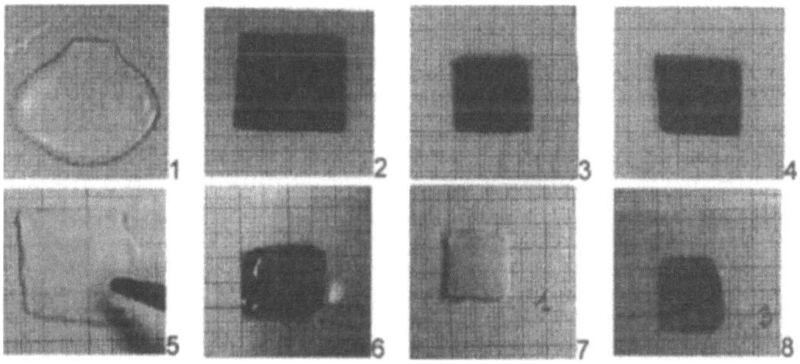

Figure 10. Photographs showing the texture and dimensions of different cast (1 to 6) and thermally molded (7, 8) film samples after a 24 h immersion at 25 °C in distilled water. Initial surface area of the samples: 1x1cm, magnification X 0.5. (1) starch/glycerol, (2) starch/glycerol/wheat alkali lignin, (3) starch/glycerol/wheat ethanol lignin, (4) starch/glycerol/hemp alkali lignin, (5 and 7) pure starch, (6 and 8) 80/20 starch/KRL.

The effect of kraft lignins on the dimensional stability of thermally molded film is not as spectacular as for cast films, since the pure starch reference itself does not significantly swell during immersion. The general higher thickness and maybe a less destructurized state of starch granules may be responsible for this higher dimensional stability of the thermally molded films. Nevertheless, the presence of kraft lignins in the thermally molded films confers other hydrophobic properties, such as a reduction of starch water solubility when 10-20% of lignins are incorporated (Figure 11). This effect is reversed at an incorporation level of 30%, where abundant lignin particles are responsible for the brittleness of the material (Baumberger, 1999).

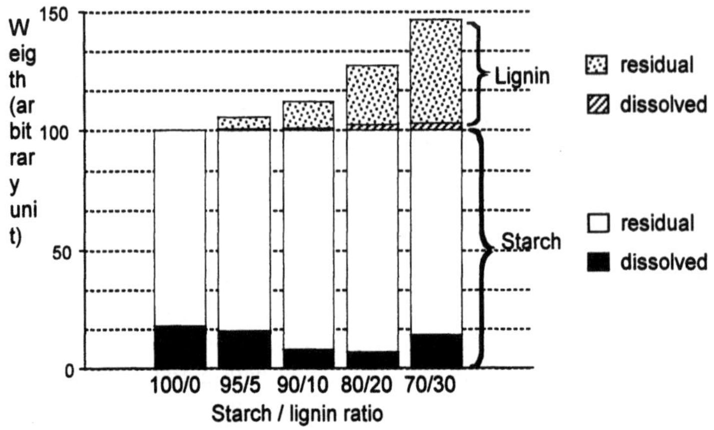

Figure 11. Water solubility (24 h at 25 °C) of WS/KRL thermally molded films. The dissolved phenolic compounds (dissolved lignin) are determined by absorbance measurements at 280 nm and the amount of dissolved starch is calculated on the basis of these measurements and of the global weight loss of the material. All the weights are expressed with respect to the initial starch weight, taken arbitrary equal to 100.

6. CONCLUSIONS

Taken together, the results described in the present chapter indicate that starch and lignins can form compatible systems with lignin acting either as filler or as extender of the starch matrix. This compatibility is favored by medium relative humidity, high amylopectin/amylose ratios and by the presence of the low molecular weight lignin components. Lignosulfonates impart a higher extensibility to starch films and indirectly contribute to a slight water sorption decrease, when used as a partial substitute for glycerol. The presence of polar sulfonic groups, likely to form hydrogen bonds with amylose and amylopectin hydroxyls, allows lignosulfonates to form an intimate blend with starch. On the other hand, non-sulfonated lignins lead to composite materials filled with high molecular weight lignin particles and plasticized by the low molecular weight phenolic fraction. The variety of molecular species constitutive of this low molecular weight fraction may result in a cumulative plasticizing effect, as observed with urea and glycols in gelatinized starch (Shogren *et al.*, 1992). The hydrophobic character of non-sulfonated lignins imparts to starch films an improved water resistance, more pronounced in the case of cast films.

Thus, starch/lignin materials not only provide model systems to investigate interaction between phenolics and polysaccharides, but they also present technical advantages that offer perspectives for utilization in the field of packaging or mulching. Though important information on the structure-property relationships have been gathered, biodegradability and permeability tests performed on thermally molded films are still lacking. The film coloration due to the presence of lignins may today be considered as a drawback towards food packaging applications, but it is not so towards agricultural applications such as mulching.

The potential of lignins to improve the functionality of polymer materials, in particular claimed by Bengs (1998) in a patent relative to biopolymer-based thermoplastic blends, goes beyong biopolymer-based systems. For example, incorporation of low amounts of lignosulfonates into polyolefins such as polypropylene or polyethylene leads to a significant increase in the elongation at break and tensile strength of the material (Simionescu *et al.*, 1996). The incorporation of over 5% of lignins in such synthetic polymers seems to require a compatibilization either through cross-linking, through the use of compatibilizers or through lignin modifications. On the other hand, mechanical properties of polypropylene films are not significantly influenced by the addition of organosolv (non sulfonated) lignins over the range 2-10% (Kosikova *et al.*, 1993). Despite the significant difference between the two matrix, a striking analogy seems thus to exist between starch/lignin and polyolefin/lignin systems, as far as thermal molding is concerned.

7. ACKNOWLEDGEMENTS

The major part of the works presented here has been supported by INRA (Institut National de la Recherche Agronomique) and by AGRICE (Agriculture pour la Chimie et l'Energie) under grant 9701021 (ADEME, France). The author would like to thank AVEBENE (Tartas, France), WESTVACO (Charleston, U.S.) and Granit (Lausanne, Suisse) for providing the different industrial lignin samples.

8. REFERENCES

Baumberger, S., Lapierre, C., Monties, B., Lourdin, D., and Colonna, B., 1997, Preparation and properties of thermally moulded and cast lignosulfonates-starch blends, *Ind. Crops. Prod.* **6**: 253-258.

Baumberger, S., Lapierre, C., and Monties, B., 1998a, Utilization of pine kraft lignin in starch composites: impact of structural heterogeneity, *J. Agric. Food Chem.* **46**: 2234-2240.

Baumberger, S., Lapierre, C., Monties, B., and Della Valle, G., 1998b, Use of kraft lignin as filler for starch films, *Polym. Degrad. Stab.* **59**: 273-277.

Baumberger, S., 1999, Obtention et caractérisation de matériaux composites amidon-lignines, Ph.D. thesis, Institut National Agronomique (France).

Baumberger, S., Lapierre, C., Lora, J. H., and Monties, B, 1999, Novel utilization of wheat straw organosolv lignins in starch composites, *Proceedings of the 10th Int. Symp. Wood Pulp. Chem.*, Yokohama (Japan), Vol I, p. 680.

Baumberger, S., Michon, C., Cuvelier, G., and Lapierre, C., 2000, Lignin utilization in starch thermoplastics: towards molecular origin of polymer compatibility, *Proceedings of the 6th Europ. Workshop Lignocell. Pulps*, Bordeaux (France), p. 121.

Bengs, H., 1998, Germain patent n°WO9902596

Callegarin, F., Gallo, J-A Q., Debeaufort, F., and Voilley, A., 1997, Lipids and biopackaging, JAOCS, 74(10): 1183-1192.

Chang, Y. P., Cheah, P. B., and Seow, C. C., 2000, Plasticizing-antiplasticizing effects of water on physical properties of tapioca starch films in the glassy state, *J. Food Sci.* **65**(3): 445-451.

Chapman, G. M., 1994, Status of technology and applications of degradable products, in: *Polymers from Agricultural Coproducts*, M. L. Fishman, R. B. Friedman and S. J. Huang, ed., ACS, Washington, pp. 29-47.

Chiellini, E. and Solaro, R., 1996, Biodegradable polymeric materials, *Adv. Mater.* **8**(4): 305-313.

Doane, W. M, 1992, USDA Research on starch-based biodegradable plastics, *Starch/Stärke*, **44**: 293-295.

Fang, Q. and Hanna, M. A., 2000, Functionnal properties of polylactic acid starch-based loose-fill packagingfoams, *Cereal Chem.* **77**(6): 779-783.

Fengel, D. and Wegener, G., 1989, *Wood: Chemistry, Ultrastructure, Reactions.* Walter de Gruyter, Berlin, pp. 613.

Fishman, M. L., Coffin, D. R., Unruh, J. J., and Ly, T., 1996. Pectin/starch/glycerol films: blends or composites? *J. Mat. Sci. Pure appl. Chem.* **A33**: 639-654.

Fritz, H. G., Aichholzer, W., Seidenst¨cker, T., and Widmann, B., 1995, Abbaubare polymerwerkstoffe auf der basis nachwachsender rohstoffe-möglichkeiten und grenzen, *Starch/stärke* **47**: 475-491.

Funke, U. and Lindhauer, G, 1994, Eigenschaften von Gießfilmen aus nativen und chemisch modifizierten Stärken, *Starch/stärke* **46**: 384-388.

Garcia, M. A., Martino, M. N., and Zaritzky, N. E., 2000, Lipid addition to improve barrier properties of edible starch-based films and coatings, *J. Food Sci.* **65**(6): 941-947.

Hänggi, U. J., 1995, Requirements on bacterial polyesters as future substitute for conventional plastics for consumer goods, *FEMS Microbiology Reviews* **16**: 213-220.

Jauregui, B., Munoz, M. E., and Santamaria, A., 1995, Dynamic viscoelastic properties of blends of poly(ethylene-co-vinyl acetate) and a modified starch, *Macromol. Chem. Phys.* **196:** 3133-3142.

Kalichevsky, M. T. and Ring, S. G, 1987, Incompatibility of amylose and amylopectin in aqueous solution, *Carbohydr. Res.* **162**: 323-328.

Ke, T. and Sun, X., 2000, Physical properties of poly(lactic) and starch composites with various blending ratios, *Cereal Chem.* **77**(6): 761-768.

Kim, M. and Pometto, A. L., 1994, Food packaging potential of some novel biodegradable starch-polyethylene, *J. Food Protection* **57**(11): 1007-1012.

Kosikowa, B., Demianova, V., and Kacurakova, M., 1993, Sulfur-free lignins as composites of polypropylene films, *J. Appl. Polym. Sci.* **47** : 1065-1073.

Krochta, J. M. and De Mulder-Johnston, C. L. C., 1996, Biodegradable polymers from agricultural products, in: *Agricultural Materials as Renewable Resources*, G. Fuller, T. A. McKeon and D. D. Bills, ed., ACS, Washington, pp. 120-140.

Leloup, V., Colonna, P., and Buléon, A., 1991, Influence of amylose amylopectin ratio on gel properties, *J. Cereal Sci.* **13**: 1-13.

Lim, S.-T. and Jane, J.-L, 1993, Preparation of water-resistant, biodegradable plastics with starch-zein mixtures in: *Carbohydrates and Carbohydrate Polymer,* M. Yalpani, ed., ATL Press, Mount Prospect, pp. 288-297.

Lourdin, D., Colonna, P., and Della Valle, G, 1995, Modes d'obtention de matériaux amylacées aux échelles laboratoire et pilote, in: *Valorisations Non Alimentaires Des Grandes Productions Agricoles,* J. Gueguen, ed., INRA, Nantes, pp. 241-248.

Lourdin, D., Bizot, H., and Colonna, P., 1997a, Correlation between static mechanical properties of starch-glycerol materials and low-temperature relaxation, *Macromol. Symp.* **114**: 179-185.

Lourdin, D., Coignard, L., Bizot, H., and Colonna, P., 1997b, Influence of equilibrium relative humidity and plasticizer content on water content and glass transition of starch materials, *Polymer* **38**: 5401-5406.

Monties, B. 1988, Preparation of dioxane lignin fractions by acidolysis, in: *Methods in enzymology,* W. A. Wood and S. T. Kellogg, ed., Academic Press, New York, pp. 31-34.

Mörck, R., Reimann, A., and Kringstad, K. P, 1988, Fractionation of kraft lignin by successive extraction with organic solvents. III. Fractionation of kraft lignin from birch, *Holzforschung* **42**: 111-116.

Muzzarelli, R. and Ilari, P., 1994, Chitosans carrying the methoxyphenyl functions typical of lignin, *Carbohydr. Polym.* **23**: 155-160.

Otey, F., Westhoff, R. P., and Doane, W. M, 1987. Starch-based blown films, *Ind. Eng. Chem. Res.* **26**: 1659-1663.

Protzman, T. F., Wagoner, J. A., and Young, A. H, 1967, Process of casting amylose films, U.S. patent n° 3,344,216.

Rivard, C., Moens, L., Roberts, K., Brigham, J., and Kelley, S, 1995. Starch esters as biodegradable plastics: effects of ester group chain length and degree of substitution on anaerobic biodegradation, *Enzyme and Microbial Technology* **17**: 848-852.

Rutlege, D. N., Barros, A. S., Vackier, M. C., Baumberger, S., and Lapierre, C., 1999, Analysis of time domain NMR and other signals, in: Advances in Magnetic Resonance in Food Science, P. S. Belton, B. P. Hills, and G. A. Webb, ed., Royal Society of Chemistry, Cambridge, pp. 203.

van Soest, J. J. G, 1996. Starch plastics: structure-property relationships, Ph.D. thesis, Utrecht University (The Netherlands).

Shogren, R. L., Swanson, C. L., and Thompson, A. R., 1992, Extrudates of cornstarch with urea and glycols: structure/mechanical property relations, *Starch/Stärke* **44**: 335-338.

Simionescu, C. I., Macoveanu, M. M., Vasile, C., Ciobanu, F., Esanu, M., Ioanid, A., Vidrascu, P., and Georgescu-Buruntea, N., 1996, Polyolefins/lignosulfonates blends, *Cellulose Chem. Technol.* **30**: 411-429.

Volperts, A., Dizhbite, T., Telysheva, G., Odermatt, J., Lehnen, R., and Faix, O., 2001, Miscibility of lignin-polymer blends as detected by differential scanning calorimetry (DSC), Proceedings of the 11th Int. Symp. Wood Pulp Chem. Nice (France), p 203-206.

Whistler, R. L., 1965, Fractionation of starch, in: *Starch: Chemistry and Technology,* R. L. Whistler and E. F. Paschall, ed., Academic Press, New York-London, pp. 331-347.

Yao, Y., Yoshioka, M., and Shiraishi, N., 1995, Rigid polyurethane foams from combined liquefaction mixtures of wood and starch, *Mokuzai Gakkaishi* **41**(7): 659-668.

Yao, Y., Yoshioka, M., and Shiraishi, N., 1994, Soluble properties of liquefied biomass prepared in organic solvents, *Mokuzai Gakkaishi* **40**(2): 176-184.

Yao, Y., Yoshioka, M., and Shiraishi, N., 1993, Combined liquefaction of wood and starch in a polyethylene glycol/glycerin blended solvent, *Mokuzai Gakkaishi* **39**(8): 930-938.

Zobel, H. F., and Stephen, A. M, 1995. Starch: structure, analysis, and application, in: *Food Polysaccharides and Their Applications,* A. M. Stephen, ed., Marcel Dekker, New York, pp. 19-66.

LIGNOSULFONIC ACID-DOPED POLYANILINE (LIGNO-PANI™) - A VERSATILE CONDUCTING POLYMER

Brian C. Berry and Tito Viswanathan*

1. INTRODUCTION

The use of renewable natural resources to develop or improve existing technologies is an important area of research to the scientific community. The availability and functionality of these resources very often make it possible to reduce the cost and improve the properties of existing products. One such natural resource that has received a great deal of attention over the years is lignin.

1.1. Lignin and Lignosulfonates

Lignin, the major non-cellulosic constituent of wood, is a complex phenolic polymer that bears a superficial resemblance to phenol-formaldehyde resins. It consists of functionalized phenylpropane units connected via alkyl and aryl ether linkages. Essentially all of the lignin commercially available is isolated as by-products from the paper industry from either the sulfite or the Kraft process. Sulfonated lignins are obtained either as spent sulfite liquor (SSL) or by sulfonation of lignin obtained from the Kraft process. The annual production of sulfonated lignins by some of the major producers is listed in Table 1.

SSL obtained from the sulfite process consists of lignosulfonates (~ 55%), sugars (30%) and other ingredients in smaller amounts. Figure 1 shows a typical monomeric unit of Kraft lignin that has been sulfomethylated at the aromatic ring and sulfonated on the aliphatic side chain. Sulfomethylation is accomplished by the reaction of the Kraft lignin with formaldehyde and sodium sulfite. The aliphatic sulfonation occurs preferentially at the benzylic position of the side chain of the phenylpropane units. Lignosulfonates are available in the form of calcium or sodium salts (Borasperse® and Ultrazine® from Westvaco, for example) and are a cheaper alternative to other forms of lignosulfonates. Some of the applications of lignosulfonates are in concrete admixtures, animal feed, oil

* Brian C. Berry and Tito Viswanathan, University of Arkansas- Little Rock, Little Rock, AR 72204.

well drilling muds, dust control, emulsion stabilizers, dye dispersants, wood preservation, and as mining aids.[1]

Table 1. Major manufacturers and their annual production of lignosulfonates (adapted from Ref.[1]).

Manufacturer	Production (tons/year)
Borregaard LignoTech	160,000
LignoTech Sweden	60,000
LignoTech USA	60,000
Flambeau Paper	60,000
Boregaard Germany	50,000
Sanyo Kokusaka	50,000
Westvaco	35,000

1.2. Inherently Conducting Polymers (ICPs)

The 2000 Nobel Prize in Chemistry was awarded to three researchers who played a pioneering role in the synthesis and characterization of inherently conducting polymers (ICPs). Some examples of ICPs are polyaniline (PANi), polythiophene and polypyrrole. A typical synthesis of polyaniline is shown in Figure 2. The product obtained is in the doped (i.e. conducting) form. Conducting polymers are of great importance because they are organic polymers exhibiting various levels of conductivity. ICPs have potential applications in a host of areas such as anti-static fabric coatings in (explosive) fuel handling garments and other electrostatic dissipative materials, EMI (Electromagnetic Interference) shielding of electronic equipment and aircraft guiding systems and the production of radar invisible (stealth) coatings.[2]

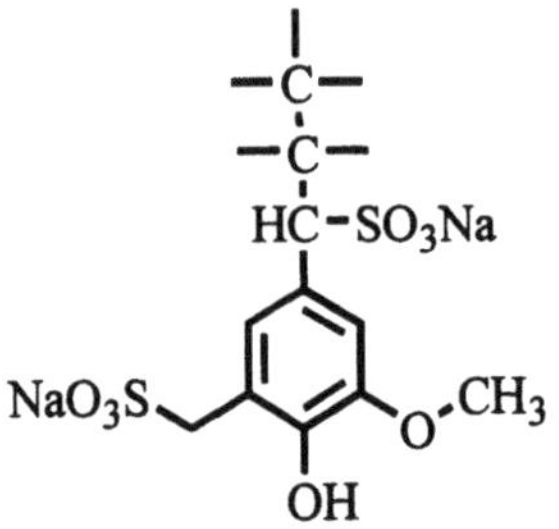

Figure 1. A typical sulfonated/sulfomethylated lignin monomer unit.

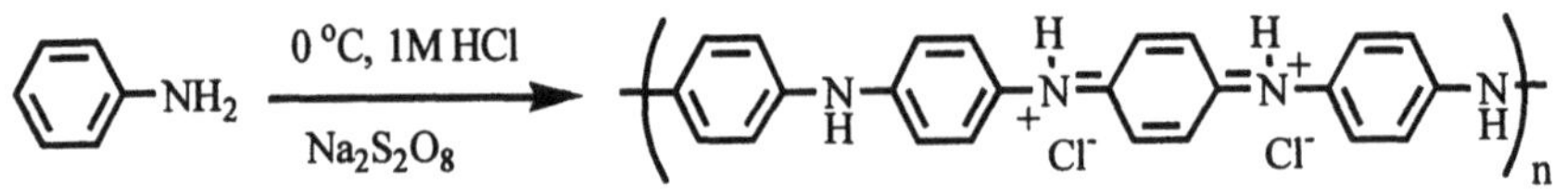

Figure 2. Typical synthesis of HCl-doped polyaniline.

Figure 3 illustrates the range of conductivity for several materials including conducting polymers. The range of conductivity for ICPs is generally in the semi-conductive range. Such conductivity may be explained based on the electronic structure of the ICP. The ICP has an extended conjugated structure and possesses a HOMO (valence band) and a LUMO (conduction band) that are separated by what is commonly referred to as a band gap. The size of this band gap determines the conductivity of the material as per the band gap theory. In the case of semi-conductors, a dopant is needed to reduce the band gap. This may be accomplished by two methods: n-type doping and p-type doping. In the former, an electron rich species is introduced which adds electrons into the backbone thereby raising the energy level of the valence band and decreasing the band gap. In the latter, more common method for ICPs, an electron deficient species such as a proton is added which serves as an electron acceptor forming a new, more easily accessible conduction band.

The electroactivity and electronic conductivity of ICPs set them apart from traditional polymers leading to many of the aforementioned applications. The realization of their potential, however, has not been achieved industrially due to the fact that ICPs (in their doped conductive form) are insoluble in water as well as most organic solvents. Therefore, much research has been concentrated in two areas: 1) increasing the solubility by modifying the polymer or by using a highly soluble dopant that is generally an acid containing a soluble polymeric anion added to the base form of PANi to make a conductive salt and 2) solution casting of thin films of the base form of polyaniline followed by a surface treatment with an acid to dope the polymer. The latter is not always industrially practical on a large scale. The former, however, has resulted in many soluble or dispersible conducting polymers.[3] Most of these processible polymers, however, are neither cost effective nor readily adaptable for a large-scale application such as corrosion prevention.

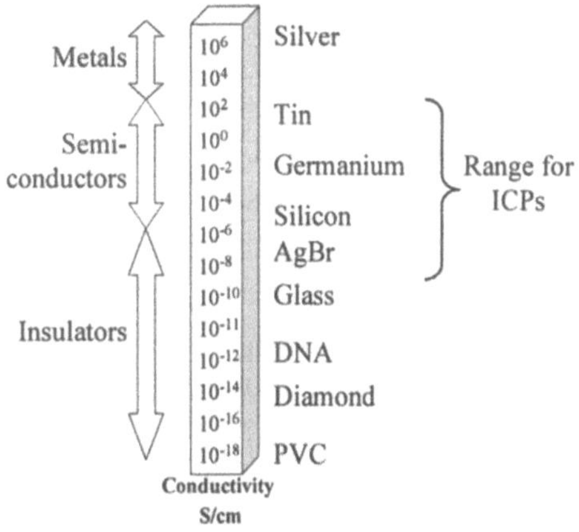

Figure 3. Range of conductivity of various materials including ICPs.

2. TEMPLATE-GUIDED POLYMERIZATION

As mentioned previously, the use of dopants that produce soluble counter ions is a common practice to improve the processibility of ICPs. Soluble polymeric acids have also been used as acid templates upon which polymerization is carried out producing a water-soluble ICP.[4, 5] Polystyrene sulfonic acid (PSSA) has been used as a template to form a water-soluble PSSA-polyaniline complex.[6] This ICP complex shows a greater thermal stability compared to HCl-doped polyaniline because the acid is not volatile and is not removed upon heating. It only becomes dedoped at pH 9 as compared to pH 4 for HCl-doped polyaniline. This indicates that the two polymers are intertwined in such a way that the electrostatic interactions between the two result in a more stable complex. The orientation of the two polymers has been described as being similar to a double helix in nature. Other polymeric acids have been used which have yielded similar results.[7]

2.1. Template-Guided Polymerization Using Lignosulfonic Acids

Lignosulfonic acids are highly soluble polymeric acids that can be used as acid templates for the polymerization of monomers such as aniline and pyrrole. This method has been patented by one of us.[8] Figure 4 is a schematic for the synthesis of polyaniline in the presence of a lignosulfonic acid. Since lignosulfonates are generally available as sodium salts, the acidic form must be produced. This may be accomplished by passing through a cation exchange resin. Of note, a method has been developed independently which does not require the use of a cation exchange resin. Once the aqueous lignosulfonate is converted to the acidic form, aniline is added to the solution. The aniline is protonated by the lignosulfonic acid forming an anilinium lignosulfonate complex. Upon addition of the oxidant, polymerization occurs to give ICP doped by the lignosulfonic acid. The resulting lignosulfonic acid-doped polyaniline (LIGNO-PANI™ – Trademark of GEOTECH Chemical Co. Tallmadge) is dispersible in water and in organic solvents such as dimethyl sulfoxide (DMSO) and N-methylpyrrolidinone (NMP).

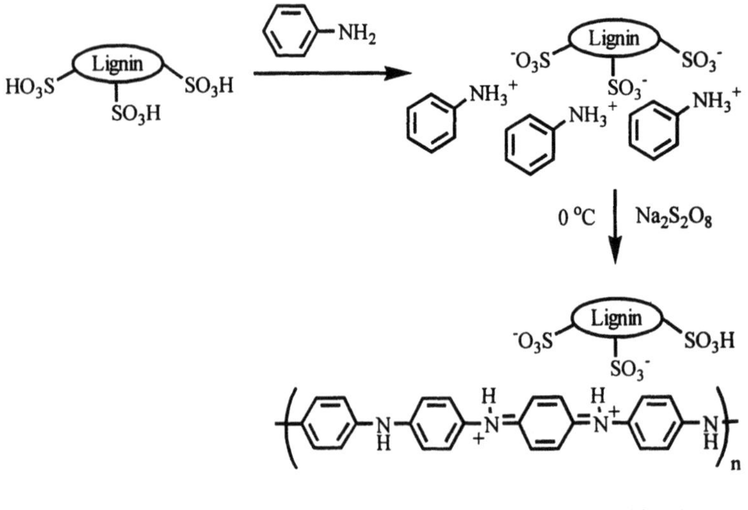

Figure 4. Template-guided polymerization of aniline using lignosulfonic acid as the dopant.

A variety of lignosulfonates are available from the companies listed in Table 1. Several of them have been used in the synthesis of LIGNO-PANI™ in our lab. The best dispersibility was obtained using a modified kraft lignin (Westvaco Reax 825E) - a sulfomethylated and ethoxylated derivative. The product is dispersible in water as well as common polar aprotic solvents such as DMSO and NMP. Varying the weight ratio of lignosulfonate to aniline as well as adding second dopants were attempted to determine the optimal ratios for dispersibility and conductivity. It was found that a ratio of 1:4 lignin:aniline was optimal for conductivity (0.1-10 Scm^{-1}) and a ratio of 1:1 optimal for dispersibility.[9]

3. APPLICATIONS OF LIGNO-PANI™

As with typical ICPs, there are numerous applications for LIGNO-PANI™. These include the previously mentioned applications as well as other novel ones that have yet to be realized. The distinct advantage of LIGNO-PANI™ in nearly all of these applications is the increased dispersibility in various solvents including water.

3.1. Corrosion Protection Using LIGNO-PANI™

Recently ICPs have received increasing attention in the field of corrosion prevention. It was first shown by DeBerry that stainless steel surfaces could be passivated with polyaniline (PANi).[10] There is considerable experimental evidence suggesting that the ICP coatings are effective in lowering the corrosion rates of metals when exposed to aqueous corrosive environments. The realization of this potential has not been achieved industrially due to the fact that ICPs (in their doped conductive form) are insoluble in water as well as most organic solvents and therefore difficult to apply using conventional methods.

ICPs are unique in that they are not only electrically conductive but also electroactive. Polyaniline in particular has three basic states that can exist as salts (protonated forms) or as bases (neutral forms). These states and the associated structures are shown in Figure 5 in their neutral forms. The pernigraniline state is the most oxidized form containing equal amounts of quinoid and benzenoid species. The emeraldine form consists of 75% benzenoid and 25% quinoid. The leuco form is the completely reduced form and consists of all benzenoid units. The underlying premise of corrosion protection using ICPs is that a protective oxide layer is produced by the electroactive ICP when in contact with the metal. This is illustrated in Figure 6. After application of the ICP, which may be a neat coating or a blend of the ICP in a resin, a passive oxide layer is formed resulting in a protective layer that remains intact and prevents further corrosion.

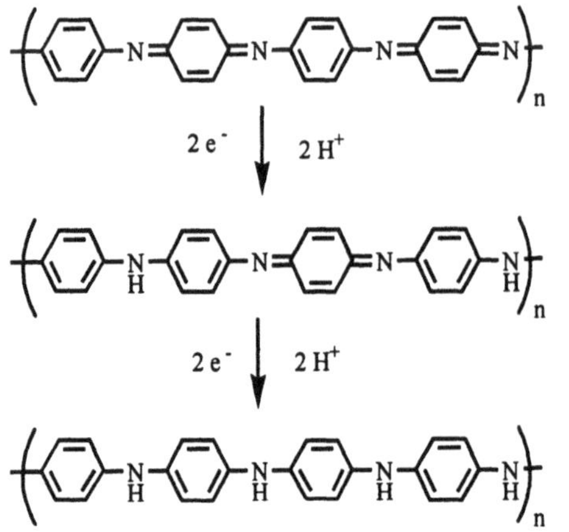

Figure 5. Polyaniline in different oxidation states.

For an ICP to protect a substrate in the manner described above, it is important that its electrochemical redox potential be more positive than that of the substrate. To confirm that LIGNO-PANI™ has a redox potential that satisfies this criterion, a series of cyclic voltammograms (CV) were performed. The CVs were performed in environments of varying pH's. The results from these experiments are shown in Figure 7. The traces are for pH values of 2.0 (A), 4.0 (B), 6.0 (C), 8.0 (D), and 10.0. These pH values were obtained using a three-component buffer prepared using a method developed by Carmody[11] containing anhydrous boric acid, citric acid monohydrate and sodium phosphate. From the CV it can be seen that there are two transitions. The transition located at the far left is the transition from the pernigraniline (i.e. most oxidized form) to the emeraldine form (i.e. the conducting form). While this transition is of lesser concern, it was studied at different scan rates and found to involve a two-electron transfer.[12] The most prominent transition is from the emeraldine to the leuco form. This reduction is associated with the formation of the passive oxide layer during the corrosion protection process and is found to involve a one-electron transfer. Interestingly, LIGNO-PANI™ remains redox active up to a pH of 9-10. This is significant since previous polyanilines were reported to lose their redox activity above pH 5.[13]

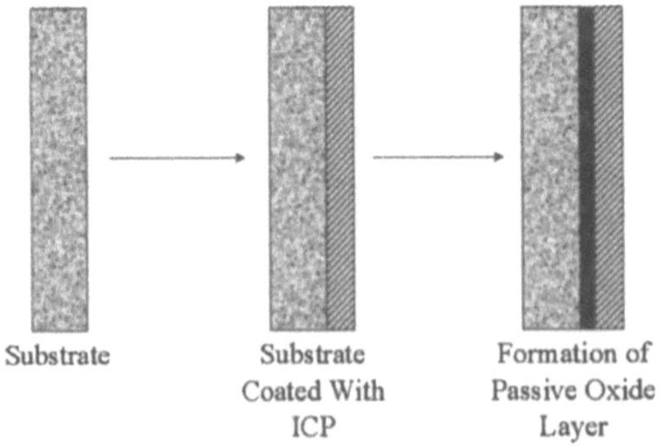

Figure 6. Mechanism of corrosion protection using ICPs.

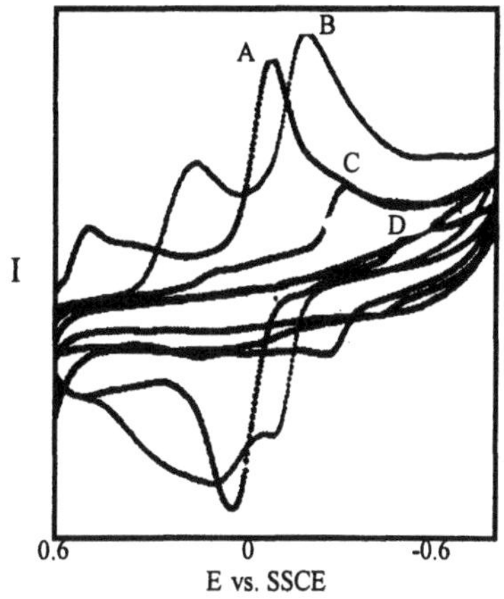

Figure 7. CV plots for pH values of 2.0 (A), 4.0 (B), 6.0 (C), 8.0 (D), and 10.0.

In order to test the effectiveness of LIGNO-PANI™ as a corrosion inhibitor, LIGNO-PANI™ was synthesized according to the template-guided procedure described earlier. The resulting product was mixed with a water-reducible resin to form a blend that was 1% LIGNO-PANI™ by dry weight. This blend was applied to the cold-rolled steel substrate panels (Q-panel) using an airbrush spraying system. The panels were then cured according to the resin manufacturer's specifications.

Since corrosion is an electrochemical phenomenon, the mechanism as well as magnitude of corrosion may be studied using electrochemical testing techniques. These techniques may include DC as well as AC techniques. In the DC case a change in the DC voltage forces a composition change of the electrical double layer (EDL) resulting in a current due to the migration of charged species to and from the EDL. This resulting current can be plotted as a function of the voltage to produce Tafel plots, linear polarization plots and cyclic polarization curves that allow for the determination of corrosion rate. A smaller current resulting from the voltage change indicates a lower corrosion rate. In AC testing commonly referred to as electrochemical impedance spectroscopy (EIS), the EDL is disturbed with a much smaller sinusoidal voltage. The resulting current will have a different magnitude as well as phase at different frequencies. The EIS data can be plotted in a variety of plots including Bode phase (phase angle as a function of frequency), Bode Magnitude of Impedance (magnitude of impedance as a function of frequency), and Nyquist (magnitude of the imaginary component of impedance as a function of the magnitude of the real component of impedance).

The coated coupons with coating compositions containing 0, 1, 5 and 10% LIGNO-PANI™ were exposed to a 3.55% solution of NaCl in a testing cell. Periodically the coupons were removed and tested over a period of 30 days. This allowed sufficient time for the corrosion rate to reach equilibrium. From the initial results, it appears that LIGNO-PANI™ inhibits corrosion by passivating the metal surface. The plates containing 1% LIGNO-PANI™ show a more noble, i.e. more positive, open circuit potential (OCP) than the water-based coating alone. Figure 8 shows the Tafel plots of the steel coupons coated with the water-based acrylic resin containing LIGNO-PANI™ (A) as well as the coupons coated with only the water-based acrylic resin (B). The LIGNO-PANI™ containing coating exhibits not only a significant shift to a higher OCP, but also a lower corrosion current which translates into a lower corrosion rate. The corrosion currents for A and B were 1.58 μA and 27.4 μA, respectively. The lower corrosion rate as well as the increased OCP for A indicates that the passive oxide layer is probably inhibiting the electron transfer between the electrolyte and the metal. The slopes of the anodic branches of the Tafel plots in Figure 8 seem to also indicate a passivation mechanism. When the slope of the anodic branch is less than that of the cathodic branch, it can be said that the corrosion process is anodically controlled indicating that the oxidation process is occurring readily.[14] This can be seen for the Tafel plot of the steel plate coated only with the water-based acrylic resin. If, however, the slope of the cathodic branch is less than that of the anodic branch, as in Figure 8, it can be said that the corrosion process is cathodically controlled. This indicates that the anodic process is no longer the determining factor in the corrosion rate. This difference can possibly be attributed to the formation of an oxide layer that impedes further oxidation processes.

EIS data also confirms that LIGNO-PANI™ significantly reduces the corrosion rate. The data shows an increase in impedance indicating a less active double layer and therefore a lower corrosion rate. Figure 9 shows the Nyquist plots for the water-based acrylic coating containing 1% LIGNO-PANI™ and the same coating containing no LIGNO-PANI™. As the plots indicate, the impedance for the control is significantly lower which results in a higher corrosion rate. Indications of a third time constant can also be seen in Bode plots (not shown) for the LIGNO-PANI™ containing plate. A third time constant was occasionally observed at very low frequencies. This time constant was difficult to study due to the low frequencies and the extended time needed to perform these tests. Low frequency testing resulted in long test times during which the electrode can change leading to inconsistencies in the data. It is believed, however, that this third time constant is actually the time constant for the double layer. The second time constant would therefore most likely be due to an oxide layer that is formed by the LIGNO-PANI™. This lends further evidence to the theory of a passive oxide layer formation.

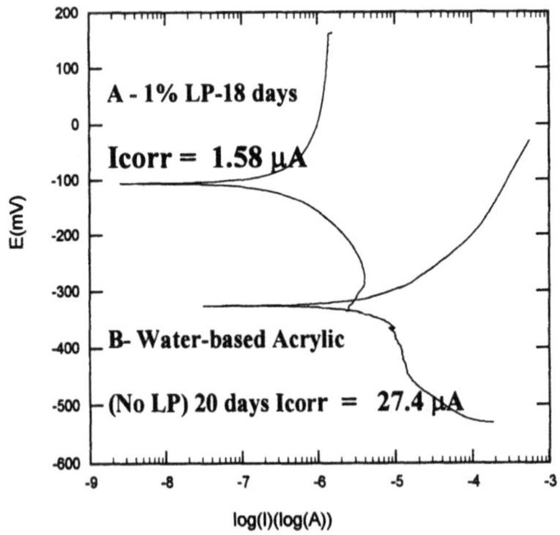

Figure 8. Tafel plots of a water based-acrylic coating containing 1% LIGNO-PANI™ (A) and no LIGNO-PANI™ (B).

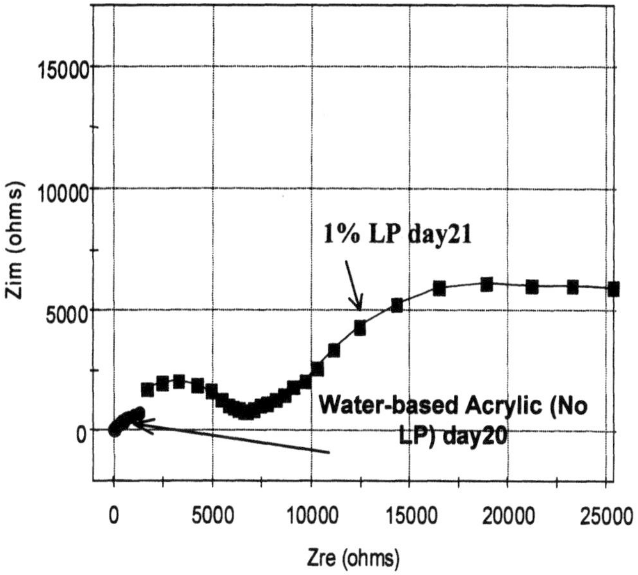

Figure 9. Nyquist plots generated using EIS for the coating containing 1% or no LIGNO-PANI™

Coupons coated with higher concentrations of LIGNO-PANI™ have also been evaluated. The 2% LIGNO-PANI™ coating performed almost as well as the 1% coating showing a reduction of 10-20 fold in the corrosion current. Another interesting phenomenon was observed with coatings containing 2% LIGNO-PANI™. Overlaid Nyquist plots in Figure 10 indicate a "correction" behavior for LIGNO-PANI™. As exposure time increases, a net decrease in the impedance is observed for all coatings. This is reasonable since infiltration of the bulk species into the pores of the coating results in corrosion of the underlying metal surface leading to a lower impedance. When examined more closely, however, the LIGNO-PANI™ containing coating indicates several disruptions in this trend that are not seen in the control plate. During coating exposure, the LIGNO-PANI™ containing coating shows decreases in impedance followed by increases in impedance. This is unexpected since the impedance would normally show no increase with exposure time. It is possible that in the case of the LIGNO-PANI™ containing coating the oxide layer is being destroyed due to infiltration of the bulk solution and that the metal surface is then repassivated by LIGNO-PANI™. This repassivation would lead to an increased impedance. The extent of this repassivation, however, decreases with time indicating that with increasing failure of the coating, the dissolution of the oxidized metal becomes so rapid that the repassivating power of LIGNO-PANI™ is no longer sufficient.

The results from the 5 and 10% coatings indicate that coating failure is accelerated by the incorporation of such a high percentage of LIGNO-PANI™. After 10 days both coatings showed significant signs of failure as well as increased corrosion currents. Leaching of the LIGNO-PANI™ by the electrolyte was also observed. On both of the plates, a significant amount of green color was found on the backside of the plate as well as in the electrolyte solution. This indicates that the LIGNO-PANI™ was possibly not cross-linked sufficiently with the resin. When the LIGNO-PANI™ was leached out, open pores probably formed in the resin allowing corrosion to occur freely on the metal surface. Figure 11 contains the Tafel plots of the coatings containing 2, 5 and 10% LIGNO-PANI™ after approximately 10 days of exposure. The coatings with 5 and 10% LIGNO-PANI™ show a lower OCP as well as an increased corrosion current compared to the one containing 2% LIGNO-PANI™. It remains to be determined whether the higher concentrations of LIGNO-PANI™ would be beneficial if the LIGNO-PANI™ were fully cross-linked with the resin.

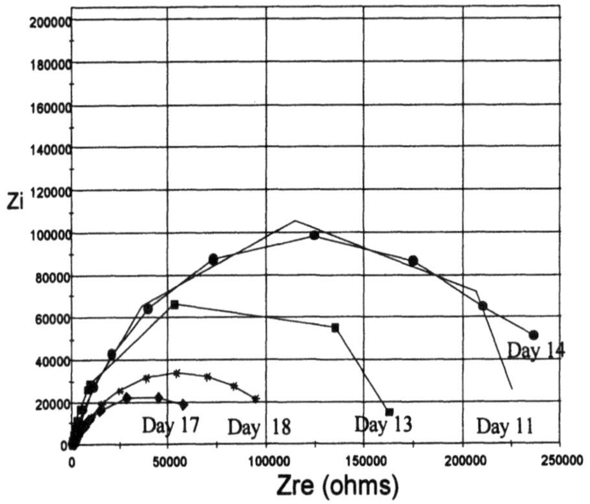

Figure 10. Nyquist plots for 2% LIGNO-PANI™ in a water-reducible acrylic coating over a period of time, indicating a repassivation mechanism.

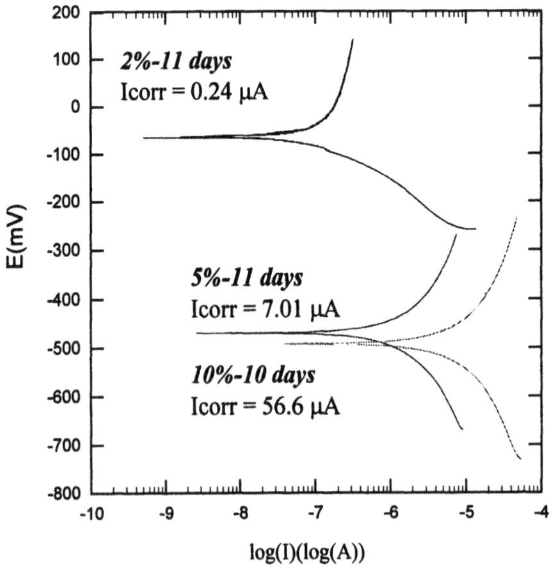

Figure 11. Tafel plots containing 2, 5 and 10% LIGNO-PANI™

In short, electrochemical data confirms that LIGNO-PANI™ is an effective corrosion inhibitor. Data indicates that at 1-2% loading levels LIGNO-PANI™ provides a 10-20 fold reduction in corrosion. There is also considerable evidence that supports the formation of a passive oxide layer.

3.2. Corrosion Protection of Metals Using LIGNO-PANI™/Active Metal Blends

The principle behind sacrificial protection is that a more active metal will act sacrificially when in contact either physically or via a conducting wire with the substrate. As the more active metal corrodes, electrons flow from the sacrificial anode to the metal substrate that is now the cathode and therefore remains in the reduced state. One of the most common cathodic protection systems used in coatings is a zinc-rich primer. This primer has inherent problems associated with the formulation of an effective coating. For the coating to be sufficiently conductive a large amount of zinc must be added. If not enough zinc is present the transfer of electrons from the zinc to the substrate will not occur. Conversely as greater amounts of zinc are added, barrier properties are decreased due to the formation of channels between the zinc particles. Increased difficulty in processing is also associated with increasing zinc content. Therefore, an alternative is needed which provides sacrificial protection but does not suffer from the disadvantages mentioned.

It has been proposed that these disadvantages might be circumvented by a blend of ICP and active metal particles.[15] Experiments on coatings containing LIGNO-PANI™ in combination with an inhibited form of aluminum (Al) have been performed. Preliminary experiments seem to indicate that the addition of Al into the coating has further beneficial results. Figure 12 shows Tafel plots of only the acrylic resin (A), the resin containing 1% LIGNO-PANI™ and 0.5% Al (B), the resin containing 1% LIGNO-PANI™ and 10% Al (C), and resin containing 1% LIGNO-PANI™ and 20% Al (D). The Tafel plots indicate a positive shift in the open circuit potential with increasing aluminum concentration except for D. The fact that D shows a decrease in OCP compared to C may indicate that in the B and C a mixture of mechanisms including cathodic protection as well as formation of an oxide layer are at work. The formation of the oxide layer would indeed result in a higher OCP. The mechanism of protection for D, however, may be cathodically dominated due to the higher loading of aluminum resulting in a lower OCP. Corrosion currents decreased as aluminum concentration increased resulting in corrosion rate values of 4.45, 0.28, 0.18 and 0.076 mils per year (mpy) for A, B, C, and D respectively. The significant decrease in corrosion rate is believed to be due to the added protection provided by the preferential oxidation of the aluminum especially in the case of D. Interestingly, values obtained early in the testing cycle indicated high corrosion rates for the aluminum containing samples. It may be possible that the instrument was actually reading the current resulting from the oxidation of the aluminum on the surface of the coating. This is possible due to the bulk conductivity of the entire coating. After two days, however, the corrosion rate decreased to below that of the resin alone. To fully understand both the true corrosion rate as well as the mechanism, another method such as the scanning vibrating electrode technique (SVET) may be needed. This would provide information about the sacrificial nature of the coating.

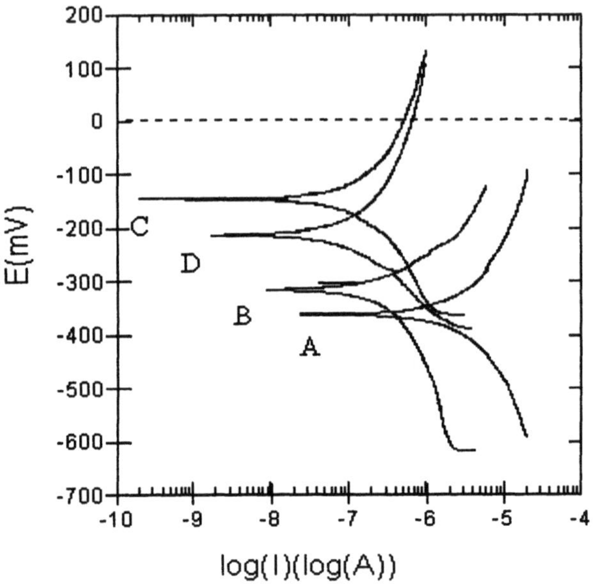

Figure 12. Tafel plots of only the acrylic resin (A) and the resins containing 1% LIGNO-PANI™ as well as 0.5% Al (B), 10% Al (C) or 20% Al (D).

Figure 13 is a photograph of two plates after exposure to salt fog for 230 hours. The panel on the left is coated with only a water-based acrylic resin while the plate on the right is coated with a water-based coating containing Catize™ (Catize™ is trademark of GEOTECH Chemical Co. Tallmadge) - a blend of ICP and metal particles (in this case LIGNO-PANI™ was the ICP of choice).[13] As shown, protection of the scribed area is evident for the Catize™ coating. It is believed that the addition of a more active metal leads to the formation of a galvanic cell that protects the steel surface similar to a zinc rich primer where the zinc acts as a sacrificial metal. The cell produces a net flow of electrons from the coating to the metal surface forcing the metal to remain in the elemental state. If the aluminum is acting as a sacrificial metal, there may be subsequent formation of a layer of aluminum oxide that should provide added barrier protection. Surface analysis is needed for further confirmation. It is presumed that even when the sacrificial metal is completely consumed, the LIGNO-PANI™ will continue to inhibit the corrosion process anodically. Therefore, all three types of protection mechanisms (anodic, cathodic and barrier), although not concurrently, are probably operating in the system.

Figure 13. Photograph of a scribed panel coated with only a water-based acrylic resin containing no Catize™ (left) and one containing a proprietary amount of Catize™ after 230 hours of exposure to salt fog.

3.3. Electrostatic Dissipative Materials

One of the advantages of LIGNO-PANI™ is that it can be easily blended with a variety of resins. This includes coating systems as well as free standing films. This is particularly advantageous in the field of electrostatic dissipation (ESD). ESD is of major importance in many fields including those dealing with sensitive electronic equipment, explosive materials, and those in which a dangerous amount of static electricity is generated. Some ESD technologies rely on substances with conductivity that is dependent upon humidity. This can lead to problems in environments with different levels of humidity. The conductivity of ICPs, however, is not dependent upon humidity. Other ESD technologies rely on conductive particles as additives such as carbon black. Blending of carbon black, however, is difficult and typically degrades the quality of the film. The polymeric nature of ICPs is advantageous in that film integrity is not degraded to such an extent as with carbon black. LIGNO-PANI™ has been blended with a variety of coatings and films. The surface resistivities of these films were measured and plotted as a function of the percent solids of LIGNO-PANI™. Surface resistivity can be measured using an electrometer and a programmable current source using two flat electrodes in contact with the surface separated by a distance equal to their width. The resulting surface resistivity is reported in ohms/square. Typical percolation plots for LIGNO-PANI™ and p-toluenesulfonic acid (p-TSA) doped LIGNO-PANI™ (synthetically different from typical LIGNO-PANI™ in that p-TSA is added during the polymerization) in a water-reducible acrylic resin are shown in Figure 14.

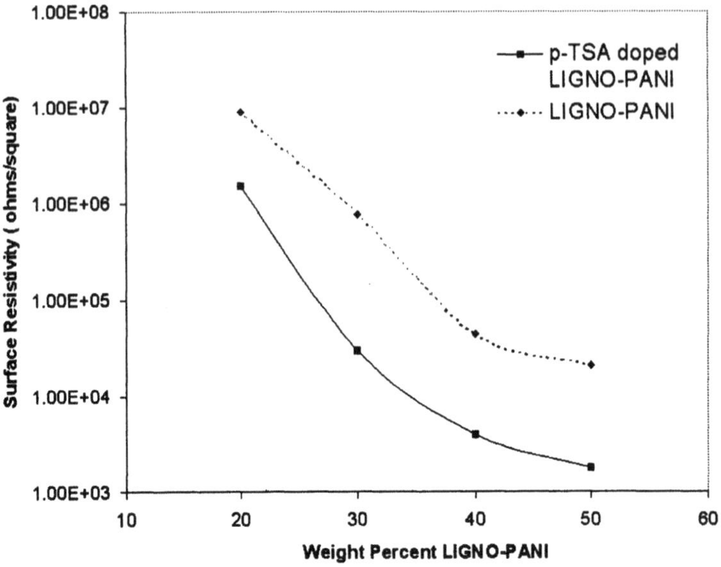

Figure 14. Percolation threshold plots for LIGNO-PANI™ and p-TSA doped LIGNO-PANI™.

It can be seen from the plots that as the concentration of the ICP is increased the resistivity changes in a non-linear fashion approaching some asymptotic point. This asymptotic point is known as the percolation threshold. For LIGNO-PANI™ in this particular acrylic resin it appears that the percolation threshold is between 40 and 50 percent although lower percolation thresholds have been seen in different coatings. Surface resistivity values applicable for ESD can be achieved with as little as 10% LIGNO-PANI™. It can also be seen from the initial slope that a wide range of resistivity values can be achieved precisely at different concentrations. If the slope were much steeper as in the case of carbon black, it would be difficult to accurately control the surface resistivity values within a certain range.

It is also important to note that the minimum surface resistivity values for the typical LIGNO-PANI™ and the p-TSA/LIGNO-PANI™ differ by a full order of magnitude. It might be assumed that this is due to conductivity differences between the two polymers. The conductivity of the p-TSA/LIGNO-PANI™, however, is actually lower when tested on a four-point probe. A possible explanation is that the p-TSA/LIGNO-PANI™ consists of polyaniline chains of greater length than a typical LIGNO-PANI™ due to the aforementioned synthetic difference. This longer chain length leads to more chain-chain interaction resulting in a more continuous network and therefore a higher conductivity.

3.4. Ferromagnetic LIGNO-PANI™ Nanocomposites

Only recently has magnetic conducting polyaniline been reported. The most common method of making magnetic polyaniline involves incorporating iron oxide crystals (such as Fe_2O_3[16] or Fe_3O_4[17]) into the conducting material. These methods, however, are difficult to use and require use of organic solvents. With the incorporation of Fe_2O_3 the magnetic nanoparticles decompose when exposed to strong acids. This is a major problem since most conducting polymers need to be doped with strong acids such as hydrochloric acid to render them conductive. Other methods include the synthesis of co-polymers that are more involved.[18] Finally, the magnetic polymers that are easily synthesized are likely not processable due to their limited solubility.

Ferromagnetic LIGNO-PANI™ (FLP) nanocomposites have been synthesized via a novel unreported procedure that utilizes the lignin grafted to the polyaniline chain. The applications of this nanocomposite include electromagnetic interference shielding,[15] sensing and actuating technologies,[16] and molecular engineering of nanomotors[19] in addition to many applications of nanocomposites exhibiting only magnetism and not conductivity.

Lignosulfonates and tanninsulfonates have been used to make water-soluble ferromagnetic species.[20] In this particular case a ferrous salt was added to a solution of the lignosulfonate and heated with alkali. Various ratios of lignin to ferrous salts as well as various pH values were tested to determine effects of various reaction conditions. Optimum results were obtained when equivalent masses of lignosulfonate and ferrous sulfate heptahydrate, $FeSO_4 \cdot 7H_2O$, were used. The solution was kept basic and heated for 2 hours. This produced a water-soluble ferromagnetic substance that retained its ferromagnetism after successive drying and redissolving.

Since LIGNO-PANI™ contains lignosulfonate that is permanently attached to the polymer chain, a similar reaction can occur. Several ferromagnetic samples have been prepared and studied. Of utmost concern is the type and size of the magnetic crystals generated. X-ray diffraction (XRD) of powdered solids has been used to determine both the crystal type and size.

XRD was performed on FLP that was synthesized using varying amounts of $FeSO_4$. As the amount of $FeSO_4$ was increased crystal type and size changed. XRD provided evidence that at low concentrations of $FeSO_4$ the nanoparticles being formed were actually α-Fe_2O_3. Figure 15 is the XRD of such a sample. The peaks occur at slightly lower angles than the card file. It was found that when the sample was removed from the auto sampler, that the level of the sample was actually lower than required. This would cause the peaks to occur at slightly lower angles. The presence of the α-Fe_2O_3 was unexpected. It is believed that all of the ferrous ions are oxidized to ferric ions. When higher concentrations of $FeSO_4$ were used, no evidence of α-Fe_2O_3 was found in the XRD pattern. Figure 16 is the XRD of a sample synthesized with a higher $FeSO_4$:lignin ratio. This XRD indicates that the only crystalline substance is Fe_3O_4 and not α-Fe_2O_3. It is possible, however, that the peaks for the α-Fe_2O_3 are buried below the noise level.

To determine the percent magnetite in each of the samples, thermogravimetric analysis (TGA) was performed on each of the samples. This effectively allowed us to perform gravimetric determinations since the organic portion of the nanocomposite completely burns away at higher temperatures. The amount of magnetite that should be in the sample based on 100% conversion of all $FeSO_4$ was also calculated. Conductivity was also determined for each of the samples. An interesting anomaly occurred in these calculations. The theoretical percent magnetite was lower than the actual percent observed. A possible explanation which has been alluded to earlier, is that some Fe_2O_3 may also be forming. The formation of the non-conducting Fe_2O_3 would serve as an insulator and could contribute to the lower conductivity of the samples.

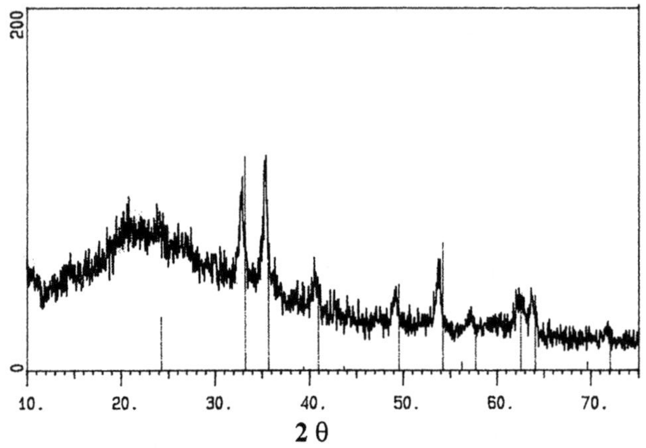

Figure 15. XRD of FLP sample containing α-Fe_2O_3.

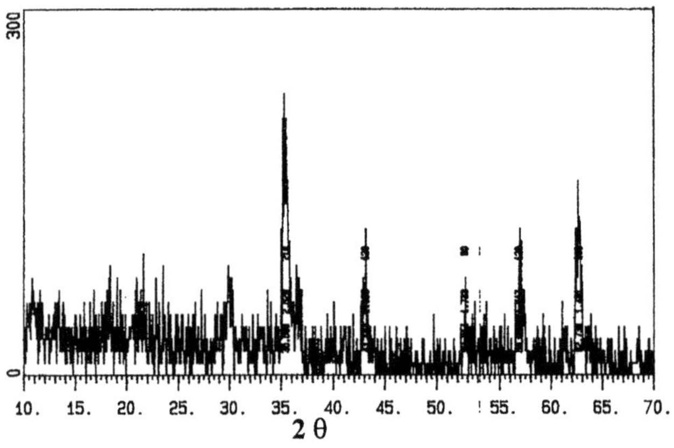

Figure 16. XRD of FLP sample containing magnetite nanoparticles.

Crystal size is also of importance. Special properties are inherent to magnetic crystals in the nanometer range.[15] Therefore, if the crystals that are formed are too large, the composite could lose desirable properties inherent to nanocomposites. Calculations of crystal size were performed using the Scherrer equation:

$$L = \lambda K / \beta \cos \theta$$

where L is the crystallite dimension in angstroms, K is the shape constant, β is the full width at half maximum (FWHM) of the peak in radians in 2θ, and θ is the angle in degrees of the peak. Instrument broadening was accounted for using a quartz standard. β in the Sherrer equation can be defined as the difference between the FWHM of the sample and the standard. Crystal sizes were determined based on the (311) peak. Crystal sizes were found to be from 33 to 45 nm for the magnetite. The crystal size for the hematite in the sample was found to be 105 nm.

Magnetic susceptibilities using a vibrating sample magnetometer or a SQUID have not yet been determined. Measurements, however, have been made using a magnetic susceptibility balance. Since magnetic susceptibility balances cannot measure susceptibilities in the ferromagnetic range directly, an alternate method was used. The samples were diluted with KBr (a diamagnetic salt) to approximately 1-10%. The samples were then tested and the values were converted to a value representative of a neat sample. This gave quantitative evidence of the comparative magnitude of the samples. Table 2 contains results from a series of samples. The highest conductivity obtained for the nanocomposites was 2 S/cm. This is quite high considering that this is a processable material. The highest conductivity observed for one of the highly ferromagnetic composites was 0.2, which is sufficient for some applications. Increasing the ratio of the ferrous salt to LIGNO-PANI™ also resulted in larger crystals. This was particularly evident when a high concentration of OH^- was present. Figure 17 is an XRD of such a sample. The narrow peaks indicate that the crystal size is larger than the previous samples and calculations indicated that they were outside the nanoscale range (i.e. >100 nm).

Table 2. Results for FLPs synthesized with various ratios of LIGNO-PANI™ to $FeSO_4{\cdot}7H_2O$ and at various pH's

Sample	LIGNO-PANI to $FeSO_4{\cdot}5H_2O$ ratio	pH	Conductivity (S/cm)	Magnetic susceptibility	Iron oxide by X-ray
1	3.75 : 1	~ 12	2.00	2.126×10^{-4}	Mixture
2	1.25 : 1	~ 12	0.50	1.686×10^{-3}	Magnetite
3	1 : 2	~ 8	0.24	1.57×10^{-5}	Goethite
4	1.9 : 1	~ 12	0.20	3.22×10^{-4}	Hematite
5	1 : 1	~ 12	0.20	1.11×10^{-3}	Hematite
6	1 : 1	~ 10	0.12	7.8×10^{-3}	Magnetite
7	1 : 2	~ 12	0.12	6.93×10^{-3}	Magnetite
8	0.7 : 1	~ 12	0.07	6.21×10^{-3}	Magnetite
9	1 : 2	~ 5	0.008	1.76×10^{-3}	Mix[a]
10	1 : 2	~ 13	0.006	1.07×10^{-2}	Pure[b]
11	0.19 : 1	~ 12	0	N/A	Magnetite
12	1 : 2	~ 12	0.001	1.17×10^{-2}	Pure[b]
13	1 : 2	~ 12	0.00039	1.28×10^{-2}	Magnetite
14	1 : 2	~ 14	0.00023	3.98×10^{-3}	Mix[a]
15	1 : 2	~ 13	0	1.00×10^{-2}	Magnetite
16	1 : 2	~ 14	0	N/A	Mix[a]

[a] mixture of goethite and magnetite.
[b] pure magnetite large crystals.

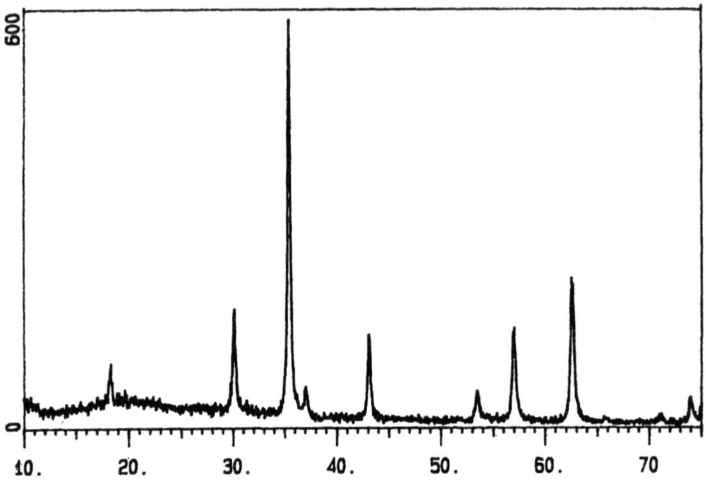

Figure 17. XRD of a LIGNO-PANI™ sample containing large crystals of magnetite.

SEM analysis confirms crystal sizes obtained from the Scherrer equation. SEM also indicates that the crystals cover the entire surface of the LIGNO-PANI™ and are needle-like in shape. Figure 18 is the SEM of the FLP. Small spikes in the nm range can be seen in the micrograph. The micrograph also confirms that the size of these crystals is in the 30-50 nm range. The crystals can be seen over the entire surface of the composite. This might suggest that the crystals are not only associated with the lignin. A Mossbauer spectrometer will allow for further characterization.

Recently, nanocomposites consisting of polyaniline grafted to tannin and magnetite have been synthesized. While magnetic studies have not been completed it appears that the values for magnetic susceptibility will be higher for these composites. As with FLP, an increase in the concentration of $FeSO_4$ leads to an increase in the concentration of the magnetite in the sample, as evaluated by TGA. XRD analysis indicates that there may also be some contamination by hematite in these samples.

Figure 18. SEM of FLP indicating the presence of needle shaped crystals in the nanometer range.

Nanocomposites consisting of magnetite and a polyaniline synthesized by template polymerization with 2-acrylamido-2-methyl-1-propanesulfonic acid (AMPS) grafted to lignin have also been synthesized. These samples appear to behave in a similar manner to the previous samples except that they seem to have better water dispersibility.

4. CONCLUSIONS

Lignin is an abundant renewable resource that is useful in a wide variety of applications. Lignosulfonic acid is particularly useful as a dopant for ICPs. The high solubility of the lignosulfonate gives an ICP that is dispersible in water, many organic solvents and a variety of resins. LIGNO-PANI™ is such an ICP that may have many more applications than typical ICPs that are not dispersible in their conducting forms.

5. ACKNOWLEDGEMENTS

We would like to acknowledge NASA Kennedy Space Center as well as NASA EPSCoR for funding.

6. REFERENCES

1. J. D. Gargulak and S.E. Lebo, in: *Lignin: Historical, Biological, and Materials Perspectives,* edited by W. G. Glasser, R. A. Northey, and T. P. Schultz (ACS Symposium Series 742, ACS, Washington, 1999), pp. 304-320.
2. H. H. Kuhn and A. D. Child, in: *Handbook of Conducting Polymers,* edited by T. A. Skotheim, R. L. Elsenbaumer, and J. R. Reynolds (Marcel Dekker Inc., New York, 1998), pp. 993-1013.
3. M. Sudhakar, P. W. Stoecker, and T. Viswanathan, in: *Recent Research Developments in Polymer Science Vol. 2 Part II,* edited by S. G. Pandali (Transworld Research Network, Trivandrum, India 1998), pp. 173-181.
4. M. Angelopoulos, N. Patel, and J. M. Shaw, Water-soluble polyanilines: properties and applications, *Mat. Res. Soc. Symp. Proc.* **328**, 173-178 (1994).
5. M. Angelopoulos, N. Patel, T. N. Seeger, and J. Gerome, *US Patent* 5,370,825 (1994).
6. L. Sun and S. C. Yang, Solution processable conducting polymer: polyaniline-polyelectrolyte complexes *Mat. Res. Soc. Symp Proc.* **328**, 209-214 (1994).
7. S. Yang, W.M. Chen, and K.S. You, The properties of polyaniline-polyelectrolyte complexes, *Synth. Met.* **84**, 77-78 (1997).
8. T. Viswanathan, *US Patent* 6,059,999 (2000).
9. M. Sudhakar, A. D. Toland, and T. Viswanathan, in: *Semiconducting Polymers,* edited by B. R. Hsieh and Y. Wei (ACS, Washington, 1999), pp. 76-87.
10. D. W. DeBerry, Modification of the electrochemical and corrosion behavior of stainless steels with an electroactive coating, *J. Electrochem. Soc.* **132(5)**, 1022-1026 (1985).
11. W. R. Carmody, An easily prepared wide range buffer series, *J. Chem. Educ.* **38**(11), 559-560 (1961).
12. B. C. Berry, A. U. Shaikh, and T. Viswanathan, pH Dependent electrochemical studies of water-dispersible lignosulfonic acid-doped polyaniline, ACS *Polymer Preprints* **41**, 327-328 (2000).
13. D. C. Trivedi, in: *Conductive Polymers: Synthesis and Electrical Properties,* edited by H. S. Nalwa (John Wiley & Sons Ltd., 1997), pp. 520-535.
14. W.-K. Lu, S. Basak, and R. L. Elsenbaumer, in: *Handbook of Conducting Polymers,* edited by T. A. Skotheim, R. L. Elsenbaumer, and J. R. Reynolds (Marcel Dekker Inc., New York, 1998) pp. 881-920.
15. T. R. Hawkins and S. R. Geer, *US Patent* 5,976,419 (1999).
16. B. Z. Tang, J. W. Y. Lam, and B. Li, Processible nanostructured materials with electrical conductivity and magnetic susceptibility: preparation and properties of maghemite/polyaniline nanocomposite films, *Chem. Mater.* **11**, 1581-1589 (1999).

17. M. Wan and J. Li, Synthesis and electrical-magnetic properties of polyaniline composites, *J. Polym. Sci. Part A.* **36**, 2799-2805 (1998).
18. J. H. Fan, M. X. Wan, and D. B. Zhu, Electrical and magnetic properties of water-soluble conducting polyaniline derivatives, *Solid State Commun.* **110**, 57-62 (1999).
19. B. Z. Tang and H. Xu, Preparation, alignment, and optical properties of soluble poly(phenylacetylene)-wrapped carbon nanotubes, *Macromolecules* **32**, 2569-2576 (1999)
20. W. S. Briggs and N. J. Kjargaard, *US Patent* 4,019,995 (1977).

POLYURETHANES CONTAINING LIGNIN

Hyoe Hatakeyama*

1. INTRODUCTION

Natural and synthetic polymers are indispensable in today's world. Recently, however, it has been found that most synthetic polymers developed from petroleum and coal are not compatible with the environment, since they cannot be included in the natural recycling system.

There are some conflicts between the convenience we enjoy today and its compatibility with nature. It is easy to say that we have to return to nature in order to solve the problems created by man-made materials. Yet, it is difficult for us to comtemplate the loss of the convenient features and materials that science has brought to us through human history.

Accordingly, in the present polymer industry, polymers that are compatible with natural circumstances are the key to the sustainable developments that can keep the rich and convenient life made possible through scientific development. For the development of environmentally compatible polymers, it is essential to understand that nature constructs a variety of materials that can be used for human life. Saccharides have already been used extensively in the food, medical and cosmetic industries. Plant materials such as cellulose, hemicelluloses and lignin are the largest organic resources. However, they are not very well used except for cellulose. Hemicelluloses have not yet been utilized. Lignin, available in large amount, over forty million tons per year in the world, is mostly burnt as fuel which only increases the amount of carbon dioxide in the environment. Thus, the importance of the utilization of lignin has repeatedly been emphasized by many people working in the scientific and industrial fields.

Lignin has been considered as a plant material that is difficult to utilize since modern pulping industries were established. However, the scientific advances have enabled us to understand the molecular features of biomaterials through modern analytical methods such as infrared spectroscopy, nuclear magnetic resonance spectroscopy, thermal and mechanical analysis, and electron microscopy. Now it is time to consider that lignin can be used as half-made raw materials for the synthesis of useful plastics and materials for human life. Major plant components, such as saccharides and lignin, contain highly reactive hydroxyl groups that can be used as reactive chemical sites. As shown in Figure

* Hyoe Hatakeyama, Fukui University of Technology, 3-6-1 Gakuen, Fukui 910-8505, Japan.

1, using such reaction sites, it is possible to convert saccharides and lignin, for example, to gels, membranes, functional polymers, engineering plastics and biodegradable polymers that are environmentally compatible.[1-10]

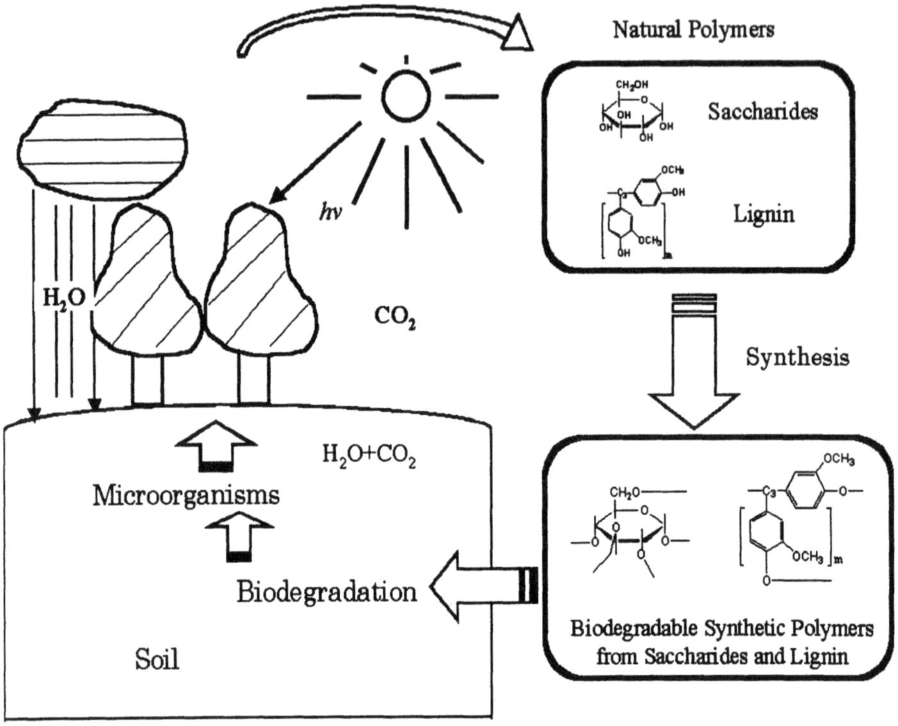

Figure 1. Recycling of environmentally compatible materials from biomass.

This paper reviews the synthesis and physical properties of polyurethanes derived from lignin. It is generally recognized that polyurethane (PU) is one of the most useful three-dimensional polymers because of the many unique features it has. For example, various forms of materials such as sheets, foams, adhesives and paints can be obtained from PU, and their properties can easily be controlled. Consequently, many attempts to use lignocelluloses as raw materials for PU synthesis have been made since natural polymers having more than two hydroxyl groups per molecule can be used as polyols for polyurethane preparation if the polyols from natural polymers can be reacted efficiently with isocyanates.

2. RAW MATERIALS

Various types of industrial lignins were used as raw materials for the preparation of lignin-based PU's in our laboratory. Industrial kraft lignin was commercially obtained from Westvaco Inc.. Hardwood solvolysis lignin (SL) was obtained as a by-product in organosolve pulping of Japanese beech (*Fagus crenata*) with aqueous cresol at 185 °C without an acid catalyst. The SL was provided by the Japan Pulp and Paper Research Institute. Alcoholysis lignin (AL) was provided by Repap Inc. Lignosulfonate (LS) was provided by Nippon Paper Industries Inc. Polyethylene glycol (PEG), polypropylene glycol (PPG) and diphenylmethane diisocyanate (MDI) were obtained commercially.

3. PREPARATION

The following methods are examples for the preparation of lignin-based PU's. Prior to making PU, polyol solutions containing lignin were prepared by dissolving lignins (KL, AL, SL and LS), in polyols such as PEG and PPG. The resulting polyol solutions were mixed with MDI and a plasticizer (PEG or PPG) at room temperature to form precured polyurethanes. The precured polyurethanes were heat-pressed to form PU sheets. To prepare PU foams, the polyol solutions were first mixed with a plasticizer, surfactant (silicone oil), and catalyst (di-^{n}butyltin dilaurate), followed by the addition of MDI. The mixture was vigorously stirred with a droplet of water that was added as a foaming agent. In the above processes, the NCO/OH ratio, the weight of starting materials and the contents of lignin (shown as Lig in the following equations), polyols such as PEG and PPG (shown as PEG in the following equations), and MDI were calculated as follows:

$$NCO/OH = M_{MDI} \times W_{MDI}/(M_{Lig} \times W_{Lig} + M_{PEG} \times W_{PEG})$$

$$W_t\,(g) = W_{Lig} + W_{PEG}$$

$$\text{Lignin content (\%)} = (W_{Lig} / W_t) \times 100$$

$$\text{PEG content (\%)} = (W_{PEG} / W_t) \times 100$$

wherein NCO/OH is the molar ratio of the isocyanate and hydroxyl groups, M_{MDI} the number of moles of isocyanate groups per gram of MDI, W_{MDI} the weight of MDI, M_{Lig} the number of moles of hydroxyl groups per gram of lignin, W_{Lig} the weight of lignin, M_{PEG} the number of moles of hydroxyl groups per gram of PEG, W_{PEG} the weight of PEG, and where W_t is the total weight of lignin and PEG in the PU system.

4. MACROMOLECULAR STRUCTURE

The chemical structure of the prepared polyurethanes is dependent on the plant raw materials. The polyurethanes consist of the core structures of lignin linked by urethane bonding. Figure 2 shows a schematic structure of lignin with urethane block copolymers having three-dimensional networks of urethane bonding that combine with the lignin components. Figure 3 shows a schematic macromolecular bonding. Accordingly, we may assume that lignin structures are essentially included in the network structure of the obtained polyurethanes. Lignin components, which are circled by dotted line, are connected with either polyoxyethylene chains (from PEG) or polyoxypropylene (from PPG) chains by urethane bonding, which are shown as solid lines.

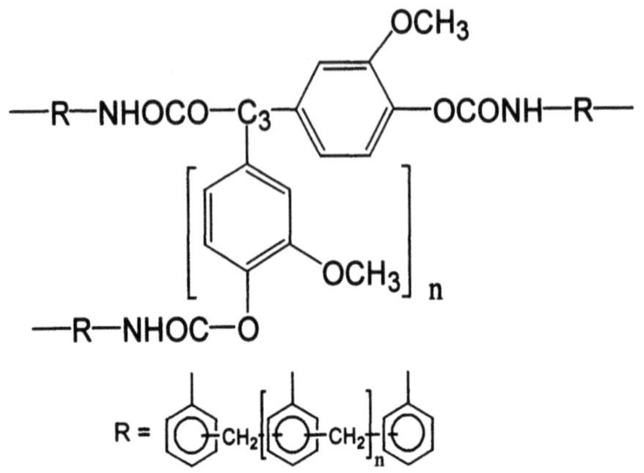

Figure 2. Schematic structure of lignin with urethane bonding.

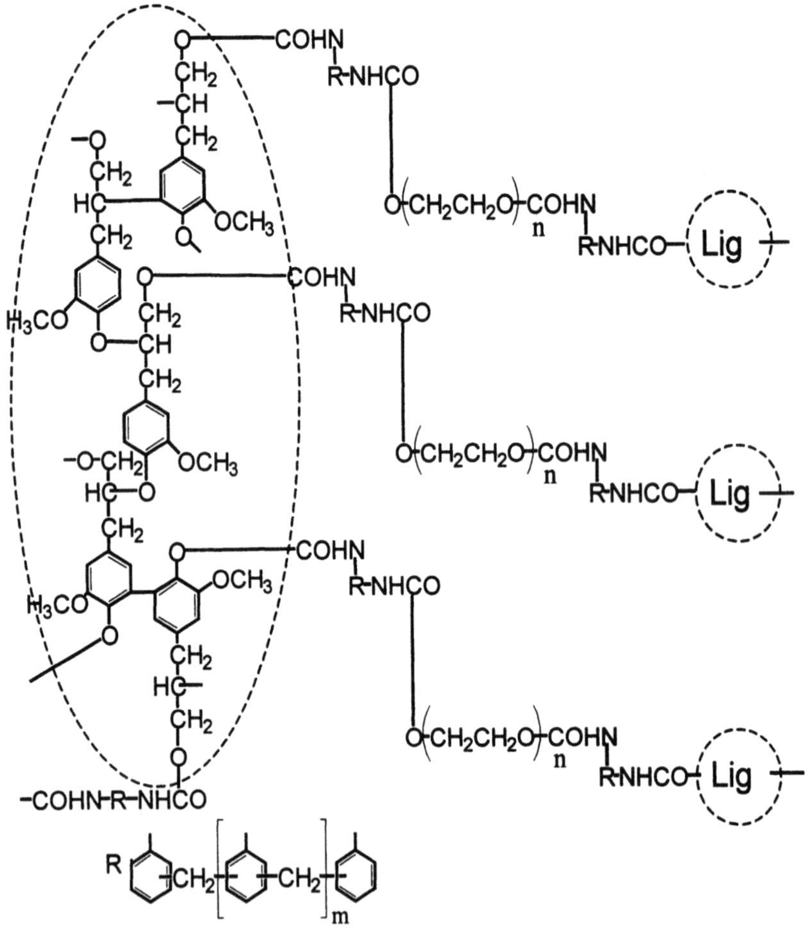

Figure 3. Schematic macromolecular network structure of the obtained polyurethanes, Lig: Lignin structure contained in the PU network.

5. CHARACTERIZATION

5.1. Differential Scanning Calorimetry (DSC)

DSC is a useful technique for studying thermal behaviour of PU derivatives, since PU's are usually cross-linked and amorphous polymers. DSC can be used to identify glass transition and melting temperatures (T_g and T_m) during heating process. Observed endotherms can be used to calculate enthalpy changes occurring with phase transitions. The following procedure is an example of DSC measurements performed at our laboratory. DSC was performed on all PU samples using a Seiko DSC 220C. The temperature program consisted of three stages; heating from 30 °C to 180 °C (holding time 5 min), quenching from 180 °C to 120 °C (holding time 5 min) and finally heating from 120 °C to 180 °C in a nitrogen atmosphere (10 ml/min). The scanning rate was 10 °C /min. The final heating run was used for the analysis. Before the measurements, the samples were dried in an air-oven at 120 °C for 2 h. T_g was defined as the temperature at the point of intersection between the tangents drawn at the point of inflection of the transition and at the flat line of the curve before the transition.

5.2. Thermogravimetry (TG)

TG is a useful technique for studying thermal degradation of PU's. In our laboratory, TG-curves of all the PU samples were recorded using a Seiko TG/DTA 220. The curves were obtained at a heating rate of 10 °C /min in a nitrogen atmosphere (30 ml/min). In order to examine the effect of atmosphere on decomposition behaviour, the TG curves were also obtained in air. The thermal degradation temperature (T_d) of these materials is defined as the temperature at the point of intersection of the tangents drawn from a point before the main decomposition step (i.e. where the curve is horizontal) and from the point of inflection of the main step.[11]

5.3. TG-Fourier Transform Infrared Spectrometry (TG-FTIR)

TG-FTIR is a method for characterizing compounds contained in evolved gases from thermally degraded samples. In our laboratory, TG-FTIR was performed using a Seiko TG 220 thermogravimeter equipped with a JASCO FTIR 7000 spectrometer. TG measurements were carried out at a heating rate of 20 °C/min. Nitrogen and air flow rates were controlled at 100 ml/min. The evolved gases during thermal degradation were simultaneously analyzed by FTIR. In order to obtain one spectrum, data of ten scans with a scan interval of 1 sec were accumulated. Each spectrum was recorded every 30 sec. The resolution of spectra was 1 cm^{-1}.

5.4. Measurements of Mechanical Properties

Mechanical properties are very important in the potential application of PU's. In our laboratory, tensile tests were performed on all PU samples using a Shimadzu Autograph AG 2000-D equipped with data processing. The tests were carried out at room temperature according to the Japanese Industrial Standard JIS K7113. The cross-head speed was 5.0 mm/min and the gauge length was 50.0 mm. The dimensions of the samples were: length 140 mm, width approx. 5 mm and thickness ca. 1 mm. An average of five measurements was done.

Compression tests were carried out using a Shimadzu Autograph AG 2000-D at room temperature. Test specimens were a rectangular solid, and the added stress was less than 10 MPa / min. Compression strength (σ) was defined as the value of the highest point of the linear part in the stress-strain curve. Static Young's modulus (E) was calculated using the initial stage of compression curves. Conditions were in accordance with the Japanese Industrial Standard (JIS Z-2101).

6. SL-BASED PU SHEETS

Figure 4 shows an example of how SL-based PU sheets were prepared. It was found that T_g markedly increased with increasing SL content, and in contrast, that the M_n of PEG only affected the T_g values slightly.[12] As reported elsewhere,[13, 14] T_g values range from 100 to 160 °C, although T_g values of lignins are dependent on measuring methods and thermal histories of the samples. The increase of T_g with increasing SL content clearly suggests that SL acts as a hard segment in the PU networks.

The NCO/OH ratio is another factor that affects T_g values of the PU's. The three-dimensional relationship between T_g values, the SL content and the NCO/OH ratio is shown in Figure 5. As seen from the figure, T_g increases with increasing NCO/OH ratio and increases markedly with increasing SL content. This suggests that the increase in the SL content and the NCO/OH ratio cause an increase in crosslinking in the PU systems. It is also recognized that the effect of the SL content is larger than that of the NCO/OH ratio.

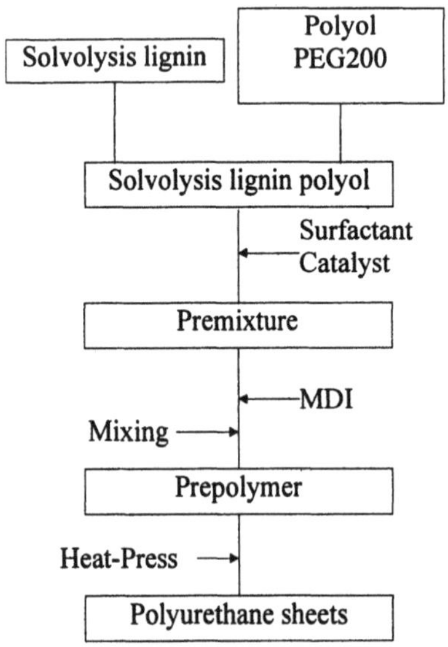

Figure 4. Scheme for the preparation of SL-based PU sheets.

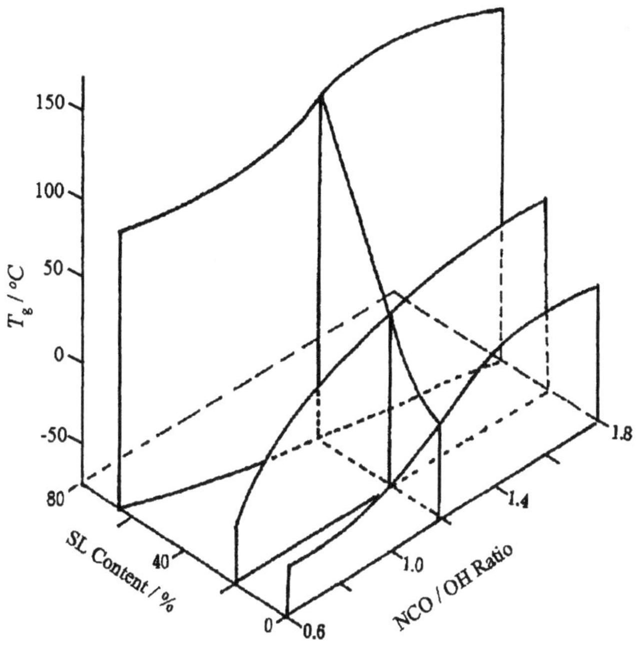

Figure 5. The three dimensional relationship between T_g, the SL content and the NCO/OH ratio.

It was also found that the T_d of the PU decreased with increasing SL content, suggesting that the decomposition of the PU occur at the lower temperature side, where the dissociation between the isocyanate groups and the phenolic OH groups in SL is dominant in the decomposition.

Figure 6 shows the three-dimensional relationship between T_d, the SL content and the NCO/OH ratio. As seen from the figure, T_d is dependent on the SL content in the PU's regardless of the NCO/OH ratio.

The above results suggest that the thermal properties of SL-based PU's can be controlled by changing the SL and PEG contents at a certain NCO/OH ratio. T_g of SL-based PU's is governed mainly by the SL content. This strongly suggests that the introduction of the rigid phenyl group into the main chain of PU's increases the rigidity of the SL-based PU's. The decomposition of the SL-based PU's is markedly dependent on the content of SL. The other factors such as the NCO/OH ratio and M_n of PEG are less significant compared with the SL content. This suggests that the dissociation between the phenolic OH groups and the isocyanate groups in PU's may be the major factor in the thermal decomposition process of the SL-based PU.[12]

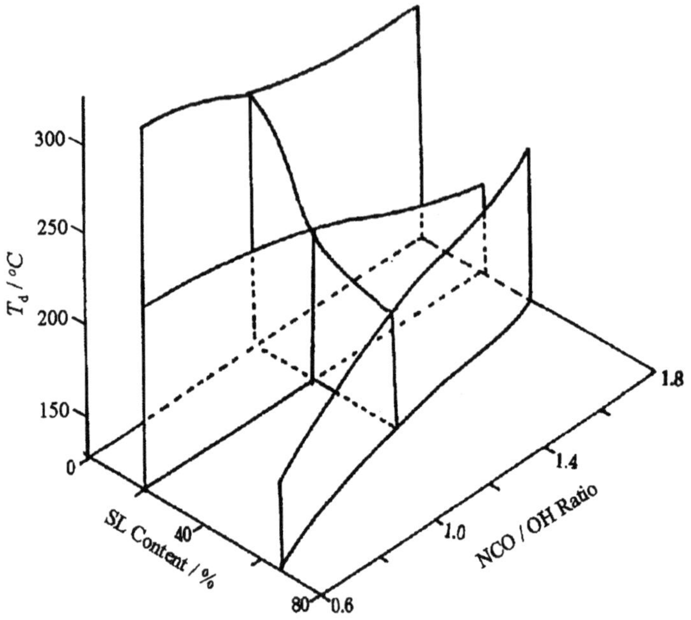

Figure 6. The three-dimensional relationship between T_d, the SL content and the NCO/OH ratio.

7. KL -BASED PU SHEETS

In order to study the influence of the molecular weight of lignin on the molecular motion of KL-based PU's, KL was fractionated by a successive precipitation method. The average molecular weights of the fractionated KL, Fractions 1 to 5, were measured by gel permeation chromatography (GPC)[15] and the results listed in Table 1.

Table 1. Number average molecular weight (M_n), weight average molecular weight (M_w), molecular weight distribution (M_w/M_n) and hydroxyl group content of fractionated KL.

Fraction	M_n x 10^{-3}	M_w x10^{-3}	M_w/M_n	Hydroxy group content
1	1.84	8.02	4.35	5.11
2	1.53	3.28	2.15	6.37
3	1.42	3.20	2.25	4.73
4	1.01	1.92	1.90	4.72
5	0.86	1.18	1.37	4.49

The changes of T_g's plotted against KL contents for the PU's from Fractions 3 and 5 are shown in Figure 7. Although the T_g values are somewhat scattered, it can be seen that the T_g values of the PU's from Fraction 5 are higher than those of PU's from Fraction 3. The number average molecular weights (M_n) of Fractions 3 and 5 are 1420 and 860, respectively. At the same time, hydroxyl group contents of Fractions 3 and 5 are almost the same. It is considered that the length of PU chains consisting of PEG and MDI between lignin molecules becomes shorter when lignin molecules having a low molecular weight are introduced in PU chain networks, at the same KL contents. The changes of T_g's were plotted against KL molar contents, in order to study the substantial effect of molecular weight of KL on the molecular motion of PU's.[15] In this case, the PU chain lengths between lignin molecules are the same when the same amounts of mole % of KL are introduced into PU molecules. This suggests that molecular weight of KL does not affect the PU molecular motion in the molecular weight region shown in Table 1.

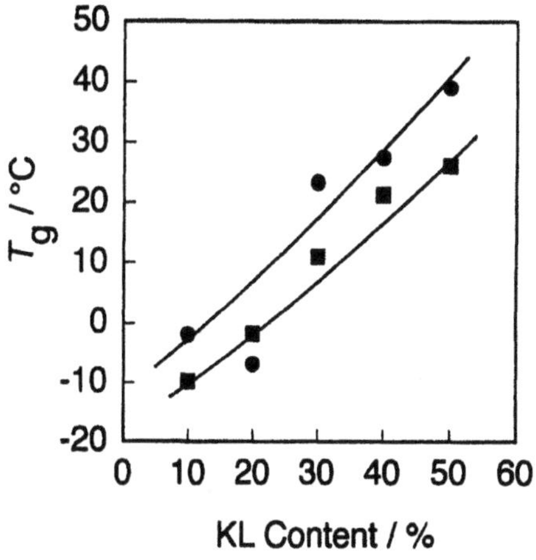

Figure 7. The relationship between glass transition temperatures (T_g's) and KL contents in PU's, Fraction 3 (■) and Fraction 5 (●).

The temperature differences between initial glass transition temperature and final glass transition temperatures (ΔT), and the heat capacity differences between glassy state and rubbery state (ΔC_p) were also calculated for the PU's. Figure 8 shows the changes of ΔT and ΔC_p plotted against KL contents for the PU's from Fractions 3 and 5. As shown in Figure 8, ΔT's increase with increasing KL contents. This suggests that the distribution of units for main chain motion becomes broad due to the introduction of KL molecules into PU chains. The difference between ΔT values of PU's from Fractions 3 and 5 is not prominent. ΔC_p's decrease with increasing KL contents, suggesting that molecular motion particularly in rubbery state is restricted at higher KL contents i.e. at higher cross-linking densities. It is known that ΔC_p decreases with increasing T_g's of ordinary polymers.[16] It seems reasonable that ΔC_p values of PU's from Fraction 5 are lower than those of PU's from Fraction 3, since T_g values of PU's from Fraction 5 are higher than those of PU's from Fraction 3.

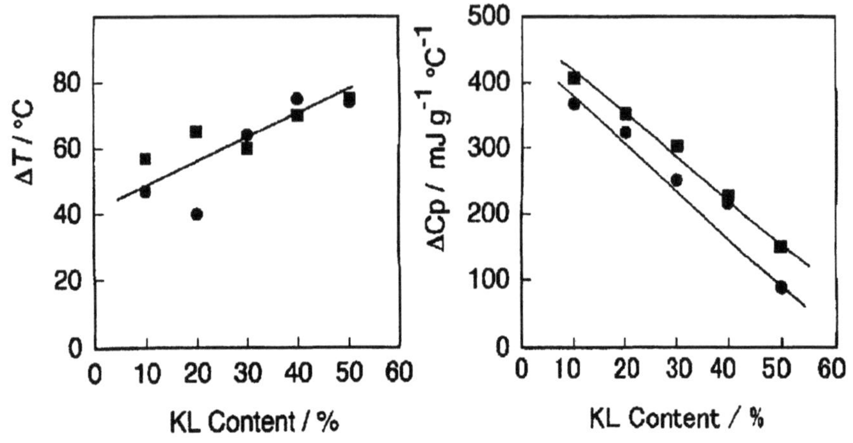

Figure 8. Changes of ΔT and ΔC_p plotted against KL content for the PU's from Fraction 3 (■) and Fraction 5 (●).

Figure 9 shows the changes of T_d's plotted against KL contents for the PU sheets from Fractions 3 and 5. T_d's decrease slightly with increasing KL contents. The same tendency has already been found in thermal degradation of PU's from cresol-solvolysis lignin (SL).[12] As shown in Figure 9, the difference between T_d's of the PU's from Fractions 3 and 5 is not prominent. This suggests that the molecular weight of KL does not affect the thermal degradation, if the molecular weight region is within a certain extent. It can be concluded that in order to change thermal properties such as T_g, ΔC_p, ΔT and T_d, it is sufficient to change the content of lignin, and unnecessary to change the molecular weight of lignin.

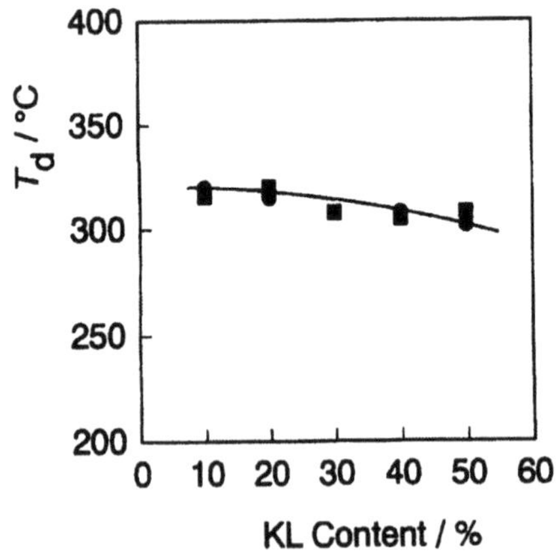

Figure 9. The relationship between thermal decomposition temperatures (T_d's) and KL contents in the PU's, Fraction 3 (■) and Fraction 5 (●).

It is also recognized that the tensile properties are markedly affected by the state of samples at the temperature where tensile tests were carried out. As shown in Figure 7, T_g values of the samples change from ca. - 15 °C to ca. 40 °C. This indicates that some samples are in the glassy state and others are in the rubbery state at 25 °C. Therefore, values for tensile properties such as elongation at break (ε), strength at break (σ), and Young's modulus (E) were plotted against T_g of each sample, in order to remove the effect of T_g on tensile properties (Figure 10). It is seen that the σ and E values are highly dependent on T_g's of samples. However, the difference between σ and E values of PU's from Fractions 3 and 5 is not prominent.

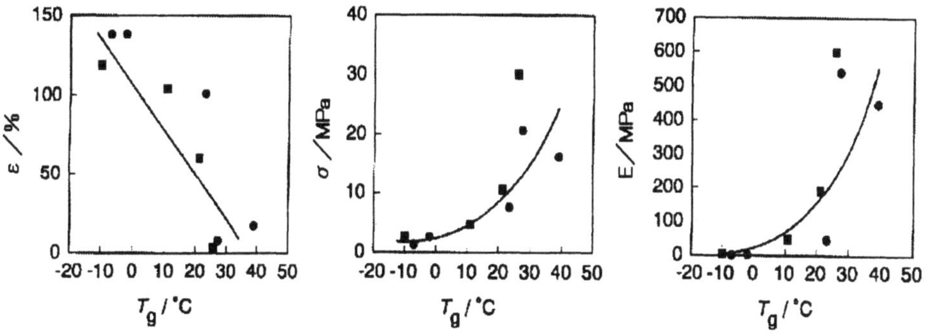

Figure 10. Changes of tensile strength at break (σ), elongation (ε) and Young's modulus (E) plotted against T_g for the PU's from Fraction 3 (■) and Fraction 5 (●).

Figure 11 shows the stacked FTIR spectra of evolved gases at various temperatures. The main peaks observed for samples were as follows: wave number (assignment): 1126 cm^{-1} (C-O-), 1260 cm^{-1} (-C(=O)-O-C-), 1517 and 1617 cm^{-1} (C=C), 1718 cm^{-1} (C=O), 2345 cm^{-1} (CO_2, NO_2), 2980 (C-H) and 3700 cm^{-1} (H_2O).

As shown in Figure 11, evolution of CO_2 gas is observed in the initial stage of thermal degradation of the PU's in the temperature range from ca. 200 to 350 °C. In this temperature range, the sample weight decreases from 100% to 70 %. Other gases are evolved in a higher temperature region from ca. 350 to 600 °C, where sample weight decreases from 70 to 30 %.

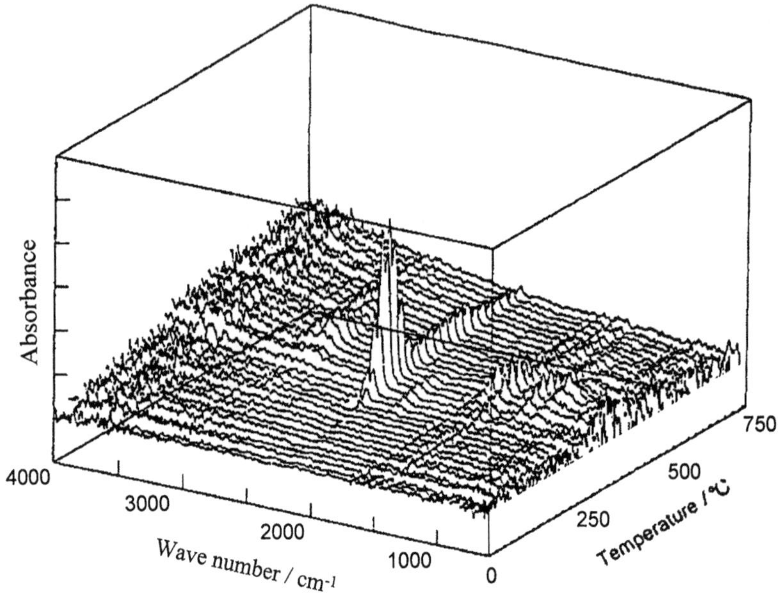

Figure 11. Stacked FTIR spectra of gases evolved during thermal degradation of PU containing 50 % KL.

It is considered that the degradation mechanism of PU's is different from that of the raw materials, since the end-groups will affect markedly the degradation mechanism. Characteristic absorption peaks can be observed for each sample and the intensities of those peaks change with temperature. Figure 12 shows the changes of intensities of the peaks for PU containing 50 % KL.

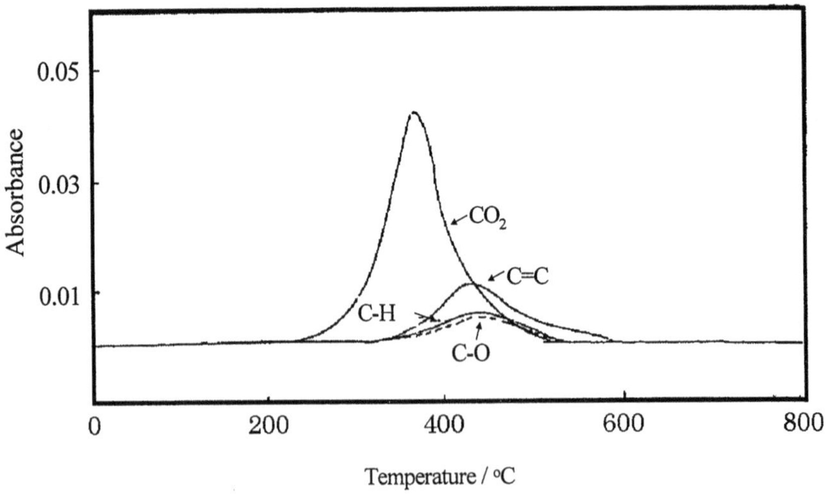

Figure 12. FTIR spectra (absorbance versus temperature) at various wave numbers for gases evolved from PU containing 50 % KL.

The evolution of CO_2 gas is one of the major obvious characteristics of PU samples. The evolution of CO_2 gas was also observed in KL.[17] However, the shape of its evolution curve is different from that of PU containing 50 % KL. In addition to this, KL shows apparent evolution of compounds with alkyl groups (1126 cm^{-1}), which can not be observed for PU containing 50 % KL. The above observation suggests that the degradation mechanism of KL in PU's is different from that of KL itself, and also that the evolution of CO_2 and compounds having -CH- groups is due to the degradation of the end-groups.

8. AL-, KL- AND LS-BASED PU FOAMS

Rigid PU foams were prepared from AL, KL and LS. It was found that their glass transition temperature (T_g) and degradation temperature (T_d) were mostly dependent on the content of AL, KL and LS. As shown in Figure 13, T_g increases with increasing content of the lignins, since phenyl groups work as hard segments in PU networks regardless of the kinds of lignin.

As shown in Figure 14, T_d decreased slightly with increasing content of lignins, since the dissociation of urethane bonds between isocyanate groups and phenolic hydroxyl groups occurs in the temperature region lower than the dissociation of urethane bonds between isocyanate groups and alcoholic hydroxyl groups.[18]

Figure 15 shows the relationship between compression strength (σ_c), modulus (E_c) and apparent density (ρ) of the PU's from AL, KL and LS. As seen from the figure, σ_c and E_c of the PU's increase with increasing ρ values. The effect of lignin contents on σ_c and E_c is apparent in the lignin contents ranging from ρ = ca. 0.05 to 0.12 g/cm^3, although it is necessary to consider that the PU's from AL and KL have ρ values around 0.05 and that the PU's from LS have ρ values from 0.1 to 0.12 g/cm^3. The above results suggest that mechanical properties of the PU's are markedly dependent on ρ values of the PU's from AL, KL and LS.

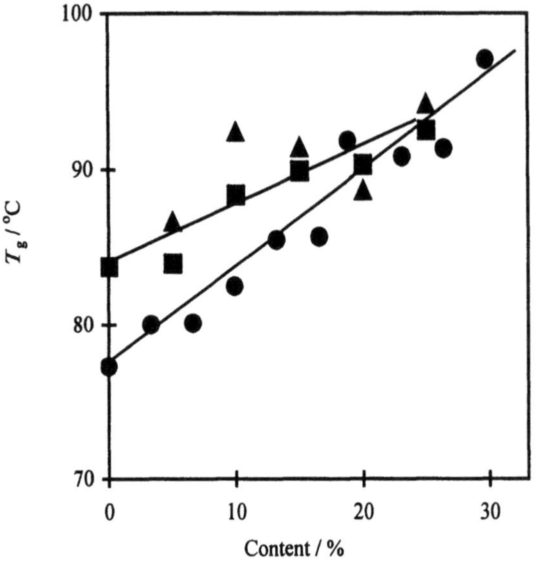

Figure 13. Relationship between glass transition temperatures (T_g's) and of AL, KL and LS content; ● LS-PU, ■ KL-PU and ▲ AL-PU.

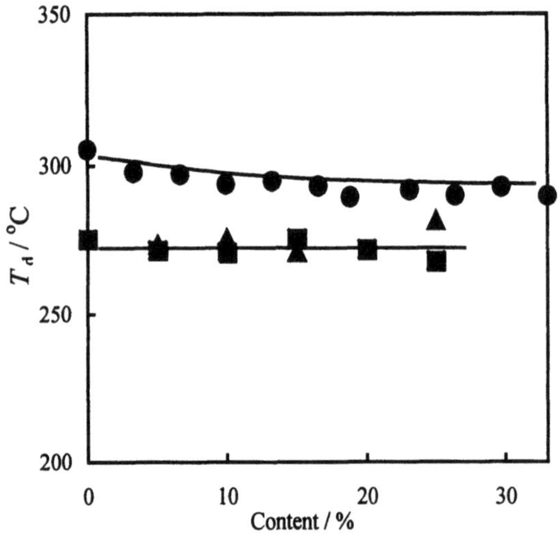

Figure 14. Relationship between thermal degradation temperatures (T_d's) and of AL, KL and LS contents; ● LS-PU, ■ KL-PU and ▲ AL-PU.

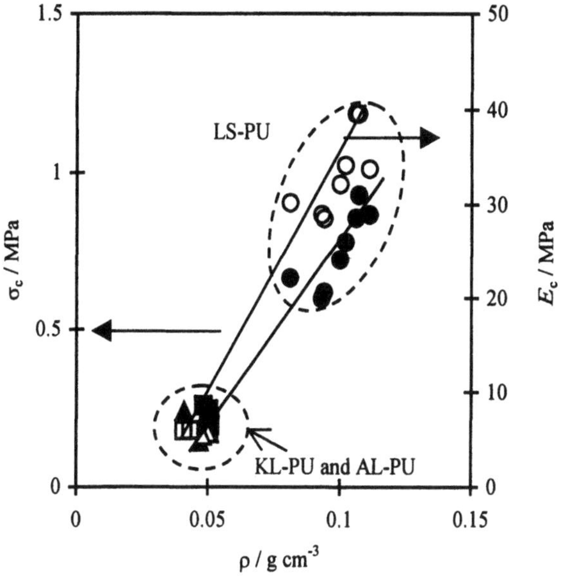

Figure 15. The relationship between compression strength (σ_c), modulus (E_c) and apparent density (ρ) of PU's from AL, KL and LS; ● σ(LS-PU) O *E*(LS-PU), ■ σ(KL-PU) □ *E*(KL-PU), ▲ σ(AL-PU) Δ *E*(AL-PU).

9. CONCLUSIONS

From the above results, we may conclude that lignins can be used as very useful components of rigid polyurethane sheets and foams. Lignins efficiently act as hard segments in the polyurethanes. It can also be concluded that the influence of lignin on the molecular motion of the polyurethanes is greater than that of other factors such as the NCO/OH ratio and molecular weight of polyols such as PEG. By introducing polyols such as PEG and PPG into the polyurethanes, it is possible to control the flexibility and rigidity of the polyurethanes. The thermal and mechanical properties of the polyurethanes can be controlled by changing the amounts of lignin and polyol.

10. ACKNOWLEDGEMENT

The author is grateful to Nippon Paper Industries Inc., Japan Pulp and Paper Research Institute, Westvaco Inc. and Repap Inc., for providing the lignin samples.

11. REFERENCES

1. V. P. Saraf and W. G. Glasser, Engineering plastics from lignin. III. structure property relationship in solution cast polyurethane films, *J. Appl. Polym. Sci.* **29,** 1831-1841 (1984).
2. V. P. Saraf W. G. Glasser, Engineering plastics from lignin. VI. structure property relationship of PEG-containing polyurethane networks, *J. Appl. Polym.* Sci. **30,** 2207-2224 (1985).
3. H. Hatakeyama, S. Hirose, K. Nakamura, and T. Hatakeyama, in: *Cellulosics: Chemical, Biochemical and Material Aspects*, edited by J. F. Kennedy, G. O. Phillips and P. A. Williams (Ellis Horwood, Chichester, 1993), pp. 524-536.

4. N. Morohoshi, S. Hirose, H. Hatakeyama, T. Tokashiki, and K. Teruya, Biodegradability of polyurethane foams derived from molasses, *Sen-i Gakkaishi* **51**, 143-149 (1995).
5. H. Hatakeyama, S. Hirose, T. Hatakeyama, K. Nakamura, K. Kobashigawa, and N. Morohoshi, Biodegradable polyurethanes from plant components, *J. Macromol. Sci. Pure Appl. Chem.* **A32**, 743-750 (1995).
6. M. J. Donnely, Polyurethanes from renewable resources. IV-properties of linear, crosslinked and segmented polymers from polytetrahydrofuran diols and their glucosides, *Polym. Int.* **37**, 297-314 (1995).
7. K. Nakamura, Y. Nishimura, P. Zetterlund, T. Hatakeyama, and H. Hatakeyama, TG-FTIR studies on biodegradable polyurethanes containing mono- and disaccharide components, *Thermochmica Acta* **282/283**, 433-441 (1996).
8. P. Zetterlund, S. Hirose, T. Hatakeyama, H. Hatakeyama, and A.-C. Albertsson, Thermal and mechanical properties of polyurethanes derived from mono- and disaccharides, *Polym. Int.* **42**, 1-8 (1997).
9. H. Hatakeyama, K. Kobashigawa, S. Hirose, and T. Hatakeyama, Synthesis and physical properties of polyurethanes from saccharide-based polycaprolactones, *Macromol. Symp.* **130**, 127-138 (1998).
10. T. Hatakeyama, T. Tokashiki, and H. Hatakeyama, Thermal properties of polyurethanes derived from molasses before and after biodegradation, *Macromol. Symp.* **130**, 139-150 (1998).
11. T. Hatakeyama and F. X. Quinn, in: *Thermal Analysis* (John Wiley & Sons, Chichester, 1994).
12. K. Nakamura, T. Hatakeyama, and H. Hatakeyama, Thermal properties of solvolysis lignin-derived polyurethanes, *Polym. Adv. Technol.* **3**, 151-155 (1992).
13. H. Hatakeyama, K. Nakamura, and T. Hatakeyama, Studies on factors affecting the molecular motion of lignin and lignin-related polystyrene derivatives, *Pulp Paper Mag. Can.* **6**, TR 105-110 (1980).
14. T. Hatakeyama, K. Nakamura, and H. Hatakeyama, Studies on heat capacity of cellulose and lignin, *Polymer* **23**, 1801-1804 (1982).
15. J. Nakano, Y. Izuta, T. Orita, H. Hatakeyama, K. kobashigawa, K. Teruya, and S. Hirose, Thermal and mechanical properties of polyurethanes derived from Fractionated Kraft lignin, *Sen-i Gakkaishi* **50**, 416-422 (1997).
16. T. Hatakeyama and H. Hatakeyama, Effect of chemical structure of amorphous polymers on heat capacity difference at glass transition temperature, *Thermochim. Acta* **267**, 249 (1995).
17. S. Hirose, K. Kobashigawa, Y. Izuta, and H. Hatakeyama, Thermal degradation of polyurethanes containing lignin studied by TG-FTIR, *Polym. Int.* **47**, 247-256 (1998).
18. J. H. Saunders and K.C. Fisch, in: *Polyurethanes, Chemistry and Technology in High Polymers, Vol. XV* (Interscience Publishers, New York, 1962), pp.103.

LIGNINS AS MACROMONOMERS FOR POLYESTERS AND POLYURETHANES

Alessandro Gandini*, Mohamed Naceur Belgacem, Zhao-Xia Guo and Suzelei Montanari

1. INTRODUCTION

Lignins from different sources, namely papermaking processes or biomass refineries vary widely in terms of structure, average molecular weight and polydispersity as a function of both the actual species used and their mode of isolation. However, they all share two common features; they are always in the form of solid particles and they all bear phenolic and aliphatic hydroxy groups, albeit in different proportions. Hence, the utilization of lignins in polymeric materials either as fillers, blends or as macromonomers can be applied in principle to all samples, provided that their specific characteristics have been carefully assessed.[1-6]

Our laboratory has been engaged for many years in the synthesis, characterization and possible applications of polymeric materials derived from renewable resources.[7] Within this broad research strategy, lignins have played an important role in the realms of both fillers and macromonomers. The studies related to the former aspect deal with the determination of the surface energy of samples of various origins[8] and with the inclusion of these materials into different matrices such as printing inks, paints and polyurethanes.[9] Specific improvements in rheological and thermal properties have been made and work is still in progress in these contexts. Other studies on this topic have been reviewed recently[2] and the use of lignins in macromolecular blends[2] is treated elsewhere in this book.

In chemical terms, the use of lignins as macromonomers has been approached systematically through the most logical and general strategy that consists of exploiting the ubiquitous presence of the OH groups in all lignins. The idea of condensing alkenyl functions into the lignin macromolecules to prepare structures which can be polymerized by chain reactions has, in our mind, major drawbacks related both to the difficulty of synthesizing mono-alkenyl lignin molecules and to the intrinsic retarding role of the phenolic moieties in free radical processes.[1, 2] It seems, therefore, more reasonable to call upon the use of both aliphatic and phenolic OH groups in polycondensation reactions with appropriate

* Alessandro Gandini, Ecole Française de Papeterie et des Industries Graphiques (INPG), BP 65, 38402 St. Martin d'Hères, France.

Chemical Modification, Properties, and Usage of Lignin
Edited by Thomas Q. Hu, Kluwer Academic/Plenum Publishers, 2002

co-monomers bearing complementary functions capable of yielding ester, urethane or ether groups. Alternatively, the same OH groups can be exploited as a source of chain extension, typically by oxypropylation and the use of the ensuing polyols as macromonomers. A considerable amount of research has been devoted to both these approaches within the last couple of decades.[2, 4, 6]

One notable exception to the use of reactions involving OH groups is the addition of lignins to typical phenol-formaldehyde resins where the chemical incorporation of lignins occurs through the classical reaction of the unsubstituted sites in their aromatic rings.[2, 4, 6]

The purpose of the present chapter is to bring up to date our contribution to the preparation of polyesters and polyurethanes based on different lignins, including either direct polycondensations or oxypropylation to convert first the solid lignin into a liquid polyol. Our working hypothesis when we started this type of investigation was that, independent of the origin and treatment associated with a given lignin, two features could be considered common to all samples; one based on structural considerations, viz. the presence of OH groups and the other based on chemical behavior, viz. the involvement of both phenolic and aliphatic homologues in the reactions envisaged.

2. LIGNINS USED IN THIS WORK

Three different samples of lignins were prepared /used throughout our studies, namely a conventional kraft lignin from *Pinus pinaster* (KL), an oxygen H_2O/acetone organosolv lignin[10] from *Picea excelsa* (OOL) and an organosolv hardwood lignin (OL) obtained according to the ALCELL process.[11] OOL was obtained under the following conditions:

Wood/white liquor = 1/10; temperature = 120 °C; O_2 pressure = 12 bars; cooking time = 5 hours.

The characterization of these different lignin samples included the determination of the relative abundance of the various OH groups, the molecular weight and the polydispersity index. Table 1 gives the data related to these parameters. Other features, such as the ash content, the percentage of methoxy groups, the residual sugars, the elemental analysis data and the detailed inspection of the FTIR and ^{1}H and ^{13}C NMR spectra were also examined.

Table 1. The relevant characteristics of the lignins used in this work

	Lignins		
	KL	OOL	OL
M_n	1220	1400	700
M_w	11880	4100	1700
M_w/M_n	9.7	2.9	2.4
Primary aliphatic OH/C9	0.39	0.55	0.23
Secondary aliphatic OH/C9	0.27	0.40	0.14
Aromatic OH	0.78	0.41	0.63
Total OH/C9	1.44	1.36	1.00
Average OH/lignin molecule	10	12	6

The use of model compounds was an integral part of some of our studies and will be dealt with in the appropriate sections.

3. POLYESTERS

This investigation called upon the KL and OOL lignins in polycondensations leading to cross-linked polyesters, because the complementary monomers used were always bifunctional reagents and all lignin macromolecules bore more than two OH groups. In preliminary studies,[12-14] the lignins were treated with either of the two diacyl dichlorides bearing a stiff aromatic bridge (terephthaloyl chloride) or a flexible oligomethylene chain (sebacoyl chloride). Reactions conducted in polar solvents at low temperatures in the presence of a proton acceptor gave high yields of cross-linked polyesters whose FTIR spectra indicated unambiguosly that both the phenolic and the aliphatic OH groups of the lignins had participated in the polyesterification reactions. Model reactions involving the same acyl chlorides and a mixture of 1,4-butanediol and hydroquinone confirmed the reaction of the two types of OH functions. Bulk polycondensations with the same reagents were also carried out in which the solvent was replaced by oligo(ethylene glycol)s (PEG) which were good solvents for the lignins used and played the complementary role of co-monomers. Whereas the polyesters obtained with the first system had properties that could only be varied by the initial monomer ratio and the type of acyl chloride, the second approach was more flexible because we could also change the proportion and the molecular weight of the macrodiol. A wide range of products was thus obtained with T_g varing from –60 to 150 °C. With the second system, it was possible to obtain the maximum yields of cross-linked polyesters when the initial [COCl]/[OH] ratio was close to unity, indicating that virtually *all* the lignin OH had indeed taken part in the polycondensation.

The continuation of this work consisted first of replacing the macrodiols with solvents possessing a similar structure, namely di- and tetraglyme. The homogeneous reactions between sebacoyl chloride and the lignins were carried out at 120 °C in the absence of a catalyst. In these systems, up to 30% by weight of lignins could be used. The yields of cross-linked polyesters were close to 90% when stoichiometric reagents were adopted. Figure 1 shows the FTIR spectrum of one such polyester. The relevant feature here is the presence of two carbonyl peaks at 1738 and 1765 cm^{-1}, arising from the entirely aliphatic and aliphatic aromatic ester functions respectively, providing again a clear proof of the involvement of both types of lignin OH in the condensation reactions.

In order to widen the scope of these novel systems, we proceeded to study a series of polyesterifications in which PEGs were introduced as the third monomers. The proportions of lignin with respect to PEG were varied between 100 and 35%, while the initial condition [COCl]/[OH] = 1 was maintained. A typical result, reflecting the behavior of all these polymerizations is described in detail: with KL/PEG_{300} = 55% and sebacoyl chloride in diglyme, a reaction conducted at 120 °C for three hours under a nitrogen atmosphere gave rise to 79% of pyridine-insoluble, cross-linked copolyester with a T_g of –42 °C.

The influence of an increasing proportion of lignin on the properties of the ensuing polymers was clearly illustrated by a corresponding increase in T_g and a better thermal stability. The latter feature is illustrated in Figure 2 which also shows that the higher the proportion of lignin introduced in the initial monomer mixture, the higher the percentage of graphitization at 700-800 °C. This is an interesting result in terms of the possible application of these materials in domains where fire resistance (intumescence) is required.

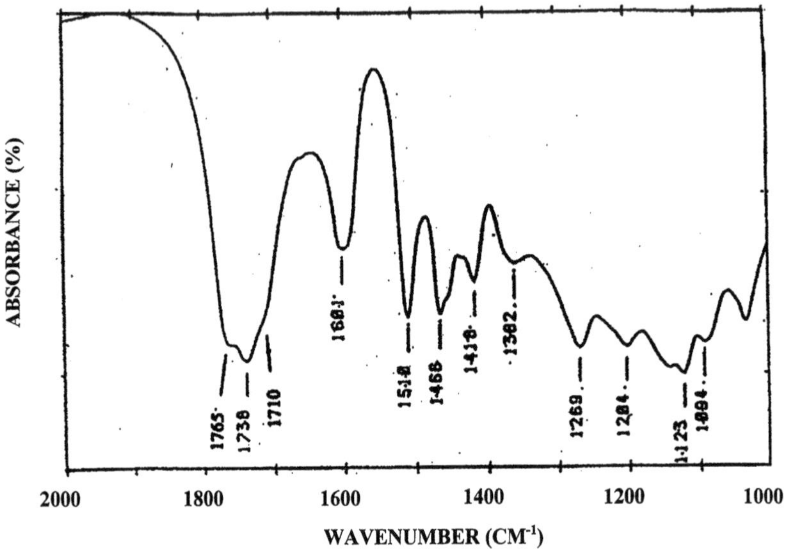

Figure 1. FTIR spectrum of KL-sebacate.

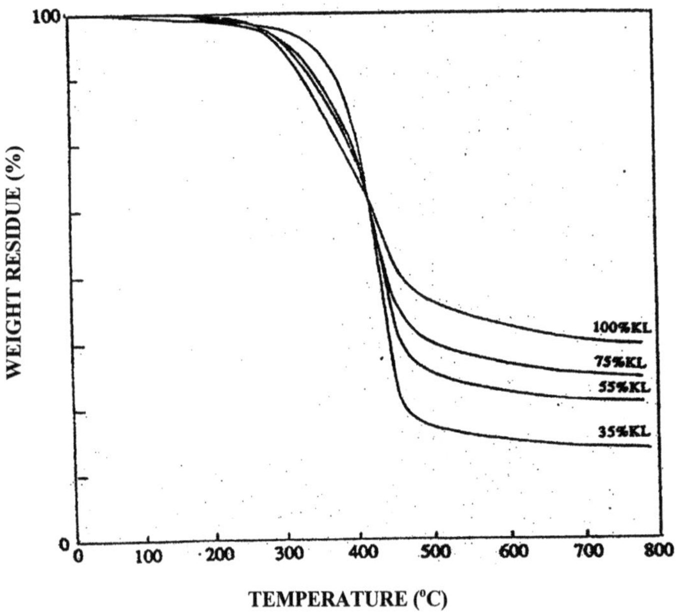

Figure 2. TGA thermograms of four KL-PEG_{300}-sebacates with different KL contents.

The dynamic mechanical properties and the morphology of these copolyesters were also examined. For this purpose, the polycondensations were carried out after pouring the reaction mixture in between two teflon plates in order to prepare membranes of different thicknesses. The reactions were prolonged for 24 h at 120 °C to ensure complete cross-linking. Figure 3 shows the DSC thermograms of three products as a function of the relative KL content for the systems involving PEG_{300}, KL and sebacoyl chloride, compared with that of the linear polyester obtained without KL.

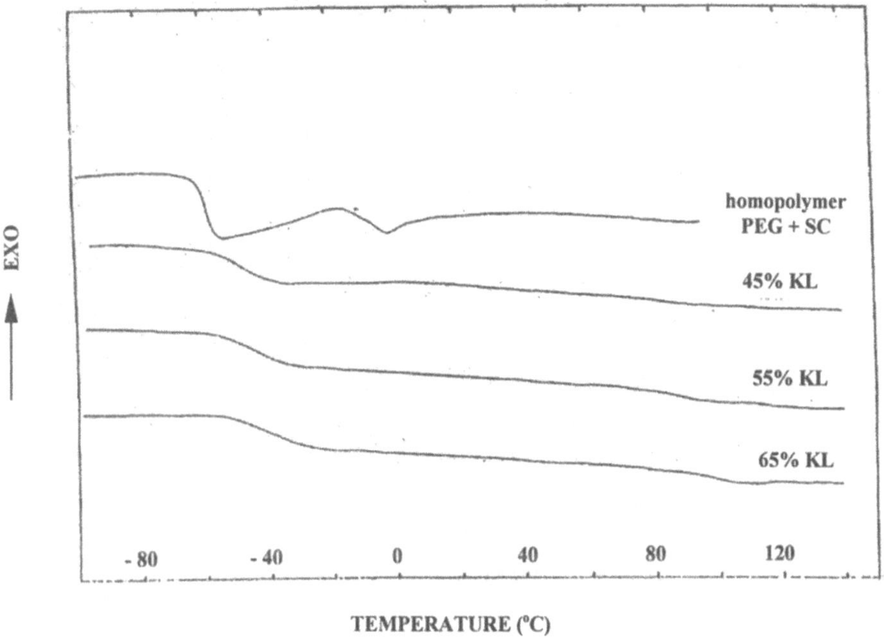

Figure 3. DSC thermograms of four KL-PEG_{300}-sebacates with different KL contents.

The variation of the storage modulus (measured at 0.1 Hz) of these copolyesters, as a function of temperature, is shown in Figure 4. The progressive increase in lignin content is reflected in the corresponding displacement in the drop in modulus (related to the glass transition) and in the corresponding increase in the modulus of the rubbery plateau. The variations of tan δ for the same samples are shown in Figure 5. The α transition is related to the glass transition, whereas the β and γ transitions are due to the localized segmental motions of the -CH_2- sequence in the sebacoyl moiety and of the -CH_2-CH_2-O- sequences in the PEG's, respectively.

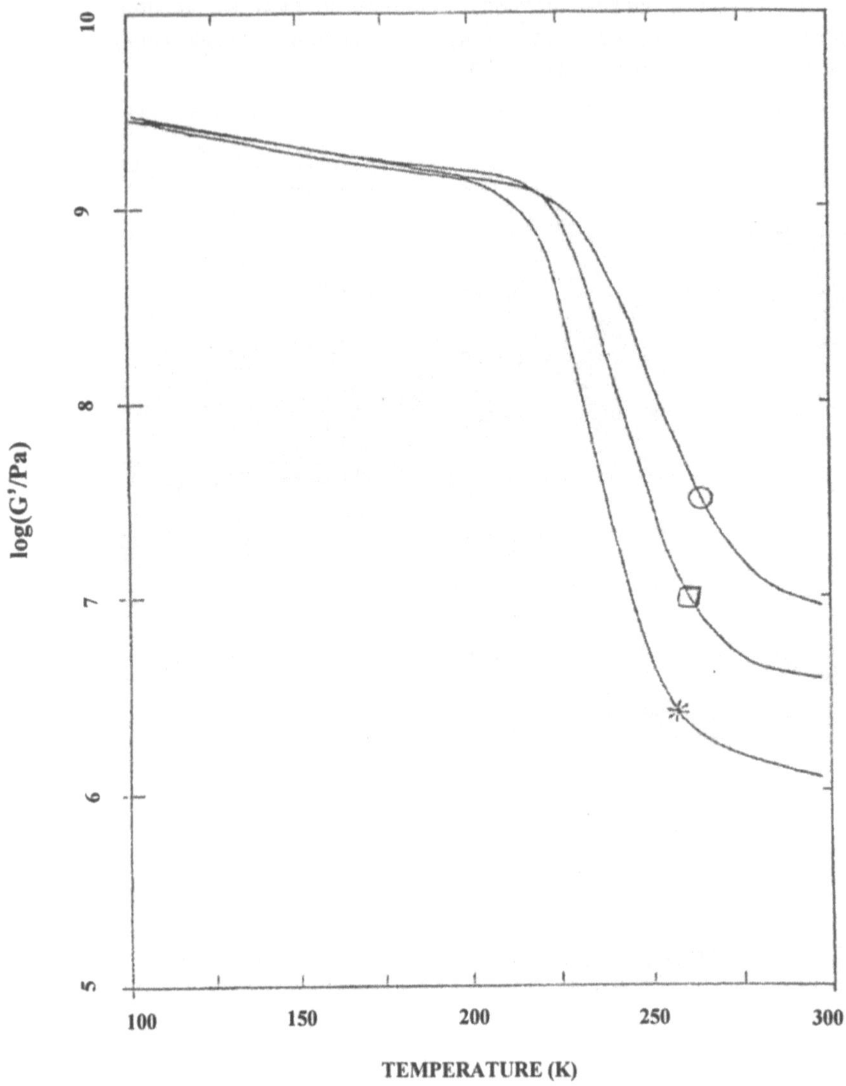

Figure 4. Storage modulus as a function of temperature of three KL-PEG$_{300}$-sebacates with different KL contents (∗) 45%; (□) 55%; (O) 65%.

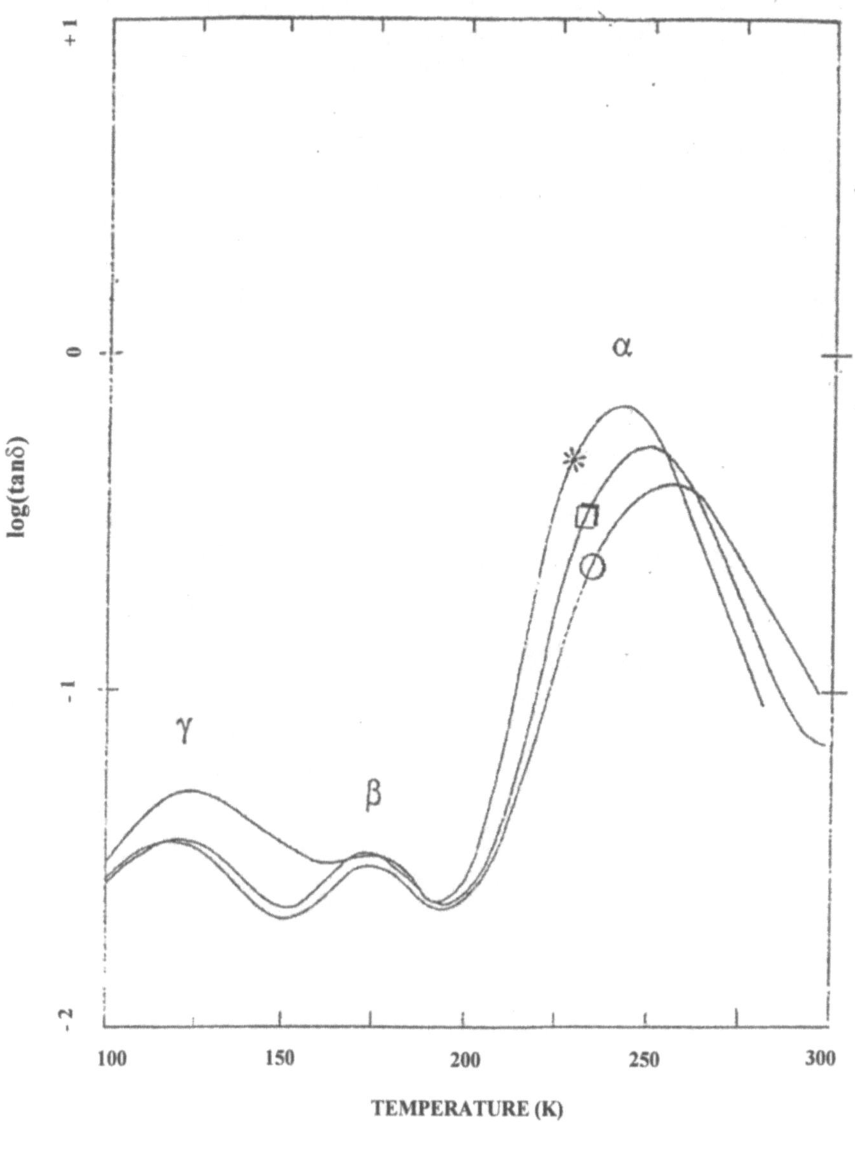

Figure 5. Variation of tan δ as a function of temperature of three KL-PEG_{300}-sebacates with different KL contents (*) 45%; (□) 55%; (O) 65%.

These membranes were fractured at liquid nitrogen temperature and the cross sections thus obtained examined by scanning electron microscopy. Figure 6 gives the morphologies of the two homogeneous materials. Figure 7 clearly shows the drastic change in the phase behavior when the lignin content exceeds about 60%. In fact, the two membranes with less lignin (Figure 6) appear fairly homogeneous, whereas with 65%, a blatant phase separation is visible (Figure 7). The latter feature is seen more dramatically in the corresponding magnified micrograph given in Figure 7b. The remarkable feature of these materials is undoubtedly related to the fact that very high proportions of lignin can be used in the construction of the networks without phase segregation.

On the basis of the results obtained with polycondensation systems using widely different conditions, it can be concluded that lignins are very adequate macromonomers for the synthesis of polyester networks. Figure 8 is a schematic view of the general structure of these networks in which the moiety bridging the ester functions is aliphatic or aromatic, according to the T_g required. Of course, a third variable consists of working at [COCl]/[OH]<1 in order to vary the cross-link density and leave some OH groups free.

The kinetic observation that the reactivity of both types of OH groups of lignins is substantially lower than that of more conventional monomeric and oligomeric structures can be attributed to steric factors. This is not in itself a problem, because it should be possible to make this difference beneficial by using lignins as latent cross-linking agents in prepolyesters bearing acidic end-groups. In other words, linear oligomers of this nature could be prepared and mixed with lignins to give thermoplastic materials capable of further reaction, leading to a gel, when heated during the processing of an object. Such systems could replace those made up of unsaturated oligoesters which are cross-linked by the radical polymerization of styrene, with the advantages of eliminating the presence of a toxic monomer and making good use of a renewable, industrial by-product.

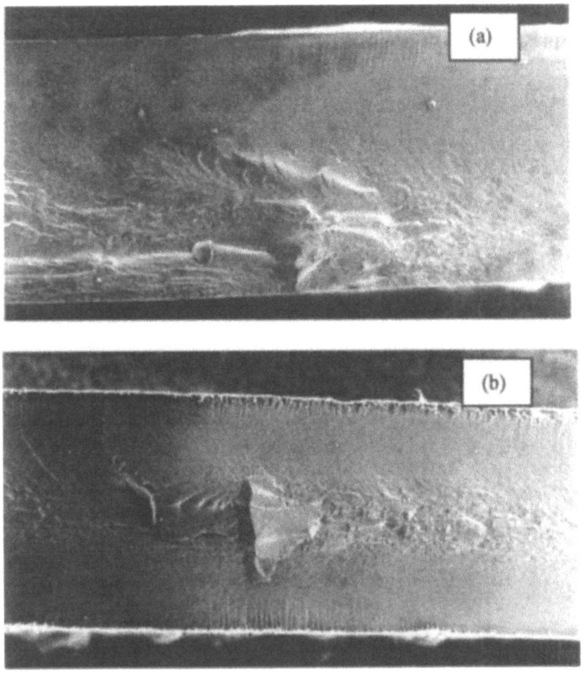

Figure 6. SEM micrographs (x60) of two KL-PEG_{300}-sebacates with different KL contents, (a) 45%, (b) 55%.

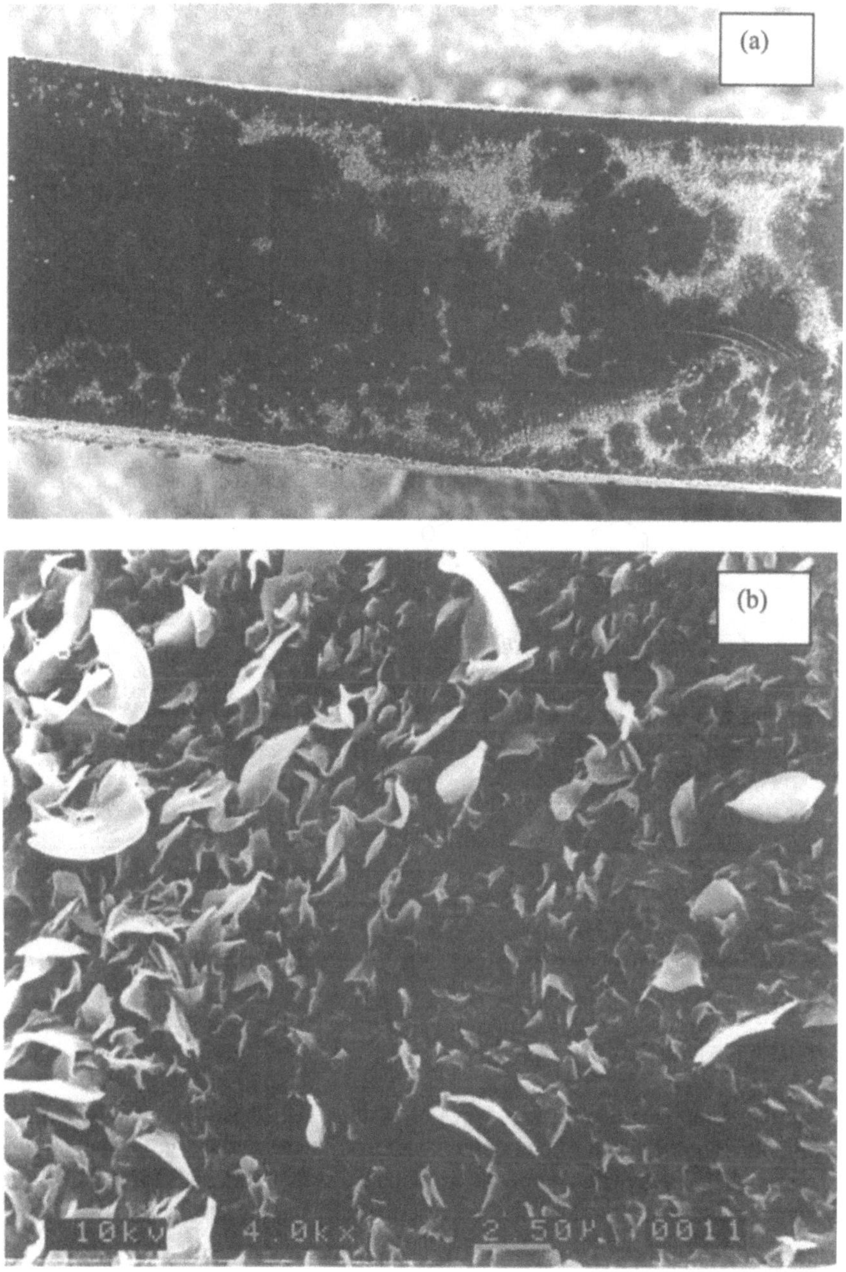

Figure 7. SEM micrographs of the KL-PEG_{300}-sebacate with 65% KL, (a) x60, (b) x4000.

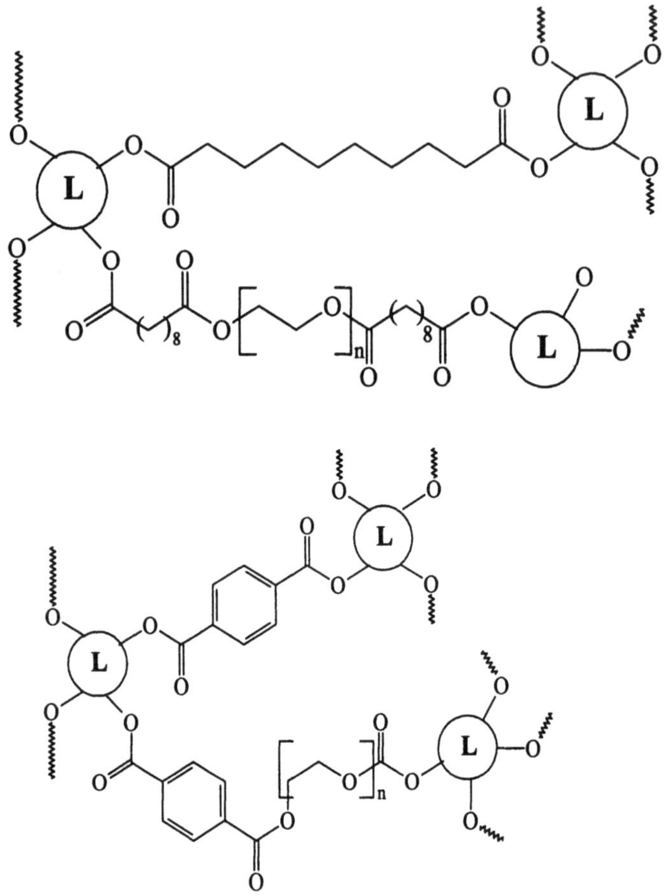

n = 5-6

Figure 8. Schematic representation of the cross-linked KL-based copolyesters described in this work.

4. POLYURETHANES

This section comprises two distinct types of systems, i.e. those involving the direct reactions of lignins (or their model compounds) with isocyanates and those in which lignins are first oxypropylated to give liquid polyols as the actual macromonomers for the synthesis of polyurethanes. Consequently, they will be described in separate sub-sections.

4.1. Direct Polycondensations

Most of the investigations in this context were carried out with OL and unless specified otherwise, the term "lignin" under this heading will mean OL. As shown in Table 1, this lignin was characterized by a relatively low molecular weight and a rather narrow dispersity. With respect to the several previous studies, carried out mostly by Glasser's group,[2, 4, 6] the present approach differed in three major aspects, namely the use of this specific lignin, the study of the behavior of lignin model compounds and the type of the isocyanates employed, which also included a model compound.

4.1.1. Model Plus Model

The three commercial compounds chosen to be the lignin model compounds, representing different characteristic structures of lignin units are shown in Scheme 1:

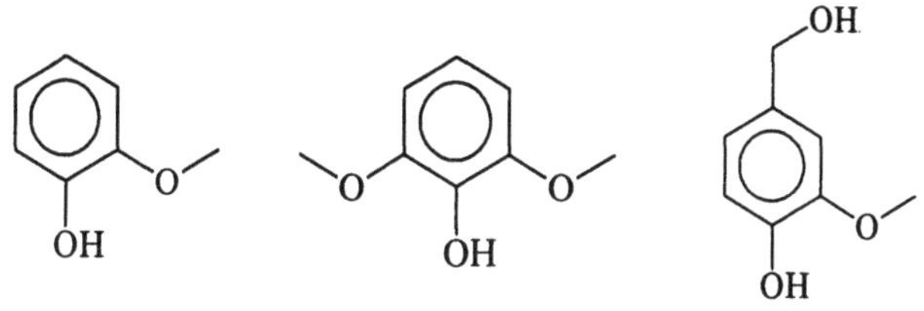

G **S** **GA**

Scheme 1. Lignin model compounds used in our studies.

Compounds **G** and **S** constitute models of phenolic guaiacyl and syringyl moieties, respectively, whereas compound **GA** was selected to study the relative reactivity of phenolic versus aliphatic OH groups in lignin.

The model isocyanate used was the "monomeric" structure (MEI) corresponding to the oligomeric mono- and di-isocyanates employed for the actual syntheses of the lignin-based polyurethanes. Its structure and synthesis are shown in scheme 2.

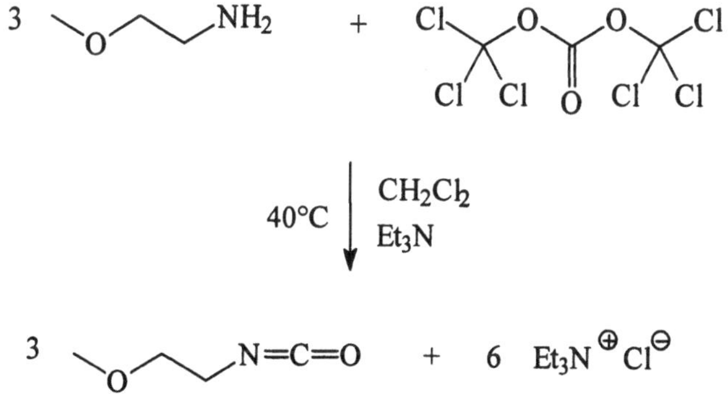

MEI

Scheme 2. Structure and synthesis of model isocyanate, MEI.

The conversion of primary amino groups of aminoethers and more generally jeffamines (see below) into NCO functions using bis(trichloromethyl) carbonate (triphosgene) was studied and optimized in our laboratory a decade ago within another context.[15] The purification of MEI was conducted by vacuum distillation and the product was thereafter stored under dry nitrogen. Its FTIR, ^{1}H and ^{13}C NMR spectra were consistent with the expected structure.

The reactions of MEI with the three lignin models were carried out under stoichiometric conditions at room temperature in THF using dibutyltin dilaurate as catalyst. They were followed by monitoring the decrease of the NCO infrared band at 2250 cm^{-1}. The products were isolated by evaporating the solvent under vacuum and characterized by FTIR and ^{13}C NMR spectroscopy. In all instances the condensations reached completion and with **GA**, the ^{13}C NMR spectrum of the product clearly showed that both the aliphatic and the phenolic OH had reacted.

It could therefore be concluded that in the absence of major steric hindrance, the typical hydroxy functions of lignin readily form urethanes with aliphatic isocyanates.

4.1.2. Model Plus Polymer

When OL and MEI were subjected to the same reaction conditions applied to the model + model systems (see *4.1.1.*), both types of OH groups reacted again, but the aliphatic ones were found to be consumed more rapidly, from the analysis of the ^{13}C NMR spectra of the products isolated as the reaction proceeded. This difference in kinetic behavior is in tune with the more pronounced acidic character of the phenolic OH compared with the aliphatic counterpart and the fact that in the NCO + OH reaction the latter plays the role of a nucleophile.

The situation was then reversed, viz. the systems examined involved the lignin model compounds and a macromolecular isocyanate (OPOI) (Scheme 3) prepared from a commercial monofunctional jeffamine using the same synthetic procedure described for the synthesis of MEI.[15]

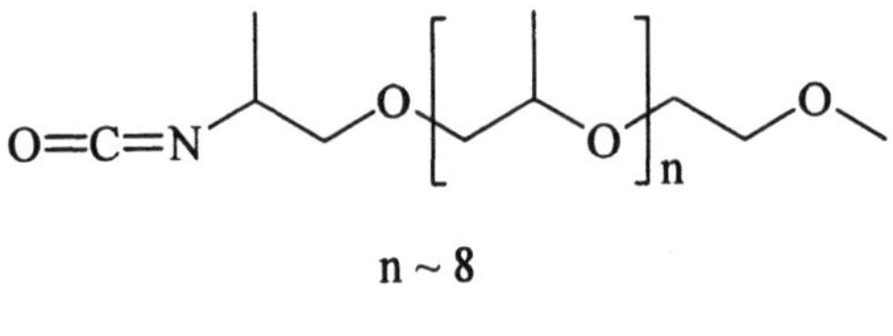

Scheme 3. Structure of a macromolecular isocyanate, OPOI.

The reactions were again conducted following the conditions described above and the products thoroughly characterized spectroscopically. Their FTIR spectra clearly indicated the formation of the expected urethanes in complete conversions, except with the product obtained from the system **GA** + OPOI which showed a residual OH peak at 3450 cm^{-1}. The ^{13}C NMR spectra of these products showed no residual OH groups in the case of **G** + OPOI and **S** + OPOI, but a substantial proportion of residual *phenolic* OH groups in the case of **GA** + OPOI. The most reasonable explanation for these results, in light of the observations gathered from the model + model and OL + model experiments, is that when the two types of OH groups were present, as in **GA**, the condensation first took place with the more reactive aliphatic OH and the steric hindrance introduced by the appended oligoether chain reduced considerably the possibility of advancement of the second reaction with the remaining phenolic OH group.

4.1.3. Macromonomer Plus Macromonomer

Having established the feasibility of synthesizing urethanes based on lignin macromonomers, we turned our attention to the preparation of actual materials by using lignin and macroisocyanates as reagents. The first set of reactions in this context was aimed at obtaining grafted lignins that would conserve their thermoplastic character, but possess various physico-chemical properties. OL was thus treated with variable amounts of OPOI in terms of the initial [OH]/[NCO] ratio, using the experimental conditions selected for the model reactions described above. Reactions were followed by FTIR spectroscopy which enabled us to monitor the decrease in the NCO peak with time, the corresponding increase in the carbonyl absorption and the gradual shift of the OH peak towards the NH counterpart. Scheme 4 shows the representative structures of the products and Table 2 summarizes their properties. Their FTIR and ^{1}H NMR spectra were completely consistent with the expected structures. The progressive increase in the grafting of oligoether "hairs" around the lignin core produced the expected increase in molecular weight and polydispersity, but also, and more importantly, a corresponding decrease in T_g with the formation of elastomeric to pasty materials and a drastic increase in solubility in solvents like ether and even water. Thus, a solid powder with little solubility in water could be readily transformed into new materials with low T_g and good solubility in water.

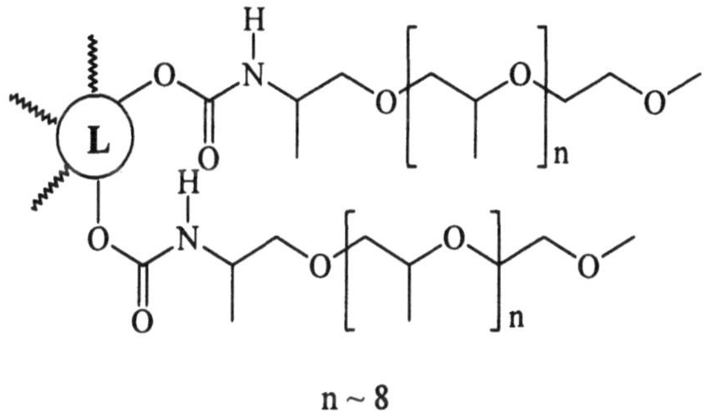

Scheme 4. Representative structures of polyurethanes prepared from OL and OPOI.

Table 2. Properties of the OL-OPOI polyurethanes

[NCO]/[OH]	M_n	M_w	M_w/M_n	T_g (°C)	Solubility (%) in Diethyl ether	Solubility (%) in Water
0	700	1700	2.4	64	12	10
0.15	1010	3760	3.4	-5	65	50
0.25	1110	5090	4.6	-35	73	65
0.4	1050	4760	4.5	-58	83	75
0.6	1140	5160	4.5	-66	92	90
0.90	1210	6730	5.6	-77	100	100

Our second approach consisted of extending the domain of the materials provided by this strategy by replacing OPOI with a bifunctional homologue prepared from the corresponding jeffamine following the synthetic protocol already described. The structure of the macrodiisocyanate, OEDI, obtained is shown in Scheme 5.

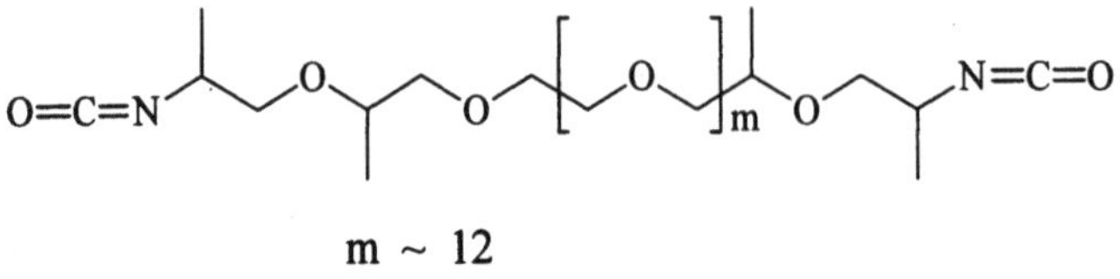

Scheme 5. Structure of the macrodiisocyanate, OEDI.

The reactions of OL with OEDI were again conducted following the conditions utilized for the model reactions, but here the product precipitated progressively as a gel with time. Again, different initial [OH]/[NCO] ratios were studied and the ensuing cross-linked polyurethanes characterized. The FTIR spectra of each network corroborated the occurrence and extent of the polycondensation reaction that produced the representative structures shown in Scheme 6.The only variable was the cross-link density and consequently the extent of internal plasticization by the frequency of the oligoether bridges.

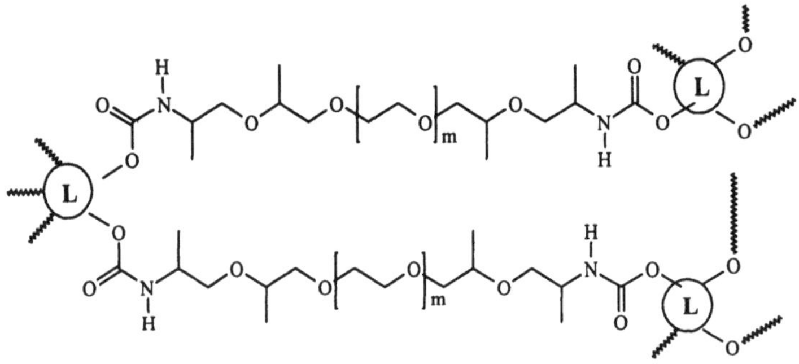

Scheme 6. Representative structures of polyurethanes prepared from OL and OEDI.

Table 3 gives the relevant properties obtained with these novel polyurethanes, as a function of the proportion of the two macromonomers used. Once again the presence of the OEDI-based sequences in these networks provided a means of reducing their T_g which behaved as elastomeric materials.

Table 3. Properties of OL-OEDI cross-linked polyurethanes

[NCO]/[OH]	Gel fraction (%)	T_g of gel
0.2	54	44
0.6	85	4
0.8	98	-15

The above investigation provided very useful information on the possibility of making polyurethanes based on unmodified lignins and data concerning the relative reactivity of the two kinds of OH groups and the role of steric effects. The specific choice of very flexible macroisocyanates led to grafted and cross-linked materials in which the stiff character of the lignin macromolecules was readily overwhelmed leading to low-T_g pastes or elastomers. Other lignin properties were also drastically modified.

4.2. Polycondensation of Oxypropylated Lignins

4.2.1. Oxypropylation

Although the results described in the previous section were interesting and satisfactory, they must be viewed as a special case within the general problem related to the preparation of polyurethanes based on lignins, because the macroisocyanates selected were particularly suitable in terms of physico-chemical compatibility with the natural polymer. In order to widen the scope of this research topic, namely to have the possibility of using conventional isocyanates, it is necessary to render lignins more reactive by deploying their OH groups and freeing them from steric and/or electronic constraints, while tranforming at the same time most of the phenolic OH groups into aliphatic ones. Hence our second approach to these materials uses oxypropylated lignins as macropolyols.

The most abundant lignins are those from kraft and sulfite processes. These are high-molecular weight branched materials and their transformation by chain-extension with propylene oxide (PO) were reported to require high temperatures and pressures and long reaction times.[16] These operating conditions also induced the self-condensation of the lignin macromolecules which gave rise to insoluble fractions. Thus, conventional lignins have not yet found industrial applications in the elaboration of polyurethane formulations, because their modification into viscous polyols is deemed too complex and energetically onerous. Since OL and OOL are low-molecular weight products soluble in common solvents, including PO itself, we thought that their oxypropylation could take place more readily than the reaction with their kraft and other homologues. In this vein, we carried out comparative experiments on the oxypropylation of different lignins under various operating conditions such as the reaction temperature, the lignin/PO ratio, and the nature and the amount of catalyst.

Typically, the chain extension of lignins was carried out in a 1-litre stainless, pressurized steel reactor heated in a silicone oil bath. The reactor was charged with the chosen lignin, the corresponding quantity of PO and the catalyst. It was then closed and heated to the desired temperature (see Tables 4 and 5) which led to the onset of the oxypropylation reaction with a corresponding increase in pressure up to a maximum value and then to a progressive decrease which reflected the gradual consumption of PO. At the end of the reaction, signaled by $P_{relative} = 0$, the products were recovered by vacuum removal of the small amounts of unreacted PO. The viscous products of these reactions were always a mixture of oxypropylated lignin and homooligomeric diols (POP) derived from the anionic polymerization of PO. This is a well-known features of all oxypropylation reactions of natural polymers which we had previously encountered with other substrates like sugar beet pulp,[17] chitosan[18] and cork.[19] The various oxypropylated lignins obtained by this procedure are abbreviated as OLOP for OL, KLOP for KL and OOLOP for OOL.

The results obtained showed clearly that OL reacts much more readily and completely with PO than the other lignins and mixtures containing up to 50% w/w of OL, with respect to PO, can be converted into liquid polyols without any unreacted residue. With KL at 10% initial lignin content, the reaction already leaves some insoluble solids (Table 6).

As expected, the viscosity of the polyols from the OL + PO systems increases with increasing OL content and at the same time the proportion of POP homopolymer decreases. The values of I_{OH} grew steadily from ~ 200 to ~ 300 as a function of the OL/PO initial ratio (Table 6). These values are close to those of commercial polyols usually employed in the manufacture of polyurethanes. The viscosity of the OL-PO polyols grew steadily from 5 to extremely high values (paste-like materials) as the initial OL/PO ratio was increased.

Different applications can be envisaged for these materials, but if they were to be used as macropolyols for polyurethanes, only those with η values lower than about 200 Pa.s would be viable, i.e. one would have to limit the initial OL/PO ratio to below about 30/70 w/w. The other more viscous products could also be employed, but would require the addition of a more fluid co-polyol.

The proportion of POP in the OLOP polyols decreased as the initial ratio OL/PO increased and practically vanished when equal weights of OL and PO were used. This is indeed the main reason for the increase in viscosity of the final products.

The polyols formed from all the other lignins tested in this study were characterized by the presence of solid residues, high proportions of POP and, as a consequence, relatively low viscosities. In other words, the actual grafting of POP chains on the lignin core molecules was much less efficient than with OL.

The catalysis by alkali metal hydroxides was much more efficient than that induced by Lewis bases. Thus, given the inevitable traces of moisture in the reaction medium (mostly in OL), it seems likely that the best way to accelerate the oxypropylation is to have free OH^- groups present in the system. Moreover, the fact that KOH in pellets and fine powder gave similar results, suggested that the catalysis took place homogeneously. However, about 5% w/w of KOH or NaOH, with respect to lignins, was the minimum amount required for an adequate reaction rate. We chose to work at 180 °C, but lower temperatures (and therefore pressures) can also be applied. In our experience with other substrates,[17-19] temperatures as low as 80-90 °C were sufficient, although this required longer reaction times. Recent results on the oxypropylation of cork showed that a preimpregnation of the substrate by a KOH alcoholic solution led to an efficient oxypropylation under mild conditions.[19]

In conclusion, the chain extension of KL, OL and OOL with PO was successful at different degrees and yielded polyols with characteristics similar to those required for industrial applications related to polyurethane production. In particular, with OL + PO, after the reaction temperature had reached 175-180 °C, the time needed for the total transformation of PO was only about 15 min. Such a reaction time is much shorter than what is required for the industrial oxypropylation of sugars like sorbitol (the precursor used in most PU systems). The difference in reactivity found between OL and OOL stems from the presence of relatively high proportions of carbonyl and carboxyl groups in the latter, which reduces its reactivity towards PO.

Table 4. Formulations related to the oxypropylation of lignins

Lignins	Experiment number	PO/Lignin (w/w)	PO/Lignin[a] (w/w)	Catalyst [C]	[C]/Lignin (% w/w)
	1	20/80	30/120	KOH	5
	2		30/120	KOH	7
KL	3		30/120	KOH	10
	4	15/85	30/170	KOH	10
	5	10/90	20/180	KOH	10
	6	20/80	20/80	KOH	5
	7		40/160	KOH	5
	8		60/240	KOH	5
OOL	9	30/70	30/70	KOH	5
	10		30/70	NaOH	5
	11		60/140	KOH	5
	12	40/60	80/120	KOH	5
	13	20/80	20/80	KOH	5
	14		30/120	KOH	1
	15		30/120	KOH	3
	16		30/120	KOH	5
	17		30/120	KOH	7
	18		30/120	KOH	10
	19		30/120	KOH[b]	5
	20		60/240	KOH[c]	5
	21		60/240	NaOH	5
OL	22		60/240	Org[d]	5
	23	30/70	60/140	KOH[b]	5
	24		60/140	KOH[c]	5
	25		60/140	NaOH	5
	26		120/280	KOH	5
	27	40/60	80/120	KOH	5
	28		160/240	KOH	5
	29		160/240	NaOH	5
	30	50/50	100/100	KOH	5
	31		150/150	KOH	5
	32		200/200	KOH	5

[a] Actual quantities in grams placed in the reactor.

[b] KOH in the form of pellets (3 mm of diameter and 1 mm of thickness).

[c] KOH in the form of very fine powder.

[d] Three organic catalysts were studied: triethylamine (TEA), pyridine and di-azobicyclooctane (DABCO).

Table 5. Operating conditions for the oxypropylation of lignins

Lignins	Experiment number	PO/Lignin (w/w)	T (°C)	Time (min)	P[a] (bar)
KL	1	20/80	185	900	10.5
	2		200	830	11.4
	3		195	810	11.2
	4	15/85	195	800	14.2
	5	10/90	179	900	15.4
OOL	6	20/80	175	500	7.1
	7		185	450	11.1
	8		179	520	18.7
	9	30/70	160	475	9.1
	10		177	485	9.4
	11		180	505	12.7
	12	40/60	190	330	11.9
OL	13	20/80	175	20	6.5
	14		180	300	11.5
	15		179	90	11.4
	16		180	19	11.6
	17		178	14	11.4
	18		170	13	11.1
	19		178	14	11.3
	20		175	20	17.7
	21		177	21	17.9
	22		185	1500	19.1
	23	30/70	175	20	8.5
	24		178	18	8.9
	25		177	25	8.4
	26		179	30	18.2
	27	40/60	179	30	11.7
	28		178	39	18.5
	29		178	44	18.2
	30	50/50	180	28	9.1
	31		180	39	13.5
	32		183	45	16.2

[a] Maximum pressure.

Table 6. Functional characteristics of hydroxypropylated lignins in our work

Lignins	Experiment number	L/PO (w/w)	η^a (Pa.s)	I_{OH}^a	POP (% w/w)	Solid residue (%)
	1		4.2	180	75	41
	2	20/80	5.0	160	72	40
KL	3		4.8	175	nd[c]	40
	4	15/85	3.2	165	nd	36
	5	10/90	2.3	148	83	33
	6		4.8	170	62	25
	7	20/80	5.1	175	60	nd
	8		5.5	177	nd	nd
OOL	9		5.2	178	48	29
	10	30/70	6.2	177	51	nd
	11		5.8	185	nd	nd
	12	40/60	4.1	165	nd	39
	13		5.2	200	39	0
	14		5.1	197	nd	0
	15		6.0	198	42	0
	16		6.3	195	nd	0
	17	20/80	5.5	208	45	0
	18		5.8	195	nd	0
	19		6.4	182	nd	0
	20		5.4	202	nd	0
	21		5.2	209	nd	0
OL	22		5.8	198	nd	0
	23		95	210	25	0
	24		92	212	22	0
	25	30/70	98	210	nd	0
	26		93	203	nd	0
	27		2860	250	20	0
	28	40/60	2920	280	nd	0
	29		2805	270	nd	0
	30		(b)	305	2	0
	31	50/50	(b)	310	nd	0
	32		(b)	320	nd	0

[a] Liquid fraction.
[b] Too high to be measured.
[c] nd = not determined.

4.2.2. Polyurethane Foams Preparation and Characterization

For this part of our work, we adopted the typical procedure concerning foams elaboration, which consisted of mixing vigorously the polyols, the filler (when needed), the surfactant, the catalysts, and the water for about 30 s (required for emulsification). The blowing agent was then added and the mixture stirred again, but only for 15 s, in order to minimize its evaporation. Finally, the isocyanate was introduced into the emulsion and the ensuing mixture again vigorously stirred for 20 s. The resulting emulsified liquid was poured in a special mould for free expansion.

Several foams involving different polyols were elaborated and summarized in Table 7, which contains the detailed formulations. The following polyols were used because of their adequate viscosity and hydroxy number:

OLOP20: $I_{OH} = 195$; (OL/PO = 20/80) OLOP30: $I_{OH} = 210$; (OL/PO = 30/70),

KLOP20: $I_{OH} = 180$; (KL/PO = 20/80) OOLOP20: $I_{OH} = 170$; (OOL/PO = 20/80), OOLOP30: $I_{OH} = 185$; (OOL/PO = 30/80).

We also called upon two commercial polyols commonly used in polyurethane foam formulations, in order to prepare reference materials, i.e. Stepanol (STP) and P140, The first was a polyester-polyol with a hydroxy number of 330 and $\eta = 8$ Pa.s, whereas the second was a polyether-polyol with a hydroxy number of 490 and $\eta = 16$ Pa.s.

Thus, a typical rigid polyurethane foam (RPU) formulation contains the following components:

1. The polyol (one or a mixture of two or more).

2. The isocyanate: here MDI, with an average NCO functionality of 2.7, which is the most commonly used reagent in RPUs, added in 10% molar excess with respect to the total OH present:

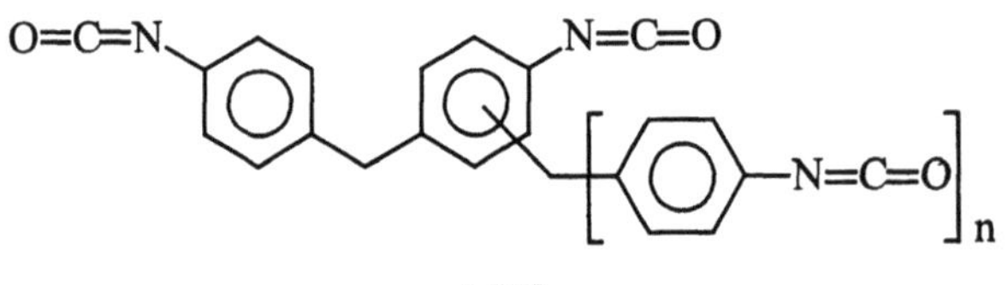

MDI

3. A blowing agent (15% w/w with respect to the total formulation) which guarantees the foam formation due to its low boiling point (evaporation induced by the initial exothermicity of the polymerization reaction): here the new HCFC **141b** agent ($C_2H_3FCl_2$: 1,1-dichloro-1-fluoroethane, bp 32 °C) is the official substitute of the banned R11:

141b

4. A surfactant, which provides the cell stabilisation: here the commercial silicone-based compound D5454, supplied by Air Product Co., systematically at 2% w/w with respect to the polyol.

5. The catalysts: here dimethylcyclohexyl amine (Polycat 8) and Niax A-1 (1 and 0.5% of these catalysts were added, respectively). Both products were supplied by Witco Chemicals:

DMCHA, (Polycat8)

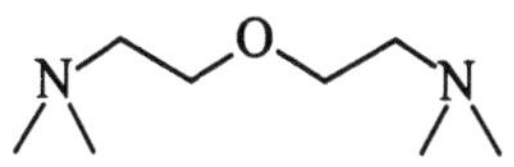

NIAX A-1, (NA)

6. Water (3 g per 100 g of total polyol), which reacted with the excess of NCO functions to give carbon dioxide, which played the same role as 141b, i.e. a blowing agent.

7. In addition to the components listed above, we also added systematically 10 g of glycerol per 90 g of the chosen polyol.

The runs summarized in Table 7 and the properties assessed for the corresponding RPUs given in Table 8, can be classified as follows:

1. Standard formulations with STP and P140 as commercial polyols. These were prepared as reference for comparison (runs 1, 2, 5, 6, 21 and 22).

2. New foams based on OLOP (3, 4, 7, 8, 23 and 24).

3. Standard foams with OL as filler (9, 10, 13 and 14).

4. New foams with OL as filler (11, 12, 15 and 16).

5. Standard foams with KL as filler (17 and 18).

6. New foams with KL as filler (19 and 20).

7. Mixed foams in which a standard polyol was mixed with OLOP (25-32).

8. Foams based on KLOP (33) and OOLOP (34 and 35).

The foams prepared from OLOP showed thermal properties and dimensional stability very similar to those measured for foams obtained from conventional polyols. They were also found to have a good resistance against natural and accelerated ageing. This suggests that OLOP could replace in a very adequate manner the polyols usually employed for the manufacturing of RPUs. KL and OOL were clearly unsuitable for RPU foam formulations, since they gave materials with very poor properties. Such results can be attributed to the fact that KLOP and OOLOP possess too high a proportion of POP. Finally, OL and KL can also be envisaged as fillers for RPU foams since they do not induce any detrimental effect on key properties, including the fact that their presence in up to 30% w/w proportions do not alter appreciably the apparent density of the foams. It seems likely that the lignin particles present in these materials had participated in the polycondensation reactions through their surface OH groups thus creating a strongly adhesive interface with the matrix. In other words, the term "filler" should be intended to mean "reactive filler".

Table 7. Formulations of RPU foams

Foam number	Polyol	Content (g)	Filler	Content (g)	MDI
1	STP	45	-	-	80.0
2	P140	45	-	-	97.0
3	OLOP20	45	-	-	65.0
4	OLOP30	45	-	-	67.0
5	STP	45	-	-	80.0
6	P140	45	-	-	97.0
7	OLOP20	45	-	-	65.0
8	OLOP30	45	-	-	67.0
9	STP	45	OL	20	80.0
10	P140	45	OL	20	97.0
11	OLOP20	45	OL	20	65.0
12	OLOP30	45	OL	20	67.0
13	STP	45	OL	30	80.0
14	P140	45	OL	30	97.0
15	OLOP20	45	OL	30	65.0
16	OLOP30	45	OL	30	67.0
17	STP	45	KL	15	80.0
18	P140	45	KL	15	97.0
19	OLOP20	45	KL	15	65.0
20	OLOP30	45	KL	15	67.0
21	STP	180	-	-	319
22	P140	180	-	-	388
23	OLOP20	180	-	-	261
24	OLOP30	180	-	-	267
25	OLOP20/STP	27/18	-	-	71.0
26	OLOP20/P140	27/18	-	-	78.0
27	OLOP20/STP	22.5/22.5	-	-	72.5
28	OLOP20/P140	22.5/22.5	-	-	81.0
29	OLOP30/STP	27/18	-	-	72.0
30	OLOP30/P140	27/18	-	-	79.0
31	OLOP30/STP	22.5/22.5	-	-	73.5
32	OLOP30/P140	22.5/22.5	-	-	82.0
33	KLOP20	45	-	-	64.0
34	OOLP20	45	-	-	68.2
35	OOLOP30	45	-	-	72.1

Table 8. Characteristics of RPU foams before and after ageing

Foam number	before ageing density (kg/m³)	before ageing λ (mW/m.K)	after ageing λ[a] (mW/m.K)	after ageing λ[b] (mW/m.K)	DS[c] (%)
1	24.8	15	17	19	+17
2	26.2	17	18	20	+17
3	17.9	20	22	24	-17
4	18.2	21	22	25	-15
5	28.2	17	24	28	+8
6	29.1	17	25	27	+7
7	30.1	19	26	29	-8
8	29.2	19	27	31	-9
9	18.2	15	17	21	+7
10	19.1	15	17	19	+6
11	18.9	15	17	20	-5
12	20.1	15	17	21	-7
13	27.5	17	19	25	+8
14	28.1	18	19	26	+8
15	27.9	17	20	25	-4
16	28.3	18	21	26	-3
17	32.5	20	28	33	+10
18	33.2	23	29	30	+11
19	35.4	24	29	30	+9
20	33.5	23	30	33	+8
21	29.2	16	20	28	+13
22	30.1	17	21	27	-8
23	29.1	18	21	28	-7
24	29.8	17	20	26	+13
25	22.2	16	21	25	+11
26	24.7	17	21	27	+10
27	25.1	16	22	29	+15
28	24.2	18	22	28	+14
29	26.1	18	21	29	+12
30	25.8	18	22	30	+14
31	26.3	17	22	29	+12
32	26.8	18	24	34	+14
33	28.2	31	nd[d]	nd[d]	+27
34	25.2	22	28	32	+13
35	26.5	24	30	36	+10

[a] After ageing at room temperature for 3 months.

[b] After accelerated ageing at 70 °C for 10 days.

[c] DS = dimensional stability.

[d] After ageing the deformation was too large and measurements became impossible; nd = not determined

5. CONCLUSIONS

It is hoped that the work presented in this chapter will contribute to convince readers, who might still be skeptical, that lignins are indeed viable macromolecular reagents (and/or fillers) in the preparation of polymeric materials and that the properties of these polycondensates can be attractive for specific domains of application. It is quite clear that much remains to be done in this field in terms of both the optimization of the systems already investigated and the search of new structures, e.g. lignin-based polyethers.

6. REFERENCES

1. Gandini, Polymers from renewable resources, in: *Comprehensive Polymer Science*, edited by S. L. Aggrawal and S. Russo, (Pergamon Press, Oxford, 1992), Supplement Volume 1, pp. 527-573.
2. A. Gandini, Les lignines et leur utilisation dans les matériaux macromoléculaires, in: *Initiation à la chimie et a la physico-chimie macromoléculaires,* (Groupe Français d'Etudes et d'Application des Polymères, Paris, 2000) Volume 13, pp. 159-195.
3. A. Gandini and M. N. Belgacem, VEGETAL BIOMASS. (2. Oligomers and Their Polymerization), in: *Polymeric Materials Encyclopedia*, edited by J. Salamone (CRC Press Inc., Boca Rota, 1996), Volume 11, pp. 8530-8541.
4. W. G. Glasser and S. Sarkanen, *Lignins: Properties and Materials* (American Chemical Society Symposium Series 397, Washington D. C., 1989).
5. W. G. Glasser, R. A. Northey, and T. P. Schultz, *Lignins: Historical, Biological and Materials Perspectives* (American Chemical Society Symposium Series 742, Washington D. C., 2000).
6. W. G. Glasser and S. S. Kelley, Lignin, in: *Encyclopedia of Polymer Science and Engineering*, edited by H. F. Mark, N. M. Bikales, C. G. Overberger and G. Menges (John Wiley & Sons, New York, 1987), Volume 8, pp. 795-852.
7. A. Gandini and M. N. Belgacem, Recent advances in the elaboration of polymeric materials from biomass components, *Polym. Intern.* **46**, 267-276 (1998).
8. N. Cordeiro, P. Aurenty, M. N. Belgacem, A. Gandini, and C. Pascoal Neto, Surface properties of suberin, *J. Colloid Interface Sci.* **187**, 498-508 (1996).
9. M. N. Belgacem, A. Blayo and, A. Gandini, Lignins as fillers in inks and polymer foams, in: *Proceedings of Eurofillers'01*(Lodz, Poland, 2001), pp. 106-107.
10. I. P. Deineko *et al.*, Chim. Drev. n° 1, 54-57 (1983); n° 5, 25-31 (1983); n° 1, 39-43 (1984); n°3, 71-75 (1984).
11. C. G. Goyal and J. H. Lora, Kinetics of delignification and lignin characteristics in autocatalysed organosolv pulping of hardwoods, in: Appita *Proceedings, 6th International Symposium on Wood and Pulping Chemistry* (Melbourne, 1991), pp. 205-212.
12. Z. X. Guo, A. Gandini, and F. Pla, Polyesters from lignin. 1. The reaction of kraft lignin with dicarboxylic acid chlorides, *Polym. Int.* **27**, 17-22 (1992).
13. Z. X. Guo and A. Gandini, Polyesters from lignin-2. The copolyesterification of kraft lignin and polyethylene glycols with dicarboxylic acid chlorides, *Eur. Polym. J.* **27** (11), 1177-1180 (1991).
14. D. Evtuguin and A. Gandini, Polyesters based on oxygen-organosolv lignin, *Acta Polymerica* **47**, 344-350 (1996).
15. J. F. Le Nest, S. Callens, A. Gandini, and M. Armand, A new polymer network for ionic conduction, *Electrochimica Acta* **37**(9), 1585-1588 (1992).
16. L. C.-F. Wu and W. G. Glasser, Engineering plastics from lignin. I. Synthesis of hydroxypropyl lignin, *J. Appl. Polym. Sci.* **29**, 1111-1123 (1984).
17. P. Velasquez-Morales, J. F. Le Nest, and A. Gandini, The modification of chitosan in view of elaborating new polymer electrolytes. 2. Polyether-based networks using oxipropylated chitosan, in: *Advances Chitin Science*, edited by A. Domard, C. Jeuniaux, R. Muzzarelli, and G. Roberts (Jaques André, Lyon, 1996), Volume 2, pp. 236-359.
18. C. Pavier and A. Gandini, Oxypropylation of sugar beet pulp. 2. Separation of the grafted pulp from the polypropylene oxide homopolymer, *Carbohydr. Polym.* **42**, 13-17 (2000).
19. M. Evtiouguina, A. M. Barros, M. N. Belgacem, C. Pascoal Neto J. J. Cruz-Pinto and, A. Gandini, New liquid polyols from cork residues – an improved oxypropylation reaction, in: *Proceedings of the 6th European Workshop on Lignocellulosics and Pulp* (Bordeaux, 2000), pp. 303-306.

LIGNIN AND ITS POLYBLENDS - A REVIEW

Dorel Feldman*

1. INTRODUCTION

The lignin present in plant tissues is referred to as native or natural lignin. During the industrial delignification of lignocellulosic materials such as wood, lignin undergoes significant structural changes; so the lignins obtained under industrial conditions, the so-called technical lignins, are not identical with the native ones in their structures.

Technical lignins constitute the by-products of the pulp and paper and the hydrolytic industries. It is estimated that more than 95% of such lignins is used as an energy source for the recovery of inorganic pulping chemicals in chemical pulp mills. The most significant and challenging problem facing lignin chemists is the utilization of the vast quantities of low-cost lignin available as by-products of the forest-using industries and, especially, the pulp and paper one.

Since its beginning the chemical pulping industry has been utilizing essentially only half of the tree or other lignocellulosic materials as pulp, and discarding the other half as a waste (particularly for the pulping of non-wood materials) or using it as a source of heat and energy. Pressures for the abatement of pollutional effluents, the need for conservation to meet expanding requirements for our raw materials, and the urge for new value-added products have been responsible for initiating research on the utilization of the other half of the lignocellulosic materials.

Lignins are natural organic materials and main constituents of woody plants (their content in woods reaches ~30%) where they play the role of lining agents and are comparable in this respect to cement in steel reinforced concrete. In trees, lignin and hemicelluloses form the matrix reinforced by cellulosic fibers. Lignins impart rigidity to the cell walls and, acting as an intercellular binder, impart to woods resistance to compression, impact and bending. They also have various functions in controlling stresses to which plants (trees) are subjected. The presence of lignin renders the woody tissue more resistant to the action of microorganisms.

Lignins are a major participant in energy flows in the biosphere. More solar energy is stored in lignin than in any other functional components in plants (if cellulose and

* Department of Building, Civil and Environmental Engineering, Concordia University, 1455 De Maisonneuve Blvd. W., Montreal, Quebec, Canada, H3G 1M8.

hemicelluloses are classified as separate components). Complementary to being a major component in plant systems for the storage of solar energy, lignin plays a major role as a composite material component in woods.

It is a commonly accepted notion that in any attempted uses of lignin as a functional component of new polymeric materials one should not try to replace synthetic components but to take new innovative approaches where the uniqueness of lignin as a macromolecular material could be exploited. In respect to the competitiveness of lignins with synthetic polymers produced from oil or coal, the new economic balances may affect the choices of raw materials and the products to be used. The use of natural materials generally has the advantages of low process cost with low energy requirement. The existing plastics, rubbers and synthetic fibers industry is highly energy intensive. In a society with a balanced use of energy one should make efforts to stay low on the enthalpy scale in meeting needs and not to increase entropy by unnecessarily breaking the covalent bonds in natural polymers.

One of the most important applications of lignin and its derivatives is in the area of plastic products and polyblends. This is due to the fact that lignin itself is a thermoplastic with a molecular weight in the range of 1,000 to 12,000 depending upon the extent of degradation and condensation during its isolation. Various studies in this area employ different types of lignin such as Kraft lignin, sulfite lignin (lignosulfonates), organosolv lignin, hydrolytic lignin, steam-exploded lignin, hydroxypropyl lignin and others.

2. POLYBLENDING

The terms polyblends and polymer alloys are used interchangeably and refer to a mixture of two or more polymers. Polyblending has become one of the most commercially important and inexpensive ways to produce new materials from readily available ones. The main aim of polyblending is the production of high-performance materials at a reduced cost. This is achieved through the choice of a large number of blending methods, the use of existing or more flexible processing equipments, and the possibilities to combine expensive polymers with inexpensive and abundant ones.

Polyblending is the physical mixing of two or more polymers in appropriate proportions to give an end product that has more desirable properties or lower cost than those of the components. The combining of two or more structurally dissimilar polymers is of commercial interest due to the potential of developing materials to meet specific functional needs. Curing time, durability, mechanical performance, or other properties can be altered to meet industrial requirements; cost reductions are possible through suitable selection of materials without sacrificing the inherent properties of the polymers.

Blending two polymers usually leads to a class of materials whose properties are due in most cases to the presence of the two components. The occurrence of synergistic effects may yield superior properties within the polyblend compared to those of the components. The formation of miscible polymer blends requires that the free energy of mixing (ΔG_{mix})

$$\Delta G_{mix} = \Delta H_{mix} - T\Delta S_{mix} < 0$$

where ΔH_{mix} and ΔS_{mix} are the enthalpy and entropy variations, respectively, and that $[\delta^2 \Delta G_{mix}/\delta\varphi^2_i]_{TP} > 0$ where φ_i is the concentration of component i. Since the entropy of mixing of large polymer molecules is quite small, miscibility usually occurs when the enthalpy of mixing is negative.

The toughness properties of immiscible blends whose components form separate phases often depend on the dimensions of the dispersed phase and on the degree of interfacial adhesion between phases. These properties are largely controlled by melt processing conditions and by the thermodynamic affinity of the blend components. Two polymers may be thermodynamically immiscible but still mechanically compatible. These compatible blends are two- or multi-phase systems that show multiple glass transition temperatures (T_g) which do not change with composition. The components do not separate, however, under mechanical stress during the life expectancy of the product.[1]

Compatibilization of polymers in blends possessing poor mechanical properties can be accomplished by adding a third component block or graft copolymer that is miscible with each of the major blend components. Reactive compatibilization, where the compatibilizer is formed by chemical reaction during melt mixing, can be an economical route for achieving a compatibilized blend product.[2-6]

The progressively clear elucidation of the role of specific interactions on the miscibility in polymer blends has been one of the most noticeable advances in this field over the last decades. Many experimental results have proved that the presence of any specific interactions in polymer blends such as hydrogen bonding, ion-ion pairing, electron-donor and electron-acceptor complexation will produce a favorable mixing enthalpy and may lead to complete miscibility between the components. This progress not only provides a practical guide for compounding miscible blends from existing polymers but also opens a new area for studying miscibility enhancement in such systems. By chemical modification of the polymers or copolymerization, some functional groups capable of forming specific interactions can be incorporated into one or both of the inherently immiscible component polymers to improve miscibility.[2] Simmons and Eisenberg[7] have studied the effect of introducing cationic and anionic groups into immiscible polymers on their miscibility. Rodriguez-Parada and Percec[8] have reported the results of miscibility enhancement via interaction between pendant electron-donor and electron-acceptor groups introduced into the component polymers by chemical modification. Among all the interactions, hydrogen bonding seems most attractive, since it has a high efficiency in improving the miscibility without affecting much the properties of the parent polymers in the blends.[2]

Cowie and Love[9] have shown that the amount of modification required will depend on how immiscible the starting components are and the relative flexibility of their molecular chains. Thus, if the components have very different solubility parameters (δp) (if we use this as a rough guide to miscibility), they can be regarded as immiscible and will require a larger number of electron-donor/electron-acceptor interactions to bridge the energy gap than in a less immiscible system where $\Delta\delta p$ is smaller. Because successful secondary bonding interactions usually require the electron-donor and electron-acceptor sites to be close enough, not all of the sites in a macromolecular chain will be involved in bonding as chain flexibility may affect their accessibility.

3. PHENOL FORMALDEHYDE - LIGNIN ADHESIVES

The use of different types of lignin in adhesive formulations derives from the very great need for low cost adhesives of reliable supply and durability mainly for wood based products such as plywood, particleboard, fiberboard and waferboard. It is desirable to obtain adhesives from renewable resources that possess the high water resistance, durability, and other characteristics of petrochemically based adhesives.

Certainly, the most investigated area has been the utilization of lignin in phenol-formaldehyde (PF) adhesives although many studies have also been done on the use of lignin in other adhesives such as urea-formaldehyde, resorcinol-formaldehyde, or melamine-formaldehyde.[10, 11] Today, however, industrial lignin can only substitute for a small portion of phenol in PF or other resins due to the low number of lignin functional groups capable of reacting with formaldehyde under alkaline-catalyzed condition.[11]

Lignins, being phenolic polymers, have long been considered for use as a phenol replacement in various polycondensation resins such as PF resin. Most of the published papers dealing with the partial replacement of phenol by lignin focused mainly on the use of the resins as adhesive or molding powder, although Pimenta and Frollini[12] have studied the use of lignin-PF resins in the production of foam. For the same purpose, Mikhanov *et al.*[13] have used hydrolytic lignin.

A study on structural wood panels in which at least 35% of the PF resin solids have been replaced with organosolv lignin supports the possibility of using this type of lignin in resins for bonding crosswise laminates (i.e. plywood).[14] Wooten *et al.*[15] found that the best way to make lignin-modified PF adhesives is by first reacting the alkaline lignin with formaldehyde to form methylol lignin. In the case of Kraft lignin a substitution of up to 60% phenol is possible in PF adhesives if factors affecting the production of such adhesives such as mixing ratios are carefully controlled.[16, 17]

The reaction rates of the alkali-catalyzed condensation of formaldehyde with phenol, Kraft lignin and steam-exploded hardwood lignin have been studied by Gardner and McGinnis[18] by monitoring the consumption of formaldehyde under various conditions. The results show that Kraft lignin has a faster rate of reaction with formaldehyde than steam-exploded lignin, and that the rate of reaction between formaldehyde and lignin is dependent on the availability of reactive sites on the lignin macromolecule. The effect of substitution levels of phenolic resol by phenol-lignin mixture on the reactivities and physical properties of the resins were evaluated by differential scanning calorimetry (DSC) analysis and by testing of the resin impregnated paper laminates. The lignin resols all showed a lower reactivity and poorer physical properties than the pure phenol resol.[19] The higher the substitution level, the lower the reactivity and the poorer the physical properties.

Kharade and Kale[20] found that the incorporation of lignin into a phenolic resin or molding powder as a filler enhanced the electrical properties. The change in the tensile strength was much less if the molding powder was made from phenol, lignin and formaldehyde. The novolak resin based on phenol alone and lignin may not be compatible, as lignin-filled resins exhibit reduced mechanical properties.

Data obtained by Piccolo *et al.*[21] using lignin from sugar cane bagasse show that during resol synthesis lignin reacts chemically with the phenol growing chain, acting as a chain extender. Molded resins, especially those prepared with 40% lignin, exhibit modulus retention at elevated temperatures. This constitutes a promising result since the presence of lignin leads to an improvement in the mechanical properties of the obtained thermoset.

Acetosolv and formacell lignins from sugar cane bagasse were used as phenolic component in resol-type resins by Benar and Schuchardt.[22] In spite of some variations in functional group content and molecular weight distribution, no differences were observed for the synthesis of resols using these two organosolv lignins. Benar *et al.*[23] have also used similar lignins from *Eucaliptus grandis* chips and concluded that up to 50% of phenol in the PF-resol can be replaced by these lignins as long as the total solid content of the resin formulation is maintained at 38%. The viscosity control becomes difficult for higher solid contents. Utilization of these lignins in the resin synthesis allows a significant reduction of reaction time. A report revealed that phenolated steam-exploded lignin based resins had intrinsic retardation in cure as compared with phenolic resin.[24] The synthesis and analysis of a phenol-formaldehyde lignin plastic have been done by Glasser *et al.*[25, 26]

Olivares *et al.*[27] have used lignin from *Pinus radiata* black liquor as a copolymer in PF resin. Methylolation, demethylolation and ultrafiltration separation of high molecular weight fractions were carried out to give lignins with different reactivity. The mechanical properties and water resistance of the particle board panels manufactured with the resins obtained from the lignins were then tested. The best resin was found to be the one composed of 18.8% ultra filtrated high molecular weight lignin, 22.9% phenol and 58.3% formaldehyde.

Mathiasson and Kubat[28] reported the results of a study exploring the possibility of using unmodified Kraft lignin as a binder in particle boards manufactured using high frequency (27 MHz) heating. The authors found that mechanical properties of the boards varied with their density. For instance, the modulus of elasticity increased from 2.5 to 6.1 GPa when density changed from 700 to 1200 kg/m^3. Similar results were obtained when Kraft lignin was replaced with thermoplastics such as vinyl chloride vinyl acetate (VC-VAc) copolymer, polyvinyl chloride (PVC), low density polyethylene (LDPE), nylon 11 and an acrylic copolymer.

The chemorheology of phenol-formaldehyde resin, PF mixed with ammonium salt lignosulfonates (ALS) filler, and PF mixed with methylolated ALS (MALS) filler has been studied by Peng and Riedl[29] in order to characterize their isothermal cure behaviors. The results show that both ALS and MALS have a lower reactivity than PF and act as diluents in the PF resin. Increasing the amount of ALS or MALS in PF increases the cure rate constant. These results are in agreement with data obtained by DSC. The efficiency of the adhesives based on phenol and lignin has been determined by Pecina *et al.*[30] according to the molar ratio of the components, the nature of components, the degree of polycondensation, the solids content and the type and amount of the hardening agents.

4. EPOXY - LIGNIN ADHESIVES

Adhesives or molding products from epoxy resins (EP) based on lignin have been known since the 1960's. In an attempt to obtain epoxy-Kraft lignin adhesives, Shiraishi [31] phenolated the lignin with bisphenol-A. The resins obtained from the phenolated lignin were found to be completely soluble in certain organic solvents including acetone. An enhanced gluability was noted for two Kraft lignins with different purity, suggesting an active participation of some sugar components. The addition of calcium carbonate also improved the gluability. The author also used BF_3-ethyl etherate, a known catalyst for introducing the phenolic groups into the lignin. The obtained EP-lignin resin was tested as adhesive for plywood with hot-pressing at 140 °C. The results showed a satisfactory waterproof performance and permitted the adoption of hot pressing times as short as 3 minutes for producing three-ply plywood panels with 6 mm thickness. Nieh and Glasser[32] synthesized an EP resin based on epichlorohydrin and hydroxypropyl lignin. A mixture of a quaternary ammonium salt and KOH was used as catalyst. The EP-lignin resin was cross-linked with diethylenetriamine, amine terminated poly(butadiene-co-acrylonitrile) and phthalic anhydride. Sol fraction swelling behavior and dynamic mechanical characteristics of cured resin were studied in relation with the curing parameters. Tomita *et al.*[33] have reported the preparation of an EP resin with ozonized lignin and diethylenetriamine or hexamethylenediamine as curing agents.

Feldman *et al.*[34-40] reported a comprehensive study on the use of EP-Kraft lignin polyblends as adhesives. The study included miscibility by DSC, dynamic mechanical analysis (DMA) and nuclear magnetic resonance (NMR) analyses; interaction between EP-polyamine network and lignin by FTIR; the effects of a silane coupling agent; and the polyblend components on the adhesive characteristics. The research was done using a commercial structural EP adhesive used for bonding steel, aluminum, concrete and other materials and a Kraft lignin. The main components of the EP adhesive were a prepolymer of diglycidyl ether of bisphenol-A and an aliphatic polyamine hardener. The initial low viscosity of the EP-hardener system facilitated the dispersion of lignin even in amounts up to 40 wt%. At 50 wt% the viscosity of the adhesive prevented a good workable mix from being made. Single-lap shear joints with aluminum substrates were used to evaluate the adhesive shear strength by tension (ASTM-1002). The polyblends with up to 30 wt% lignin showed a considerable improvement in adhesive joint shear strength in comparison with EP adhesive. The highest value was obtained at a lignin loading of 20 wt%; at 40 wt% loading the shear strength was 16% lower than that of control EP. The cost of the polyblend with 20 wt% Kraft lignin that showed an adhesive shear strength of 178% of the control EP was estimated to be only 81% of the cost of the commercial EP adhesive. The DSC data show that when the lignin content in EP is 20 wt% the polyblend exhibits a single T_g, indicating a mono phasic system. At a lignin content of 25 wt% or more the polyblends exhibit two T_g's. The first one is close to the T_g of EP with a tendency to increase as the lignin load increases (Figure 1), indicating the epoxy rich phase. The second T_g is in the proximity of the T_g of lignin, indicating the lignin rich phase.[36, 39]

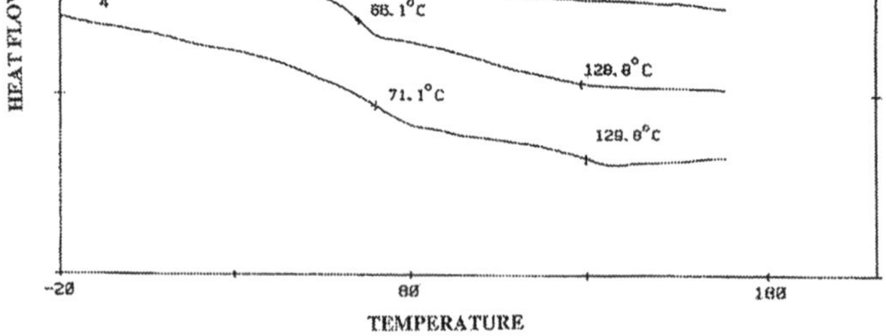

Figure 1. DSC diagrams for EP and EP - lignin polyblends (run 1) (Reprinted with permission from ref. 39. Copyright John Wiley & Sons, Inc. 1991).

Some tanδ diagrams obtained by using DMA method are displayed in Figure 2. Since the polyblend with 20 wt% lignin exhibits a single T_g, characteristic for a mono phasic system, it can be considered miscible. For the polyblend with 40 wt% lignin, a second peak on the tanδ curve, located in the temperature region characteristic of lignin motion, indicates immiscibility. These results explain the decrease in shear strength of the mixtures with more than 20 wt% lignin.

From the shape of the proton spin - lattice relaxation - time curves (calculated from solid-state NMR spectra) showing $T_{1\rho}H$ as a function of lignin content, it seems reasonable to assume that a stoichiometric reaction has occurred at a blend loading of 20 wt% lignin.[39] These findings not only reveal the basic morphology of these new polyblends, but also emphasize their potential applications in various fields since miscible systems usually can provide superior mechanical properties. The FTIR data clearly indicate the feasibility of a reaction between lignin and the polyamine hardener in an EP-lignin system through their carbonyl and amine groups, respectively, and that this can occur only after curing at relative elevated temperatures.

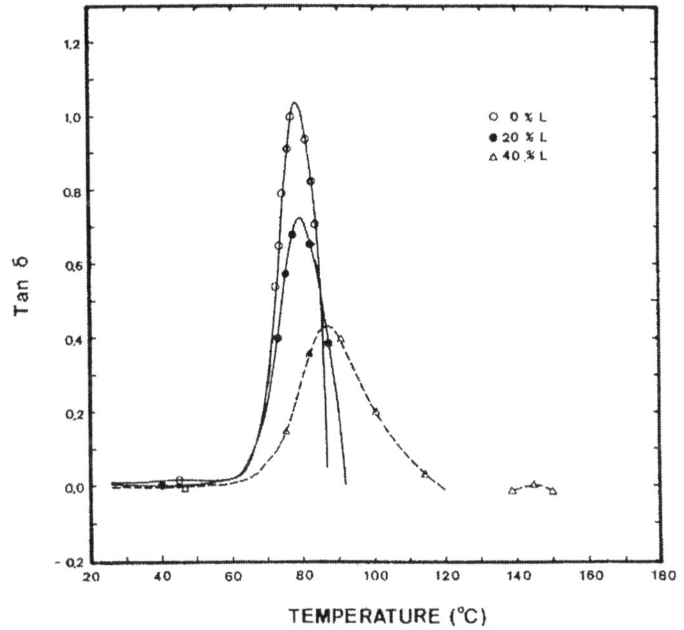

Figure 2. tanδ vs. temperature for EP, EP-lignin polyblends: specimens mounted in vertical mode; heating rate 5 °C/min (reprinted with permission from ref. 39. Copyright John Wiley & Sons, Inc. 1991).

Quantitative data concerning the reactivity of lignin towards the polyamine hardener have been obtained using a chemical titration technique. Kinetic data indicate that during the cross-linking process, the reaction between lignin and polyamine can not compete with alphatic primary and secondary unaccelerated amine cures of diglycidyl ether of bisphenol

A (DGEBA) resins.[34] FTIR data support the speculation that lignin bonds to the EP-polyamine network via unreacted amine groups of the hardener.[36]

The addition of a silane coupling agent (2 wt%) leads to an improvement of adhesion between the EP-lignin adhesive and inorganic substrates and of its durability.[37] Surface study of the contact angle values proves that the wettability is improved by adding lignin and organo-silane to EP adhesive. Such a polyblend has much higher values of adhesion than the EP adhesive.

Studies of the influence of lignin type on the adhesive properties showed that hardwood commercial lignins (Tomlinite, Eucalin, steam-exploded lignin) impart a better adhesivity to EP-lignin polyblends than the softwood lignin (Indulin). Tomlinite Kraft lignin gives the highest adhesive joint shear strength to the EP with 20 wt% of the lignin. The improvement can be related to the molecular weight of the lignin. The addition of 20 wt% mineral filler does not affect significantly the adhesive joint shear strength of EP-Kraft lignin adhesive.

Besides EP resin and lignin, a third polymeric component such as PVC or phenoxy resin was found to contribute to the improvement of adhesive properties.[38] The durability tests showed that EP-lignin adhesives had better weatherability and photostability than the control EP adhesive.

EP-thiolignin resins were prepared by Ito and Shiraishi[41] by dissolving the thiolignin in bisphenol A, followed by glycidylation. Phenolation of the thiolignin, achieved through its dissolution in bisphenol A, was found to improve both the dry and waterproof gluability of the EP-thiolignin adhesives. The cross-linking density of the cured polyblend was greater at a higher degree of phenolation. The purity of thiolignin did not have significant influence on the gluability of the prepared adhesives, but the catalysts used for phenolation and for glycidylation and the addition of calcium carbonate as a filler had tremendous influence on the gluability.

5. OTHER ADHESIVE AND SEALANTS WITH LIGNIN

Feldman and Baskaran[42] have prepared a three-component adhesive sealant with improved mechanical properties by blending a sealant based on poly(dimethyl siloxane), PDS, with a suspension type PVC and Kraft lignin. A silicone sealant was selected because of its good blending properties. The sealant was tested on California redwood treated with two different primers. Data showed that by adding a small amount of lignin, the modulus of the sealant based on PDS + 10% PVC was improved. The application of commercial primers to the wood substrate was found to significantly improve the adhesion of PDS + 10% PVC and lignin. A polyblend has also been prepared from an acrylic (terpolymer) sealant and Kraft lignin.[43] This polyblend can be easily applied to glass, masonry, metal, wood, porcelain or other typical building materials. DSC diagrams clearly show the incompatibility of the acrylic sealant with Kraft lignin, the latter component being used in amounts up to 20 wt%. The interpretation of IR spectra of this polyblend is difficult due to the interference of the mineral filler contained in the sealant. However, the insignificant changes in the acrylic spectrum after addition of lignin suggest that there is no interaction between the two components. The improvement of mechanical properties of the sealant may be due to the infusion of lignin which acts as a reinforcing filler. The addition of lignin up to 15 wt% appears beneficial in improving mechanical properties and durability of the acrylic sealant.

Glasser *et al.*[44, 45] have formulated hydroxyalkylated Kraft, organosolv, steam-exploded and sulphuric acid lignin samples with cross-linking agents and used them as wood adhesives in emulsion and solvent-based systems. The same authors have also undertaken a series of studies related to the formation of polyurethane films prepared from chain-extended hydroxypropyl lignin.[46, 47]

For polyblends from lignin and polyurethane sealant, Feldman *et al.*[48] have confirmed the immiscibility of the lignin with the sealant by DSC analysis and CP-MAS NMR. Furthermore, the interaction potential of various commercial lignins has been shown by surface analysis to decrease in the following order: Tomlinite > Indulin AT > Eucalin.[49-51] The addition of lignin to the polyurethane sealant, however, increases the modulus of elasticity of the formulations investigated.[52] Feldman *et al.*[53] have also obtained interesting data by blending a polyurethane sealant with maleic anhydride grafted lignin. Recently, Gandini *et al.*[54] have reported polyurethanes based on oxygen-organosolv lignin.

6. POLYOLEFIN LIGNIN POLYBLENDS

The aim of adding lignin to polyolefins is to improve their mechanical properties, thermal and light stability or to provide biodegradability. As early as 1983, Lyubeshkina[55] showed the potential of dry sulphate lignin in the filling and modification of thermoplastics to obtain frost-resistant polypropylene compositions and reprocessable polymeric material wastes. Recently, Simionescu *et al.*[56] have obtained blends of polyolefins and lignosulfonates with good mechanical and thermal properties and demonstrated that lignosulfonates could impart biodegradation susceptibility to isotactic polypropylene (PP). Vasile *et al.*[57] have carried out a comprehensive study on the compatibility between isotactic PP and epoxy modified lignosulfonates.

Similar to other polymers, polyolefins need to be thermally stabilized. Several types of thermal stabilizers - organic compounds containing phenolic hydroxyl groups, aromatic amines, organic sulphur or phosphines have been widely used in the polymer industry. Lindberg *et al.*[58] have shown that lignosulphonate can provide thermal stabilization to low-density polyethylene (PE) when the blending of PE with lignosulphonate is done at temperatures over 190 °C. The increasing amount of lignosulphonate after a slight cross-linking may contribute, in the opinion of these researchers, to the increase of branching and thus a more ideal flow behavior. The improvement of thermal stability is dependent on the quality of the lignosulphonate and on the number of the sulphonate groups. The stabilization effect can be explained by the formation of the lignin phenoxy radicals capable of inhibiting PE degradation by coupling to its macroradicals. The PE thermally stabilized by lignosulfonate and the one stabilized by commercial thermal stabilizers show no difference in their mechanical properties and electrical resistance.

The results obtained by Ait-Kadi *et al.*[59] on lignin - high density polyethylene (HDPE) blends show that it is possible to obtain composites with high lignin content (up to 50%) having very good mechanical properties and low cost. In their study lignosulphonate was modified by grafting ethylene in hexane using a Ziegler-Natta catalyst system. The improvements in mechanical properties upon grafting as well as the morphological differences between the grafted and non-grafted lignin composites are an indication of enhanced interactions at the interface, which means better adhesion, dispersion and wetting.

Model predictions were found to be in logically good agreement with the basic concepts of particulate composites in general and with those introduced through catalytic grafting in particular.

Using prehydrolysis lignin and organosolv lignin in blends for PP films, Kosikova *et al.*[60-62] found from rheological data that such lignins in amount up to 10 wt% had a good compatibility with PP. Thermal and structural data indicated chemical interaction between lignin and PP via cross-linking. Due to their ability to absorb UV light, such PP-lignin films could potentially be used in the food packing industry.

Kubat and Stromvall[63] studied the polyblends made from polyethylenes (low and high density) and lignosulphonates of varying purity up to a maximum lignin content of 70%. The mechanical properties of the injection molded materials from these polyblends showed an increase in modulus (stiffness) with increasing lignosulphonate content. The changes in the breaking strength were relatively small, whilst the elongation at rupture and the impact strength were reduced. When blended with polystyrene, the lignosulphonate with the lowest purity yielded a material transparent to light.

Chodak *et al.*[64] investigated the effect of adding a small amount of lignin or a lignin derivative on the cross-linking reaction of polypropylene initiated by peroxides and on the thermooxidation of the cross-linked product. Unmodified oak lignin was found to be the only additive affecting the cross-linking process significantly. An increase of thermooxidation stability was observed in samples containing lignin or most of its derivatives, but the effect was not significant.

Rosch and Mulhaupt[65] have used organosolv lignin as a filler for PP and poly(ethylene-co-vinyl acetate) -EVA. They measured the mechanical properties of the polyblends such as Young's modulus, yield stress, fracture stress and elongation at break of the blends at various lignin content (0 to 30 wt%) and found pronounced matrix reinforcement for both of the lignin filled thermoplastics with the increase of the lignin content. For example, the Young's modulus of the EVA - 30% wt% lignin polyblend doubled that of the EVA while the high elongation at break was retained, due to the better interfacial adhesion in the EVA - lignin polyblend. Recently, Rozman *et al.*[66] have used lignin in composites made from coconut fibers and PP. Their study demonstrates that such lignin composites possess higher flexural strength than and similar tensile properties to the control composites.

7. PVC - LIGNIN POLYBLENDS

Due to the commercial value of poly(vinyl chloride), PVC, there are already established blends of PVC with other common or less common polymers. The practical benefits of these blends are: improved processibility and impact strength, and increased T_g. Because of the abundance of lignin, its functions in plants and its functional groups, many studies have been done to examine the properties of lignin - PVC blends. The functional groups of PVC and lignin are such that proton-donor/proton-acceptor interactions can readily occur (e.g. between the carbonyl group of lignin and α-hydrogen of PVC or between the hydroxyl group of lignin and the chlorine atom of PVC). According to Hofmann,[67] such interactions likely contribute to the miscibility or partial miscibility of PVC with lignin.

Polyblends of an unplasticized PVC filled with TiO_2 (up to 10 wt%) and organosolv lignin (up to 10 wt%) were obtained by melt mixing and compression molding.[68] The mechanical (mainly tensile strength) properties of the polyblends were evaluated and their morphology studied by DMA, DSC and SEM. The tensile strength increased with an increase of lignin content without a major impact on elongation up to 7.5 wt% lignin for formulations with and without TiO_2. Above this content the tensile strength still increased but the elongation decreased. Rheometric data show that the lignin present does not affect the processibility of the PVC-lignin polyblends.[68] SEM and DSC results indicate the immiscibility of these two polymers. The DSC diagrams show the presence of two endothermic relaxation peaks in the T_g region. The fact that they are not perfectly separated and that the relaxation energy of the principal peak, which corresponds to PVC molecular relaxation is higher in the polyblends than in PVC control is an indication of the likely miscibility of PVC and lignin in the polyblends. DMA results show that the presence of lignin leads to the breaking of the intermolecular bonds within PVC to certain extent. This is supported by the decrease in the latent heat of melting of the PVC crystallites determined by DSC. It is known that such crystallites act as ties of cross-links in PVC.

The fact that the polyblends toughness is not affected by the decrease of the number of crystallites, as demonstrated by the storage moduli and the tensile strength, can be explained by a reduction of the free volume of the polymers due to their polar-polar interaction. Although the PVC-lignin polyblends are not miscible or only partially miscible, their morphology (with up to 7.5 wt% lignin) does not affect their performance, as the mechanical properties have indicated.[68, 69]

In polyblends with PVC, commercial lignins obtained through various techniques prove to behave differently, although the mechanical properties of the polyblends, as well as IR data suggest the same interaction between some of their functional groups. It has been established that this interaction decreases in the following order: Indulin (a softwood Kraft lignin) > Alcell (an organosolv lignin) > HWPSL (hardwood partially soluble lignin) > SWLS (softwood lignosulfonate) and HWLS (hardwood lignosulfonate). After artificial weathering (UV exposure, moderate temperature and humidity) yield and breaking strength of the polyblends are, in most cases, better than those of PVC (Table 1).[70] A major drawback of blending PVC with these lignins is the diminished impact strength and poor weatherability of the polyblends. A slight decomposition of lignins during the high temperature processing of the rigid PVC can partially explain their detrimental effect on these properties. A tentative rating of technical lignins on their positive influence on PVC properties would place softwood lignins above the hardwood ones.

Table 1. The influence of artificial weathering on the yield and breaking strength (reprinted with permission from ref. 70. Copyright John Wiley & Sons, Inc. 1997)

Type of polyblend	Yield strength change	Breaking strength change
PVC control	+2%	-2%
PVC-Indulin	+8%	+1%
PVC - Alcell	+8%	-5%
PVC - SWLS	+7%	+2%
PVC -HWPSL	+7%	Brittle fracture
PVC - HWLS	+1%	Brittle fracture

Recently, Feldman *et al.*[71] reported an investigation into the use of plasticizers for reducing the degree of association between lignin molecules to overcome their adverse effects on the mechanical properties of its polyblends with a VC-VAc (vinyl chloride-vinyl acetate) copolymer containing 9.3 wt% VAc. Polyblends prepared with different plasticizers, an organosolv lignin and the VC-VAc copolymer were studied by thermal analysis and mechanical testing. Variations in the mechanical properties of these polyblends were found to correlate well with the plasticizer efficiency and lignin dispersion in the case of organosolv lignin.

Galimov *et al.*[72] have studied the influence of hydrolysis lignin on the properties of plasticized PVC sheets. At a low lignin loading a sharp fall in the effective viscosity of the melts of PVC blends was observed over a wide range of temperatures. It is known that when a small amount of lignin is added in plasticized formulations, an even dispersion takes place. Arrangement ordering and an increase of chain mobility were also observed in the direction of shear deformation due to partial rupture of the connecting chains while the melt was flowing; this led to the decrease of the viscosity. The increase of the viscosity with an increase in hydrolysis lignin content is explained by the immobilisation of the volume of the polymer matrix by the lignin surface and by the increasing interaction between PVC and lignin. The authors found that the presence of hydrolysis lignin in PVC sheets led to an improvement in their technological and operating characteristics.

8. OTHER POLYBLENDS WITH LIGNIN

The amounts of lignin incorporated into polymeric materials are in general limited by the brittleness provided by lignin or by phase separation due to immiscibility. However, Sarkanen *et al.*[73] have obtained for the first time polyblends of poly(vinyl acetate), PVAc, containing 85 wt% Kraft lignin and two plasticizers; such polyblends showed promising mechanical properties. The tensile strength and modulus of the thermoplastic formulation increased linearly with the molecular weight of the Kraft lignin. Melt-flow index measurement suggested that such Kraft lignin based polyblends could be successfully extrusion molded. Li and Sarkanen[74] have also obtained alkylated 100% Kraft lignin-based polymeric materials with tensile properties virtually identical to those of polystyrene.

Polymer blends of hydroxypropyl cellulose (HPC) and organosolv lignin were prepared by Rials and Glasser.[75] HPC and the lignin showed partial miscibility as demonstrated by a single T_g up to a composition of 40 wt% lignin, above which phase separation was detected. No evidence of hydrogen bonding or other types of association was gathered by FTIR. Young's modulus and tensile strength increased by about 140% and 150%, respectively, when the lignin content was increased to 15% for some molded products. Ultimate elongation decreased sharply with the amount of lignin rising to 15%. The development of a fibrillar supermolecular morphology was clearly seen for the molded blends by SEM. Glasser *et al.*[76] have also prepared thermoplastic blends of several biodegradable polymers such as cellulose acetate butyrate (CAB), a starch-caprolactone copolymer/blend (SCC) and poly(hydroxybutyrate) with organosolv lignin esters (acetate, butyrate, hexanoate and laurate). Studies of these polyblends showed widely different levels of interaction between the components. Blends of the lignin acetate and lignin butyrate with CAB exhibited a high level of compatibility that was lost when the acyl substituent increased in size. The addition of unmodified lignin to poly(hydroxybutyrate) and starch-caprolactone copolymer greatly retarded crystallization and lowered the melting temperature. The addition of lignin led to a significant increase of modulus at room temperature due to an increased crystallinity of the lignin blends.

Baumberger *et al.*[77, 78] have obtained films containing 0-30% of commercial Kraft lignin mixed with wheat starch. After extrusion the films were thermally molded. The authors found that Kraft lignin could be incorporated into starch films up to a concentration of 30% with a significant decrease of extensibility at higher water content. The low molecular weight lignin fractions were most probably responsible for a partial compatibility between the two polymers in cast films. The presence of lignin reduced the overall water affinity of the thermally molded films.

The preparation of polyblends from poly(vinyl alcohol), PVA (with 80, 87, and 98% hydrolysis degree) and alkaline lignin from sugar cane bagasse was carried out by Corradini *et al.*[79] DSC results showed miscibility of the PVA-80-lignin and PVA-87-lignin systems and immiscibility of the PVA-98-lignin. Apparent activation energy of pyrolysis (E_a) was evaluated from TGA for individual polymers and their blends. In the case of PVA, E_a increased with increasing acetyl groups content. E_a value of the blends suggested a dependence of thermal stability on the hydrolysis degree of PVA.

As a polyol, lignin can be used as a macromonomer for the synthesis of cross-linked polyesters. Gandini *et al.*[80, 81] obtained copolyesters based on Kraft lignin with sebacoyl chloride and terephthaloyl chloride. Ethylene glycol was used as a co-monomer and a solvent for the Kraft lignin. The polyesters showed elastomer properties at room temperature due to the presence of segments arising from polyethylene glycol. The esterification with only sebacoyl chloride led to a duroplast due to the polyfunctionality of lignin and cross-linking.

In his comprehensive review, Faix[82] has concluded that engineering plastics and thermosetting resins will be the fields of utilization for lignin and its derivatives with high production potentials in the future.

9. RUBBER - LIGNIN POLYBLENDS

Literature data show that lignin can be used in rubber materials as a filler, pigment, modifier or extender. Most of the studies have been done with the aim to use lignin as a filler. As early as 1947 lignin was introduced as a reinforcing filler in natural and synthetic rubber latices.[83, 84] The first reported practical test of the use of Kraft lignin in tires was performed in 1964 where half of the carbon black was substituted by Kraft lignin, resulting in an improvement in the abrasion index of the rubber. Natural rubber, polybutadiene, nitrile rubber (NBR), neoprene (polychloroprene), styrene-butadiene copolymer rubber (SBR) in the form of latexes are compatible with lignin; even in high amounts lignin does not affect the stability of such latexes. Furthermore, sugar cane bagasse lignin has been found as an efficient thermal stabilizer for polybutadiene rubber, capable of substituting phenol stabilizers commonly used for this purpose.[85] It has also been reported that rubbers reinforced with Kraft lignin are lighter than those filled with carbon black or silica and that the presence of lignin improves their ozone resistance.

In principle, lignin filled rubber can be obtained through the following two ways: (1) the coprecipitation of lignin and rubber from a lignin-rubber latex mixture, and (2) by introducing hydrated lignin into rubber with a mixing equipment. Lyubenskina[55] has shown that an important feature in the preparation of lignin-rubber systems was the introduction of lignin into the latex, since its addition into dry rubber did not lead to the reinforcement. The reason for this is the binding tendency of lignin particles to one another through hydrogen bonds, which leads to the growth of its particles and prevents their dispersion during the rolling operation of the filled rubber.

It has also been demonstrated that besides surface activation properties and other factors, lignin particle size (usually 100-200μm for technical lignins) is also very important in the reinforcement of rubbers, and this effect increases significantly when lignin is precipitated in the presence of sodium silicate. As the result of coagulation, fine particles and thus a rubber with low stress parameters for 300% elongation and a high hardness are formed. The lignin-filled vulcanizates, thus, exhibit a lower resistance to wear than carbon black filled rubber. The mechanism for the reinforcement of rubbers with lignin is complex and has not been completely illustrated. For Kraft lignin, apart from its physical characteristics such as surface activity and particle size, interaction of the hydroxyl groups of lignin with the π electrons in the double bonds of the rubber has been found to be important.[86]

Kraft lignin (density = 1.3 g/cm^3) compares very favorably with carbon black, a common, low density (1.81 g/cm^3) reinforcing filler for many rubber products.[87] From a technical standpoint of view, lignin can compete well with carbon blacks in rubber reinforcement. For example, lignin can provide a higher tensile strength than carbon black for styrene-butadiene copolymer rubber (SBR) (Table 2).[88] Abrasion resistance is very important for some of the rubber products, especially for tires. It has been found that for lignin-reinforced SBR the abrasion resistance is governed not by failure characteristics but by the viscoelastic properties of this complex polyblend. Particle size affects the failure characteristics more than the viscoelasticity, and according to Falkehag,[88] no correlation can be established between the lignin particle size and the abrasion resistance of the rubber.

Table 2. Physical properties of lignin and carbon black reinforced SBR at 68 parts of lignin per 100 parts of rubber (Oil-extended SBR) (reprinted with permission from ref. 88. Copyright John Wiley & Sons, Inc. 1975)

Property	Lignin	HAF[a] carbon black	ISAF[b] carbon black
Modulus (Mpa)	3.6	4.2	5.1
Tensile strength (Mpa)	21.7	17.5	20.5
Elongation (%)	720	720	750
Hardness (Shore)	54	56	61
Tear resistance (Mpa)	2.4	2.1	2.3
Corrected Pico Abrasion	86	91	114

[a] High abrasion furnace.
[b] Intermediate super abrasion furnace.

10. PROSPECTS OF LIGNIN IN POLYMER SYSTEMS

Research into the utilization of various technical lignins is among the most challenging and promising areas in natural polymer science. The problems facing their utilization in polymer systems (i.e. the drawbacks of lignin) include variability, inhomogeneity (polydispersity), low solubility, high thermal glass transition temperature and a complex chemical and macromolecular structure posing difficulties especially for their quality control. It is well known, for instance, that lignins from the steam explosion and organosolv processes are different from their counterparts obtained from the Kraft and sulfite pulping processes.

It is not surprising that outside the pulp and paper industry technical lignin has been used mainly as a cheap extender and macromolecular substitute for end products that involve surface-active and dispersing characteristics. Cross-linking lignin with suitable resins such as phenol-formaldehyde, resorcinol-formaldehyde and isocyanates, or using lignin in blends with linear synthetic polymers capable of functional group interaction, seems to offer a promising avenue for using larger quantities of lignin in thermosetting network systems or in thermoplastics. In order to arrive at such applications, research must focus on finding new ways to circumvent the drawbacks of lignin.

Lindberg *et al.*[89] has summarized the positive and promoting factors in the chemical utilization of lignin as follows:

- readily available in huge amounts
- high energy content owing to its aromatic nuclei
- its reactive groups on the carbon backbone can be used for a wide range of substitution and addition reactions
- good compatibility with several important basic chemicals
- excellent colloidal and rheological properties especially in the case of lignosulphonic acid
- good absorbent and ion exchange and adhesive properties
- a direct source of various kinds of phenolic and aromatic compounds
- if disposed of as waste, black liquor can be a serious pollution hazard.

11. CONCLUSIONS

Lignin is a renewable, non-toxic, commercially available and low cost natural resource that has a real potential of being used as a basic raw material for the polymer industry. In spite of many years of developmental effort, this potential has not yet been fully realized. Some of the difficulties encountered in the production of useful polymeric end products arise from the fact that the processes by which lignin is separated from wood alter its structures in such a fashion as to reduce its chemical reactivity. Recent developments in the lignin - synthetic polymer and in lignin - modified cellulose systems have shown an increased awareness of both the inherent limitations and the potential of the reactivity of lignin, and of the vast prospects of its application in thermoplastics, thermosets, adhesives, sealants, molding powders, rubbers, foams, films and many others.

12. REFERENCES

1. J. W. Barlow, Polymer blends for engineering applications, *Anales de la Asociacion Quimica Argentina* **84**(1), 87-93 (1996).
2. X. Cao, M. Jiang, and T. Yu, Controlable specific interaction and miscibility in polymer blends; hydrogen bonding and morphology, *Makromol. Chem.* **190**, 117-128 (1989).

3. H. A. Schneider, The Gordon-Taylor equation. Additivity and interaction in compatible polymer blends, *Makromol. Chem.* **189**, 1941-1955 (1988).
4. K. A. Solen and M. C. Kuchar, Chemical compatibility of polymeric materials, *Chem. Eng. Edu.* (Spring), 94-98 (1990).
5. M. Ratzsh and G. Handel, Interactions between polymers, *Makromol. Chem. Macromol. Symp.* **38**, 81-98 (1990).
6. N. C. Liu and W. E. Baker, Reactive polymers for blend compatibilization, *Ad. Polym. Technol.* **11**(4), 249-262 (1992).
7. A. Simmons and A. Eisenberg, Miscibility enhancement in ionomeric blends, *Polym. Prepr.* (ACS Div. Polym. Chem) **27**(1), 341 (1986).
8. J. M. Rodriguez-Parada and V. Percec, Interchain electron-donor-acceptor complexes - a model to study polymer-polymer miscibility, *Macromolecules* **19**, 55-64 (1986).
9. J. M. G. Cowie and C. Love, The use of molecular recognition to obtain selective blending in polymer systems, *Polym.* **42**, 4783-4789 (2001).
10. E. Roffael and B. Dix, Lignin and lignosulfonate in non-conventional bonding - an overview, *Holz als Roh -und Werkstoff* **49**, 199-205 (1991).
11. A. Pizzi, F-A. Cameron, and G. H. van der Klashorst, Soda bagasse lignin adhesives for particle board, in: *Adhesives from Renewable Resources,* edited by R. W. Hemingway, A. H. Conner, and S. J. Branham (ACS Symposium Series 385, Washington, 1987), pp. 82-95.
12. M. J. de A. Pimenta and E. Frollini, Lignin utilization as "macromonomer" in the synthesis of phenolic type resins, *Anais. Assoc. Bras. Quim.* **46**(1), 43-49, (1997).
13. S. A. Mikhanov, V. M. Golubev, and E. S. Bilimova, Use of lignin in the production of phenol-formaldehyde foam plastic, *Intl. Polym. Sci. Technol.* **14**(6), 74 (1987).
14. P. M. Cook and T. Sellers Jr., Organosolv lignin-modified phenolic resin, in: *Lignin, Properties and Materials*, edited by W. G. Glasser and S. Sarkanen (ACS Symposium Series 397, Washington, 1989), pp. 324-333.
15. A. L. Wooten, T. Sellers Jr., and P. M. Tahir, Reaction of formaldehyde with lignin, *For. Prod. J.* **38**(6), 45-46 (1988).
16. H. Pecina, G. Kuhne, Z. Bernaczyk, and O. Wienhaus, Lignin-Phenol-Bindemittel fur die Holzwerkstoffherstellung, *Holz als Roh- und Werkstoff* **49**, 391-397 (1991).
17. H. Pecina, G. Kuhne, Z. Bernaczyk, and O. Wienhaus, Lignin-Phenol-Bindmittel fur die Holzwerkstoffherstelung *Holz als Roh- und Werkstoff* **50**, 407-409 (1992).
18. D. J. Gardner and G. D. McGinnis, Comparison of the reaction rates of the alkali-catalyzed addition of formaldehyde to phenol and selected lignins, *J. Wood Chem. Technol.* **8**(2), 261-288 (1988).
19. R. E. Ysbrandy, R. D. Sanderson, and G. F. R. Gerisher, DSC thermal analysis of phenol and phenol-lignin extended resols and their physical behaviour in paper laminates, *Das Papier* **46**(2), 62-67 (1991).
20. A. Y. Kharade and D. D. Kale, Effect of lignin on phenolic novolak resins and molding powder, *Eur. Polym. J.* **34**(2), 201-205 (1998).
21. R. S. G. Piccolo, F. Santos, and E. Frollini, Sugar cane bagasse lignin in resol-type resin: alternative application for lignin-phenol-formaldehyde resins, *J. Macromol. Sci. Pure. Appl. Chem. A* **34**(1), 153-164 (1997).
22. P. Benar and U. Schuchardt, Organosolv lignins from sugar cane bagasse as component in resols, *Fifth European Workshop on Lignocellulosics and Pulp* (Aveiro Portugal, August 30 - September 2, 1998), pp. 45-48.
23. P. Benar, A. G. Goncavles, D. Mandelli, and U. Shuchardt, Eucalyptus organosolv lignins: study of the hydroxymethylation and use of resols, *Bioresource Technol.* **68**,11-16, (1999).
24. H. K. Ono and K. Sudo, Wood adhesives from phenolysis lignin. A way to use lignin from steam-explosion process, in: *Lignin Properties and Materials,* edited by W, G. Glasser and S. Sarkanen (ACS Symposium Series 397, Washington, 1989), pp. 334-346.
25. P. C. Muller and W. G. Glasser, Engineering plastics from lignin, VIII. Phenolic resin, Prepolymer synthesis and analysis, *J. Adhesion* **17**,157-174 (1984).
26. P. C. Muller, S. S. Kelly, and W. G. Glasser, Engineering plastics from lignin, IX. Phenolic resin, synthesis and analysis, *J. Adhesion* **17**, 185-206, (1984).
27. M. Olivares, J. A. Guzman, A. Natho, and A. Saavedra, Kraft lignin utilization in adhesives, *Wood Sci. Technol.* **22**, 157-165 (1988).
28. A. Mathiasson and D. G. Kubat, Lignin binder in particle boards using high frequency heating. *Holz als Roh-und Wekstoff* **52**, 9-18 (1994).
29. W. Peng and B. Riedl, The chemorheology of phenol-formaldehyde thermoset resin and mixtures of the resin with lignin fillers, *Polymer* **35**(6), 1280-1286 (1994).
30. H. Pecina, Z. Bernaczyk, O. Wienhaus, and G. Kuhne, Lignin-Phenol-Bindemittel fur die Holzwerkstoffherstellung, Die Reaktivitat von Lignin-Phenol-Bindemitteln, *Holz als Roh- und Werkstoff* **52**(1), 1-5 (1994).

31. N. Shiraishi, Recent progress in wood dissolution and adhesives from Kraft lignin, in: *Lignin Properties and Materials*, edited by W. G. Glasser and S. Sarkanen (ACS Symposium Series 397, Washington, 1989), pp. 488-495.
32. W. L. S. Nieh and W. G. Glasser, Lignin epoxide: synthesis and characterization, in: *Lignin Properties and Materials*, edited by W. G. Glasser and S. Sarkanen (ACS Symposium Series, 397, Washington, 1989), pp.506-514.
33. B. Tomita, K. Kurozomi, A. Takemura, and S. Hosoya, Ozonide lignin-epoxy resins; synthesis and use, in: *Lignin Properties and Materials*, edited by W. G. Glasser and S. Sarkanen (ACS Symposium Series 397, Washington, 1989), pp. 496-505.
34. D. Feldman and D. Banu, Kinetic data on the curing of an epoxy polymer in the presence of lignin, *J. Polym. Sci.* **26**(4), 973-983 (1988).
35. D. Feldman and M. Khoury, Epoxy-lignin polyblends II: Mechanical behavior and weathering, *J. Adhesion Sci. Technol.* **2**(2), 107-116 (1988).
36. D. Feldman, D. Banu and M. Khoury, Epoxy-lignin polyblends III: Thermal properties and infrared analysis, *J. Appl. Polym. Sci.* **37**, 877-887 (1989).
37. J. Wang and D. Feldman, Effects of organo-silanes on the adhesive properties of epoxy-lignin polyblends, *J. Adhesive Sci. Technol.* **5**(7), 565-577 (1991).
38. J. Wang, D. Banu and D. Feldman, Epoxy-lignin polyblends: effects of various components on adhesive properties, *J. Adhesive Sci. Technol.* **6**(5), 587-598 (1992).
39. D. Feldman, D. Banu, A. Natansohn, and J. Wang, Structure-properties relations of thermally cured epoxy-lignin polyblends, *J. Appl. Polym. Sci.* **42**,1537-1550 (1991).
40. D. Feldman, D. Banu, M. Lacasse, J. Wang, and C. Luchian, Lignin and its polyblends, *J. Macromol. Sci. Pure Applied Chem. A* **32**(849), 1613-1619 (1995).
41. H. Ito and N. Shiraishi, Epoxy resin adhesives from thiolignin, *Mokuzay Gakkaishi* **33**(5), 393-399 (1987).
42. D. Feldman and A. Baskaran, The effect of adding lignin to poly(dimethyl-siloxane) - poly(vinyl chloride) blends, *J. Adhesion* **27**, 231-243 (1989).
43. D. Feldman, M. Lacasse, and D. Banu, Contribution to the modification of an acrylic sealant with lignin, *J. Polym. Mater.* **5**, 131-139 (1988).
44. W. G. Glasser, C. Barnett, T. Rials, and S. Kelley, Synthesis and characterization of several different hydroxyalkyl lignin derivatives, *Proc. Int. Symp. Wood Pulping Chem.* (Tsukuba, Japan) **3**, 89-94 (1983).
45. W. G. Glasser, C. A. Barnett, T. G. Rials, and V. P. Saraf, Engineering plastics from lignin II: characterization of hydroxyalkyl lignin derivatives, *J. Appl. Polym. Sci.* **29**,1815-1830 (1984).
46. V. P. Saraf and W. G. Glasser, Engineering plastics from lignin III: Structure-property relationships in solution cast polyurethane films, *J. Appl. Polym. Sci.***29**, 1831-1841 (1984).
47. T. G. Rials and W. G. Glasser, Engineering plastics from lignin IV: Effects of cross-link density on polyurethane film properties - variations in NCO:OH ratio, *Holzforcshung* **38**(4), 191-199 (1984).
48. A. Natansohn, M. Lacasse, D. Banu, and D. Feldman, CP-Mass NMR spectra of polyurethane - lignin blends, *J. Appl. Polym. Sci.* **40**, 899-904 (1989).
49. D. Feldman and M. Lacasse, Polymer-filler interaction in polyurethane - Kraft lignin polyblends, *J. Appl. Polym. Sci.* **51**, 701-709 (1994).
50. D. Feldman, M. Lacasse. and R. St. J. Manley, Polyurethane based sealant modified by blending with Kraft lignin, *J. Appl. Polym. Sci.* **35**, 247-257 (1988).
51. D. Feldman and M. Lacasse, Swelling characteristics of lignin filled polyurethane sealant, *J. Adhesion Sci. Technol.* **8**(5), 472-484 (1994).
52. D. Feldman and M. Lacasse, Mechanical characteristics of sealants based on polyurethane - lignin polyblends, *J. Adhesion Sci. Technol.* **8**(9), 957-969 (1994).
53. D. Feldman, C. Luchian, D. Banu, and M. Lacasse, Polyurethane - maleic anhydride grafted lignin polyblends, *Cellulose Chem. Technol.* **25**, 163-180 (1991).
54. D. V. Evtuguin, J. P. Andreolety, and A. Gandini, Polyurethanes based on oxygen organosolv lignin, *Eur. Polym. J.* **34**(8), 1163-1169 (1998).
55. E. G. Lyubeshkina, Lignins as components of polymeric composite materials, *Russian Chem. Rev.* **52**(7), 1196-1224 (1983).
56. C. I. Simionescu, M. M. Macoveanu, C. Vasile, F. Ciobanu, M. Esanu, A.I oanid, P. Vidrascu, and N. Buruntea, Polyolefins/lignosulfonates blends, *Cellulose Chem. Technol.* **30**, 411-420 (1996).
57. C. Vasile, M. Downey, B. Wong, M. M. Macoveanu, M. .Pascu, J. H. Choi, C. Sung, and W. Baker, Isotactic polypropylene/epoxy -modified lignin blends, *Cellulose Chem. Technol.* **32**, 61-88 (1998).
58. H. Levon, J. Huhtala, B. Malm, and J. J. Lindberg, Improvement of the thermal stabilization of polyethylene with lignosulphonate, *Polymer* **28**(April), 745-750 (1987).

59. S. Casenave, A. Ait-Kadi, and B. Riedl, Mechanical behavior of highly filled lignin - polyethylene composites made by catalytic grafting, *Can. J. Chem. Eng.* **74**(April), 308-315 (1996).
60. V. Demianova and B. Kosikova, Lignin utilization in polyolefin blends, in: *Lignocellulosics Science, Technology, Development and Use*, edited by J. F. Kennedy, G. O. Phillips, and P. A. Williams (Elis Hardwood, Chichester, 1992) pp. 827-831.
61. B. Kosikova, M. Kacurakova, and V. Demianova, Photooxidation of the composite lignin/polypropylene films, *Chem. Papers* **42**(2), 132-136 (1993).
62. B. Kosikova, V. Demianova, and M. Kacurakova, Sulfur-free lignins as composites of polypropylene films, *J. Appl. Polym. Sci.* **47**, 1065-1073 (1993).
63. J. Kubat and H. E. Stromvall, Properties of injection molded lignin-filled polyethylene and polystyrene, *Plast. Rubber Process. Appl.* **3**, 111-118 (1983).
64. I. Chodak, R. Brezny, and L. Rychla, Blends of polypropylene with lignin. I: Influence of a lignin addition on cross-linking and thermooxidation stability of polypropylene, *Chem. Papers* **40**(4), 461-470 (1986).
65. J. Rosch and R. Mulhaupt, Mechanical properties of organosolv - lignin filled thermoplastics, *Polym. Bulletin* **32**, 361-365 (1994).
66. H. D. Rozman, K. W. Tan, R. N. Kumar, A. Abubakar, Z. A. M. Ishak, and H. Ismail, The effect of lignin as a compatibilizer on the physical properties of coconut fiber - polypropylene composites, *Eur.Polym. J.* **36**(7), 1483-1494 (2000).
67. H. G. Hofmann, Polymer blend modification of PVC in: *Polymer Blends and Mixtures*, edited by D. J. Walsh, J. S. Higgins and A. Maconnachie (Martinus Nijhoff, Dordrecht, 1985) pp. 117-118.
68. D. Feldman, D. Banu, and S. El-Raghi, Poly(vinyl chloride) - lignin blends for outdoor application in building. *J. Macromol. Sci. Pure Appl. Chem. A* **31**(5), 555-571 (1994).
69. D. Feldman, D. Banu, J. Lora and S. El-Raghi, Rigid poly(vinyl chloride) - organosolv lignin blends for applications in building, *J. Appl. Polym. Sci.* **61**, 2119-2128 (1996).
70. D. Feldman and D. Banu, Contribution to the study of rigid PVC polyblends with different lignins, *J. Appl. Polym. Sci.* **66**, 1731-1744 (1997).
71. D. Feldman, D. Banu, J. Campanelli, and H. Zhu, PVC-plasticized lignin polyblends, *Poly Millennial 2000 International Conference* (Waikoloa, Hawaii), December 9-13, 2000.
72. E. P. Galimov, V. P. Dmitriev, and R. K. Nizamov, Film materials based on PVC and hydrolysis lignin, *Intern. Polym. Sci. Technol.* **18**(8), 55-56 (1991).
73. Y. Li, J. Milnar, and S. Sarkanen, The first 85% Kraft lignin - based thermoplastics, *J. Polym. Sci. Part B, Polym. Physics* **35**, 1899-1910 (1997).
74. Y. Li and S. Sarkanen, Thermoplastics with very high lignin contents, in: *Lignin: Historical, Biological, and Materials Perspectives,* edited by W. G. Glasser, R. A. Northey and T. P. Schultz (ACS Symposium Series 742, Washington, 2000) pp. 351-366.
75. T. G. Rials and W. G. Glasser, Multiphase materials with lignin. IV Blends of Hydroxypropyl cellulose with lignin, *J. Appl. Polym. Sci.* **37**, 2399-2415 (1989).
76. I. Ghosh, R. K. Jain, and W. G. Glasser, Blends of biodegradable thermoplastics with lignin esters, in: *Lignin: Historical, Biological and Materials Perspectives,* edited by W. G. Glasser, R. A. Northey, and T. P. Schultz (ACS Symposium Series 742, Washington, 2000) pp. 331-350.
77. S. Baumberger, C. Lapierre, B. Monties, D. Lourdin and P. Colonna, Preparation and properties of thermally molded and cast lignosulfonates - starch blends, *Ind. Crop. Prod.* **6**, 253-258 (1997).
78. S. Baumberger, C. Lapierre, B. Monties, and G.Della Valle, Use of Kraft lignin as filler for starch film, *Polym. Degrad. Stab.* **59**, 273-277 (1998).
79. E. Corradini, E. A. G. Pineda, and A. A. W. Hechenleitner, Lignin - poly(vinyl alcohol) blends studied by thermal analysis, *Polym. Degrad. Stab.* **66**(2), 199-208 (1999).
80. Z. X. Guo, A. Gandini, and F. Pla, Polyesters from lignin.1. The reaction of Kraft lignin with dicarboxylic acid chlorides, *Polym. Intern.* **27**, 17-22 (1992).
81. Z. X. Guo and A. Gandini, Polyesters from lignin. 2. The copolyesterification of Kraft lignin and polyethylene glycols with dicarboxylic acid clhorides, *Eur. Polym. J.* **27**(11), 177-1180 (1991).
82. O. Faix, New aspects of lignin utilization in large amounts, *Das Papier* **12**, 733-739 (1992).
83. B. B. Boonstra, Fillers: carbon black and non black, in: *Rubber Technology, 2nd Edition,* edited by M. Morton (Van Nostrand Reinhold Co., New York, 1973), p. 54.
84. B. B. Boonstra, Reinforcement by fillers, in: *Rubber Technology and Manufacture, 2nd Edition,* edited by C. M. Blow and C. Hepburn (Butterworths, London, 1982), p. 269.
85. M .A. De Paoli and L. T. Furlan, Sugar cane bagasse lignin as a stabilizer for rubbers: Part II - butadiene rubber, *Polym. Degrad. Stab.* **13**, 129-138 (1985).
86. Yu. G. Kuzmina, *The Manufacture and Applications of Filled Compositions Based on Styrene-Polyolefin*

Copolymers (Niitekhim Publisher, Moscow, 1977), p. 43.
87. D. N. Simmons, Non black fillers and colouring materials, in: *Rubber Technology and Manufacture, 2nd Edition,* edited by C. M. Blow and C. Hepburn (Butterworths, London, 1982) p. 219.
88. S. I. Falkehag, Lignin in materials, *Applied Polymer Symposium* **28**, 247-257 (1975).
89. J. J. Lindberg, T. A. Kuusela, and K.Levon, Specialty Polymers from Lignin,in: *Lignin Properties and Materials,* edited by W. G. Glasser and S. Sarkanen (ACS Symposium Series 397, Washington, 1989), pp. 190-204.

ARBOFORM® – A THERMOPLASTIC, PROCESSABLE MATERIAL FROM LIGNIN AND NATURAL FIBERS

Helmut Nägele, Jürgen Pfitzer, Edgar Nägele, Emilia R. Inone, Norbert Eisenreich, Wilhelm Eckl and Peter Eyerer*

1. ABSTRACT

The natural polymer lignin is a by-product of the pulp and paper industry and every year approximately 50 million tons are generated in chemical pulp mills worldwide. So far the majority of them has been directly supplied to a thermal use for the power supply of the chemical pulp mills. This polymer lignin is the main component of a new class of engineering structural materials and parts for equipment in industrial applications using only renewable resources. ARBOFORM®, a material of this class, consists of lignin, natural fibers for reinforcement and natural additives supporting processing and properties. Although it exhibits wood-like properties, it can be processed like a thermoplastic material and used for engineering products. The mixing and compounding of granules of this material are based on standard technologies of polymer engineering. They can be injection molded and pressed like a thermoplastic raw material. The production of component parts and pressed plates from this material takes place at lower temperatures and the resulting parts show a lower shrinkage than those made from synthetic plastics. The mechanical behavior, however, is wood-like and the measured properties lie in a range of those of polyamide. Some examples of possible applications for mass consumer goods and industrial equipment, which are currently under detailed investigation, are discussed.

2. INTRODUCTION

With the world production of more than 100 million tons of polymers from crude oil carbon-containing substances are brought to daylight and released into the atmosphere after their utilization; in the case of incineration - in the form of carbon dioxide (CO_2).

* Helmut Nägele, Jurgen Pfitzer and Edgar Nägele, Tecnaro GmbH, Am Goldberg 2, D-99817 Eisenach, Germany. Emilia R. Inone, Norbert Eisenreich, Wilhelm Eckl and Peter Eyerer, Fraunhofer Institut für Chemische Technologie, Joseph-v.-Fraunhofer-Str. 7, D-76327 Pfinztal, Germany.

The accumulation of this gas in our atmosphere is responsible for the greenhouse effect. Though a higher level of CO_2 in the biosphere may promote a faster growing of plants, new research results indicate that the world CO_2 production is now seven times higher than the biosphere can accommodate.[1-4]

In contrast, materials made from renewable resources are offering a new way to environmentally friendly products in terms of CO_2 accumulations. Production of materials from renewable resources may meet the needs of both society and industry if costs induced for their waste disposal can be reduced by, for example, developing useful, value-added products from the waste materials. Being from renewable resources, the materials are produced only to the extent that the biological raw materials are regenerated.[3, 5]

The starting point for a new class of thermoplastic materials from renewable resources is lignin, a natural polymer that is a main constituent of biomass, in trees and other woody plants with an amount of up to 30%. Lignin, produced in plants by photosynthesis, is the second most abundant natural polymer after cellulose and builds up a three dimensional network around the cellulose fibers.[6-8] Lignin imparts to wood rigidity and resistance to impact, compression, and bending, while cellulose fibers provide the tensile strength property. This makes wood a material possessing combined impact and tensile stress resistance. In wood-based materials, such attributes could be further adjusted, depending on the composition, to provide desired properties for a product.[9-12]

The need for the development of such a new class of thermoplastic materials arises from modern concepts of product design, which require new approaches in material development. Predominant issues are variability in properties, low density for light weight structures, low costs of raw materials and their processing, and an environmental friendliness in all stages of the product life cycle including an acceptable way for disposal at the end of life. Currently used materials, especially engineering plastics, do not fully address these issues. Wood, a traditional engineering material, is still competitive due to its high resistance against impact and thermal loading, and its attractive styling attributes. Disadvantages of wood-based materials are the high costs, splintering on breaks and environmentally harmful glues when shaping complex geometries via layered constructions. Plastics and their composites show a broad range of properties that can meet different needs of products of various industrial branches. However, their inherent disadvantages are softening at higher temperatures, shrinkage on processing and environmental problems with recycling, especially for multi-component blends, composites and sandwich structures. Additional driving forces for the development of such a new class of materials are a zero-waste strategy in industrial production and product responsibility on a longer term, both of which are necessary in a market requiring both ecological and economic advantages for a successful commercialization.

3. A LIGNIN MATERIAL FOR INJECTION MOLDING AND PRESSING[13-24]

3.1. Lignin

Chemical pulp mills have long produced vast waste during wood pulp productions. These waste streams, which contain lignin, hemicelluloses and a broad spectrum of other chemicals from the pulping process, are usually burned for their energy value with only a small portion being converted to a variety of valuable chemicals such as vanillin, dimethyl sulfoxide and formic acid useful in various industry branches.

As a by-product of the pulp and paper industry million tons of lignin are generated every year in chemical pulp mills worldwide. Many investigations have been made to search new application fields of lignin and new technologies to process and transform this by-product to a valuable raw material.[1, 6]

For the pulp and paper industry, the most profitable use of lignin would be as a component of finished paper. Indeed, new high-yield pulping processes, combined with new refining and bleaching processes, have led to a significant increase in the use of lignin as a component of paper products. In recent years additional incentives to exploit the more efficient use of the lignin-containing wastes from the chemical pulping and wood-using industries have also been established. Sulfite liquors and its sulfonated lignin components (lignosulfonates) still represent the most widely-used waste product of the pulp and paper industry. The relatively large-scale usage of lignosulfonates is based on their physical and chemical properties. The dispersing properties of the lignosulfonates, combined with their ability to sequester metallic ions, have contributed to their successful use as additives in the preparation of oil-well drilling mud. Other application fields of lignosulfonates include road binders, animal feed pelletizers, cement products, industrial cleaners, trace element chelation, leather, tanning, metallic ion complexation for agricultural use, and boiler-scale inhibition. The expected complete utilization of these products in the future has resulted in continued physical and chemical studies of lignosulfonates in solution.

There exist also a variety of methods to extract native (non-modified) lignin totally or partially from lignocellulose biomass. These methods use organic solvents like methanol (Organosolv process) or even pure water (steam explosion process, Aquasolv process[32-35]) but need special separation techniques to recover the lignin.

Alkaline (both kraft and soda) pulping processes are the most dominant methods for the production of chemical wood pulps. The economics of these processes require the burning of the lignin in the spent liquor both for recovering the pulping chemicals and for providing the energy. As a result, the kraft and soda pulp mills in the United States burn some million tons of lignin annually. Although this transformation of lignin into energy is an economical method for the disposal of the spent pulping liquor, it is not an ideal way to dispose such a potentially valuable raw material.

3.2. Plastic Sector

Plastics are usually defined according to their hardening processes. Those that yield solid materials by simple cooling of a polymer melt (a physical process) and soften while being heated are called *thermoplastics.* Injection molding is a plastic manufacturing process where heated molten plastic is forced into a mold cavity under pressure. After the cavity is filled with the plastic melt, it is cooled and the plastic melt becomes a solid material resulting in a desired component part.

Injection molding is the most important process for producing moldings from thermoplastics, elastomers, and thermosets. The significance of injection molding is due to its ability to manufacture complex molding geometries in a single stage with a high degree of reproducibility. A finishing operation (removing the runner, deflashing) is seldom necessary. Because of the high degree of automation possible, this molding process is also suitable for mass production operations. Apart from its use for commodity articles, this technique is increasingly being used to produce technical moldings (e.g., compact discs and automobile bumpers). The size of the molding can vary from a few tenths of a gram to several kilograms.

Lignin seems to be a versatile material in polymer application. It can be used in formulations of phenol-formaldeyde and epoxy resins or in a rubber matrix as a reinforcing agent.

3.3. ARBOFORM®

We considered a new generation of a plastic-like and wood-like biomaterial appropriate for industrial engineering applications to be the one that can be processed like a thermoplastic by injection molding. In order to develop such a material, model moldings designed for various industrial branches and aimed at substituting plastics and wood in the selected applications should be developed and tested for their usability.[24, 30-32] The material has to consist only of natural renewable raw constituents with the main components being lignin, natural fibers and additives. To avoid man-made chemicals (e.g. sulphur in lignosulfonates) in the material, a native (Aquasolv) lignin[33-35] could be used in the future. Detailed aims in our development are:

- Materials and injection-molded model component parts should possess properties of an engineering plastic, verified in standard property test procedures of the materials

- Only natural components without man-made chemicals and additives should be used in the materials

- Processing at low temperatures of injection molding (< 160 °C), mechanical properties similar to those of polyamide, thermally stable at least till 95 °C or even 105 °C

- The materials can be used to obtain high quality engineering thermoplastics reinforced with natural fibers for application in industrial goods of mass consumer, electronic, automotive industry, etc.

The need for the development of such a new class of materials arises from an analysis of the chances a material will have if it is to compete with existing plastics and wooden component parts on a commercial scale. Ecological aspects are an important factor but can be considered only in combination with economic advantages.

A material of this type and made of lignin that can be processed as a thermoplastic material has been developed at the Fraunhofer-Institut für Chemische Technologie together with the company Tecnaro GmbH. The trade name of this material is ARBOFORM® (Latin: *arbor* – the tree) and it is the resulting mixture of lignin, natural fibers and natural additives.

When heated, the mixture can be processed to molded parts or panels on conventional plastic processing machines and molds in the same way as a synthetic thermoplastic. It is a natural fiber-reinforced material with lignin being the matrix or the binder component.

This lignin-based material has an interesting advantage compared to a wood material: it is processible as a plastic material on a standard, injection molding machine, while its appearance and some of its properties are similar to wood. Lignin from various processes and sources can be used to make such a material; the choice of the lignin depends on the application field of the product. For application where the sulphur content is not important, a high content of lignosulfonates can be used. For application where no man-made chemicals are allowed, native lignin obtained by a chemical-free process can be used. Of course, some additives are also necessary, similar to the situation for plastics and especially engineering plastics for technical applications. However, the additives for ARBOFORM® are also of natural origin with the function of plasticizers or dyes.

3.4. Natural Fibers

Natural fibers are currently of high importance for application in structural materials and engineering applications because light weight is a future demand as in cars and transporting units (see below) in order to minimize fuel consumption. Although they are not as strong as glass fibers, natural fibers show many advantages over glass fibers. For example, the relative low density of lignocellulose fibers (~ 1.5 g/cm^3) compared to that of glass fibers (~ 2.5 g/cm^3) enables light weight applications of natural fibers together with synthetic or natural polymers (Table 1).

Table 1. Properties of polypropylene (PP) and PP reinforced with glass or wood fibers

Material	Density g/cm^3	Tensile strength MPa	Tensile on break %	Bending modulus GPa
PP[a]	0.905	26	> 50	> 50
PP + Glass[b]	1.14	80	7	7
PP + Wood fibers[c]	1.039	26	5.4	5.4

[a] Stamylan P 412MN10 (DSM).
[b] Stamylan P 63G1030 (30% glass fibers).
[c] North Wood Plastics PP (40% wood fibers).

The general classification of natural fibers is mineral (asbestos), animal (silk and mammalian fibers) and vegetable fibers. Fibrous materials act to reinforce a matrix material by transferring the stress under an applied load from the weaker matrix to the much stronger fibers. Polymers (such as lignin in ARBOFORM®) serve as matrices.

Production of injection-molded parts reinforced by vegetable fibers is conceivable in the automotive sector. They may reduce weight and costs in the products. The typical application of vegetable fibers and synthetic polymers in the automotive sector may be for dashboard and seat. In ARBOFORM® both the matrix and the reinforced component are renewable raw materials and the processing does not require a modification of the current production procedure. Its application could be the component part itself or as a carrier.

In general, the vegetable fibers are classified either as long and multiple-celled, or as short and single-celled. They include:

- Bast fibers (or stem fibers), e.g. jute, flax, sunn hemp and ramie
- Leaf fibers, e.g. sisal, henequen, abaca
- Seed- and fruit-hair fibers, e.g. coir, cotton and kapok.

The various vegetable fibers differ from one another in properties. For cordage uses, for example, the leaf fibers are stronger, coarser, more stretchy and more resistant compared to the bast fibers.

Different kinds of fibers such as flax, hemp, sisal, as well as wood pulp fibers can be used alone or in combination as the reinforced component for ARBOFORM®. The type of the fibers used depends on the application field and the desired properties of the end product.

The interfacial adhesion between the matrix and the reinforced material contributes to the different properties in a product. There are different degrees of adhesion between the matrix and the reinforcing material. For synthetic polymers, it is common to add either a bonding agent or a compatibilizer or to chemically modify (treat) the fiber surface to strengthen the bonding of the reinforcing material to the matrix and consequently improve the mechanical properties. ARBOFORM® does not require any bonding agents or compatibilizers as the natural bonding of lignin and the lignocellulose fibers is restored in some way during processing. Certain combinations of different types of fibers (origin and processing), however, may have a positive influence on the properties.

4. FORMULATIONS AND PRODUCTION OF ARBOFORM® GRANULES

4.1. Granules

In the processing of plastics, granules are needed to obtain injection-molded parts. The standard and most important method to produce plastic granules is the extrusion method. It involves normally a heating and a friction processes. The extruders have single or twin screws which transport, compact, melt (or soften), mix, shape and finally bring out the melt which then cools down and solidifies. Granules are generated by cutting the extruded strings into particles. For the preparation of the ARBOFORM® granules a special technique is used, which allows its preparation without inducing heat stress to the natural components, and differs from the preparation of plastic granules in the standard extrusion process. This is important for all the natural components, particularly for the natural fibers because any thermal stress will lead to the decomposition of the natural fibers and a significant decrease in their reinforcing abilities.

The process to produce lignin-containing granules influences substantially the mechanical properties of the injection-molded parts. The injection-molded parts of ARBOFORM® from granules prepared using such a special technique have better mechanical properties, especially higher tensile strength than those from granules prepared by a standard extrusion process.

The granules prepared using this special, non-heating technique have pea sizes and they are appropriate for utilization on standard injection molding machines of plastics processing. They possess the following characteristics:

- Unique formulation and distinct wood appearance
- Different kinds of recipes are available with various contents of lignin, natural fibers and additives
- Each recipe targeted to a different application field and its composition can be made for component parts of defined mechanical and thermal properties
- It is also possible to obtain colored granules with the use of natural dyes.

The fibers used are of the size of short fibers by the definition used in the plastic sector. Picture 1 shows some of the granules made from ARBOFORM® with a formulation of 30% lignosulfonates, 60% wood pulp fibers and 10% additives. The additives used are also derived from renewable raw materials and include fillers, colorants, antioxidants and plasticizers.

4.2. Fibers Content

In the formulation of ARBOFORM® there is a strong dependence of the mechanical properties of the resulting component parts on the fiber content (Figures 1-4). Figure 1 shows a nearly linear dependence of the tensile strength on the content of fibers in ARBOFORM®. In analogy the same is true for the dependence of the bending stress on the fiber content (Figure 2).

Picture 1. ARBOFORM® granules with a formulation of 30% lignosulfonates, 60% wood pulp fibers and 10% additives.

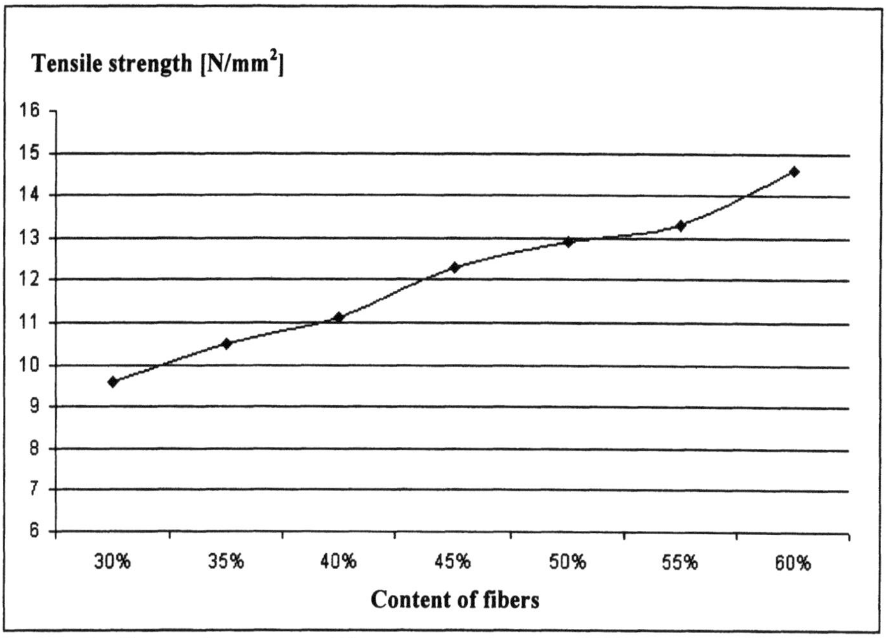

Figrue 1. Tensile strength vs. content of fibers.

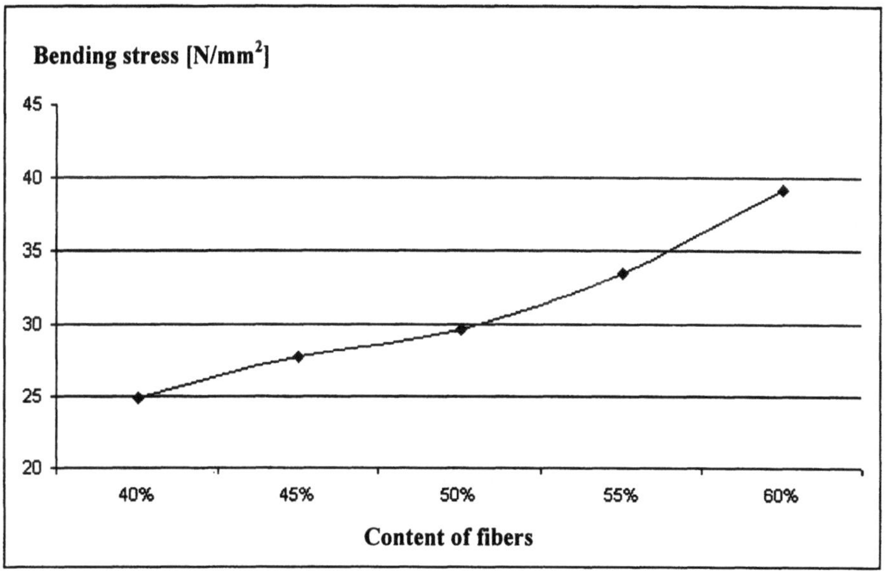

Figure 2. Bending stress vs. content of fibers.

The dependence of the indentation hardness (Figure 3) and the impact strength (Figure 4) on the content of fibers, however, is less pronounced.

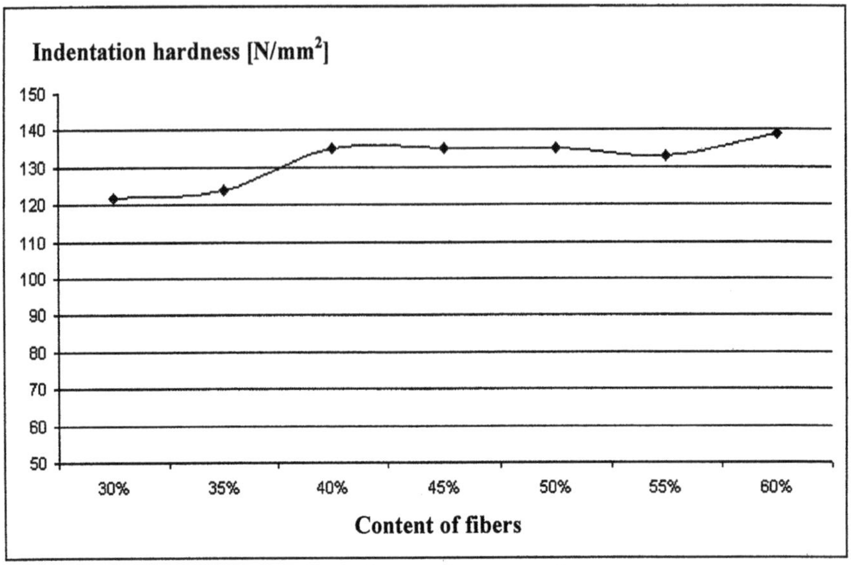

Figure 3. Dependence of the indentation hardness on the content of fibers.

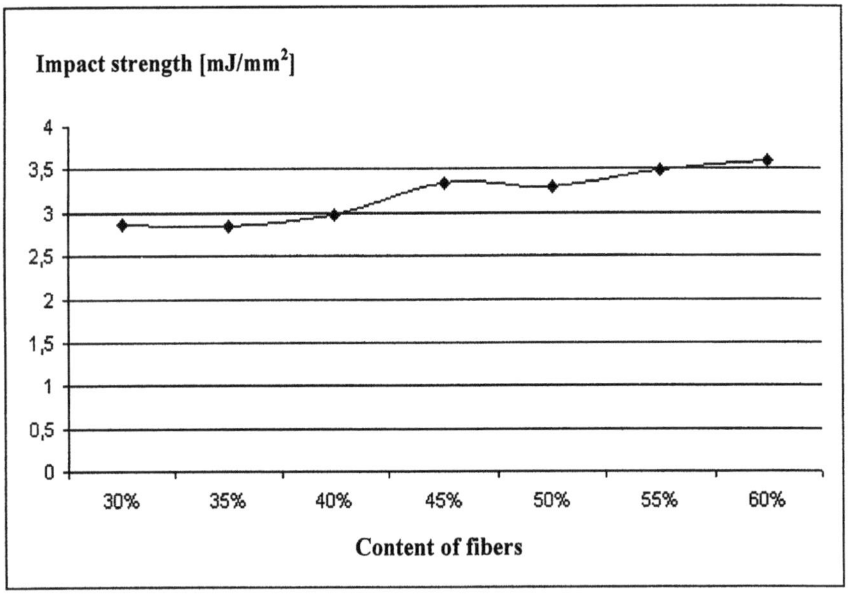

Figure 4. Dependence of the impact strength on the content of fibers.

A comparison of the stress-strain diagram of ARBOFORM® to that of low-density polyethylene (PE-LD) shows interesting characteristics. For ARBOFORM® a high Young's modulus is observed at low deformations but no visco-elastic range on linear strain loading. This means that ARBOFORM® behaves wood-like, the tensile tension coincides with the tension on break and is typical for materials of high brittleness. As it is well known, this is not a disadvantage. A designer normally uses tensile tension derived in the linear elastic range (Hook's law) for the constructed component parts.

5. PROCESSING OF ARBOFORM®

5.1. Injection Molding

The injection molding of ARBOFORM® granules can be carried out on standard injection molding machines. The normal screws for thermoplastics can be used. The temperature can be raised from 100 °C at Zone 1 to 170 °C at Zone 3 and the nozzle can have a temperature between 155 and 170 °C. The process runs with an open nozzle (2.5 – 4.0 mm). The back pressure must be as high as necessary for a smooth rotation of the screw. The injection pressure should be as high as possible (150 MPa) and so is the injection speed. The holding pressure should be about 30% of the machine's capacity and the related time only 0.5 – 2 seconds. The cooling time must be about 20% longer than for a normal thermoplastic material.

The injection molding tools must be constructed for the serial production of specially selected materials. When component parts are developed, available tools from plastic injection molding of the same parts could be used as a first approach to trials. In many cases this would give adequate results of an appropriate quality. However, for serial production and development in complex cases specially adapted tools have to be developed. Important issues to be taken into account are:

- The shrinkage on injection molding is very low and does not exceed 0.3%
- Drafts should be foreseen for de-molding which are useful to achieve fast working cycles
- Application of ribs that are not too thin is required because of the low shrinkage of ARBOFORM®
- Hot channel nozzles work only in limited cases.

Picture 2 shows a watch frame made from injection molding of ARBOFORM® granules on a standard injection molding machine.

Picture 2. Frame of a watch made from ARBOFORM®.

5.2. Pressing Plates

In the process of pressing plates and panels ARBOFORM® can be treated on a standard press for thermoplastics with special formulations. The pressing can occur with a pressure of 500 N/cm^2 (5 MPa) and a temperature of 170 °C.

The produced plates can be used for various indoor applications and for furniture constructions. All the material data are in agreement with the Norm EN 312-3 required for plates in such applications (Table 2). In Picture 3 a pressed plate made of ARBOFORM® is shown which is also a carrier for a thin veneer.

Table 2. Typical properties of a pressed plate[a] made of ARBOFORM®

Properties	Unit	ARBOFORM®-Plate	Norm EN 312-3
Thickness	mm	5.2	> 4 - 6
Size	cm	34 x 34	---
Density	kg/cm^3	1.4	---
Bending stress	N/mm^2	52.8	15
Modulus of elasticity in tension	N/mm^2	8610	1950
Perpendicular tensile strength	N/mm^2	0.55	0.45

[a] Formulation: 30% lignosulfonates, 60% wood pulp fibers and 10% additives.

Picture 3. ARBOFORM® as a carrier for a noble wood veneer.

5.3. Treating ARBOFORM®

Because of its wood-like behavior it is possible to treat an injection-molded or pressed part made from ARBOFORM® similar to wood, after the production step. It has been shown that ARBOFORM® is a material that can be easily treated by mill cutting and circular sawing.[21] For mill cutting the required power to cut ARBOFORM® is lesser than that needed to cut a Medium Density Fiberboard plate (MDF-plate) and slightly more than that needed to cut a chip board. For circular sawing two times of the cutting power of a MDF-plate is required. ARBOFORM® produces parts that are processable in the dry state. In contrast to a normal thermoplastic material the surface of parts made from ARBOFORM® does not melt during mill cutting or circular sawing.

6. ARBOFORM® - ECOLOGICAL AND TECHNICAL ASPECTS [23-29]

6.1. Ecological

One of the advantages of the lignin–containing ARBOFORM® is its processability on standard, injection molding machines, allowing its possible acceptance in the plastic sector for mass production of consumer goods. Its low processing temperature is another major advantage. Lower temperature translates to energy and resources saving.

Other advantages include good environmental aspects such as wood-like closed CO_2 loop and biodegradability. The global CO_2 balance is affected significantly by burning fossil energy carriers, but ARBOFORM® has the same biodegradability properties as wood and it can be burned with neutrality to CO_2 accumulation. ARBOFORM® can potentially substitute synthetic plastics that are based on petrochemicals, the resources of which will likely be exhausted in the near future. Materials with the same engineering properties as synthetic polymers but possessing good biodegradability with consequently low costs of waste disposal have become more and more desirable in the society and also in the industry. The use of natural resources especially from agriculture or forestry sector would also bring the following benefits to their cultivators:

- Alternative non-food application of agricultural and forestry products such as the natural fibers and better use of by-products (lignin) open new market opportunities

- Raw material derived from renewable resources does not require waste treatment.

After their life time component parts made from ARBOFORM® can be incinerated because they consist of 100% renewable materials, meaning that only the same amount of CO_2 is emitted to the atmosphere to be used by the plants to produce the same amount of the materials. Therefore, the CO_2 loop is closed and there is no accumulation of new CO_2 in the atmosphere and no net greenhouse effect. The greenhouse effect is a natural process by which various gases act like the glass of a greenhouse, trapping heat near the Earth's surface. These gases, commonly called greenhouse gases, are naturally occurring or man-made and include carbon dioxide, methane, water vapor, and other trace chemicals.

6.2. Technical

Besides the positive environmental aspects, ARBOFORM® also has substantial technical advantages over synthetic engineering polymers as a highly valuable material.

Its modulus of elasticity in tension (Table 3) is above that of common polymers such as polyethylene (PE), polypropylene (PP) and polystyrene (PS) and in the region of polyamide 66 (PA 66). ARBOFORM® has a high stiffness and is ready for use in products setting high standards (Picture 2). Since ARBOFORM® consists of materials from wood (lignin and cellulose fibers), it has similar mechanical, thermal, acoustical and other attributes to natural wood (Tables 2 and 3).

Wood and wooden veneers have found several applications in the automotive sector, especially in car interior parts. Wood is a versatile material but its complex forms are cumbersome and very expensive to produce. In a highly competitive automotive sector, all the materials used need to be of top quality and low costs. Wooden veneers, particularly those possessing wood appearance but based on a carrier feasible to be produced on a large scale on an injection molding machine, are of high interest for usage in a car. A difficult manipulation and a costly technology are necessary to combine synthetic polymers with wooden veneers. However, it is easy to use ARBOFORM® as a carrier for wood veneers (Picture 3).

Materials used as a support for wood veneers in car interiors must be able to withstand high thermal stress with regard to temperature fluctuations and maximum temperatures. Synthetic plastics cannot be used directly as a support for wood veneer parts because of the big differences of the expansion coefficients of the two materials: the carrier and the wooden veneer (Table 3). The higher linear expansion coefficient of the plastics compared to the wood veneer will cause the occurrence of cracks and its peeling off from the carrier when the temperature is increased. ARBOFORM® is an excellent alternative to be a carrier for wooden veneers because it has very similar expansion coefficient to wood (Table 3). Typical properties of an injection-molded part made of ARBOFORM® are summarized in Table 4.

Table 3. Comparison of various properties of ARBOFORM® with wood and plastics

Properties	ARBOFORM®	PE (LD, HD)	PP, not reinforced	PS	PA 66, not reinforced	Beech (cross)
Tension at break [N/mm^2]	14 - 22	8 - 30	30 - 40	45 - 65	65	7
Modulus of elasticity in tension [N/mm^2]	2000 - 7000	50 - 500	600 - 1700	1200 - 3300	2000	1500
Linear expansion coefficient [1/K] / 10^{-6}	10 - 50	170 - 200	100 - 200	70	80	45
Decrease in size, %	0.1 - 0.3	2 - 3	2 - 3	1 - 3	1 - 3	----
Vicat-Temperature, °C	80	40 - 70	90	80 - 90	250	---

Table 4. Typical properties of an injection-molded part made of ARBOFORM®

Mechanical Properties	Norm	Unit	Value range *
Tensile strength	DIN 53 455	N/mm^2	14 - 22
Elongation at break	DIN 53 455	%	0.3 - 0.7
Modulus of elasticity in tension	DIN 53 457	N/mm^2	2000 - 7000
Modulus of elasticity in bending	EN ISO 178	N/mm^2	2000 - 7000
Bending stress	EN ISO 178	N/mm^2	23 - 42
Impact strength (Charpy)	EN ISO 179	kJ/m^2	2 - 5
Indentation hardness	DIN 53 456	N/mm^2	75 - 7140
Hardness	DIN 53 505	Shore D	50 - 80
Thermical properties	Norm	Unit	Value range*
Linear expansion coefficient	DIN 53 752	1/ °C	$1 \cdot 10^{-5}$ - $5 \cdot 10^{-5}$
Vicat-Temperature	DIN 53 460	°C	80
Martens-Temperature	DIN 53462	°C	54
Heat conductivity	DIN 52 612	W/(m·K)	0.384
Glow wire test	DIN EN 60669-1	---	650 °C, passed
Electrical properties	Norm	Unit	Value range*
Conductivity, surface	DIN 53 482	G Ω	5
Conductivity, passage	DIN 53 482	G Ω	3
Other properties	Norm	Unit	Value range*
Decrease in size	---	%	0.1 - 0.3
Density (in molded part)	---	g/cm^3	1.3 - 1.4
Water content	---	%	2 - 8
Migration of certain elements	DIN EN 71-3	---	Requirements passed
Resistance to saliva and sweat	DIN 53 160	---	Requirements passed
Behaviour in fire	DIN 4102-B2	---	Requirements passed

* Depending on the formulation.

7. APPLICATION POSSIBILITIES

ARBOFORM® is a highly valuable thermoplastic material combining the attributes of natural wood and the processability of a thermoplastic. Parts made this material are easily produced by the injection molding or pressing technology and can further be processed by mill cutting and circular sawing. It can be used to make various products and parts in many industry branches.

For the automobile industry, this lignin composite can be used as a carrier for wooden veneers or decoration for the front-end parts.

For the housing industry, it can be used to make furniture with special and complex design as well as garden accessories and flooring. The shrinkage and swelling of this material are lower than those of plastics. Sawing, mill cutting and drilling of the material is the same as for wood. In addition, the surface is glossier than materials like MDF or FPY and it is also possible to varnish directly without veneering with wood.

For the electronic industry, this lignin-based material can be used to make products such as computer mouse, bass reflex tubes, covers for computers or TV sets and loud speakers with special design.

For the consumer industry, it can be used to make musical instruments, toys, cases for watches and clocks, necklace with wood aspects, button, ring, etc.

8. CONCLUDING REMARKS

The most attractive feature of the ARBOFORM® type material is that it can be embedded into a new alternative modern factory concept for the production of plastic-like and/or wood-like parts for various equipment and components of mass consumer, electronic, automotive industry, starting from a bulk extraction of renewable raw materials.[33-36] This concept is illustrated in Figure 5. With good efficiency and high environmental friendliness high-pressure hydrolysis (HPH) on a technical level could provide partially the raw material with an option of expanding the spectrum of extracts and substances. It can also provide additional constituents used in a bio-composite. The dyes or additives could be extracted under mild conditions or the cellulose from pulps used as a reinforcement agent. The large-scale application of a biomass conversion process will only be economically feasible if the main components can be used to make top-quality engineering products in an economic way. Therefore, a careful analysis of the quality of the derived components prepared preferably in the absence of chemicals has to be included and attractive applications of the bio-composites demonstrated. ARBOFORM® was a step ahead in this direction and an encouraging approach. It has good economic and ecological prospects because there is a long-term solution on the availability of raw materials and the prices for the plastic-like materials in engineering applications are estimated in a range of engineering plastics. Such a price has already been reached with low production quantities at a pilot production.

The development of ARBOFORM® demonstrates that engineering materials can be obtained economically and entirely from natural raw materials in a unique process under conditions of current technologies of pre-treatment, extraction and processing.

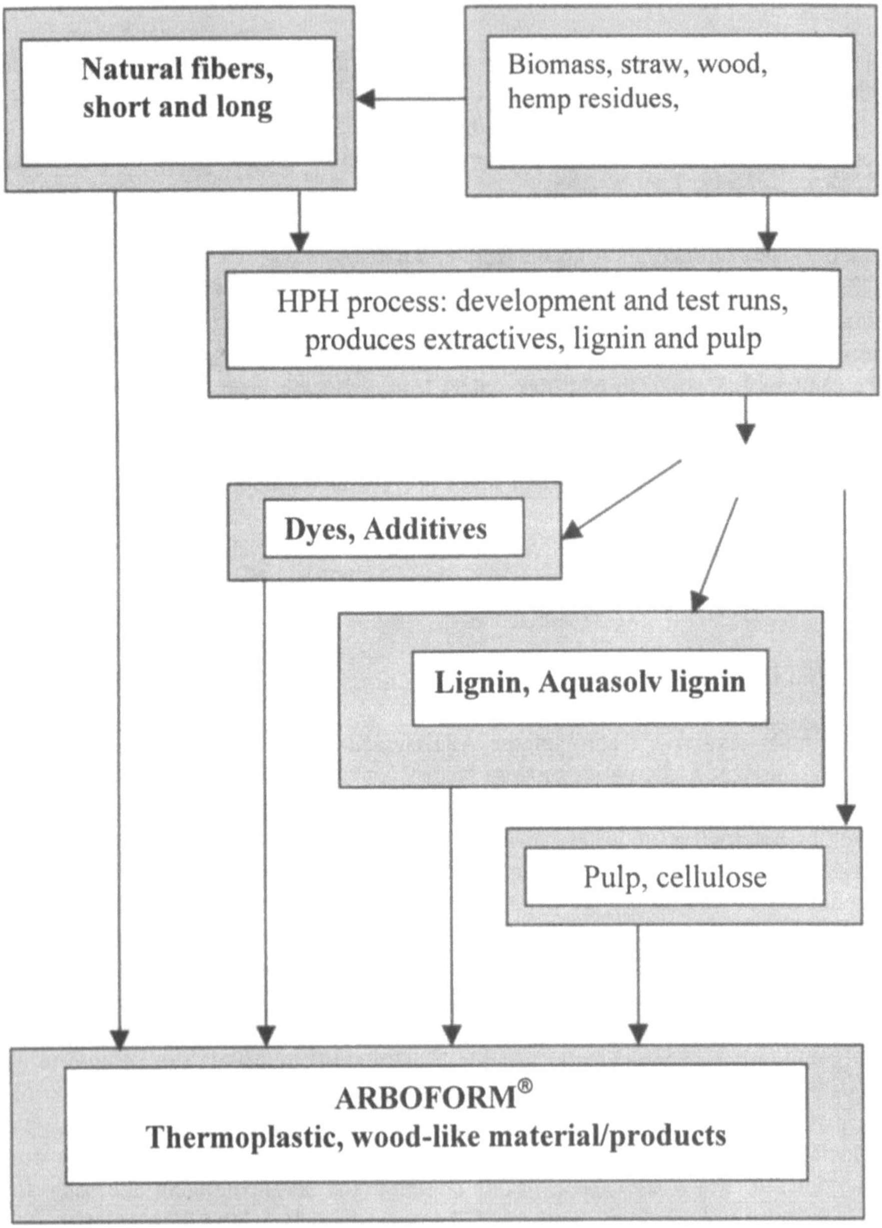

Figure 5. Key elements of a process to generate plastic-like and wood-like products from biomass - in a regional approach.

9. REFERENCES

1. A. Gandini and N. Belgacem, Recent advances in the elaboration of polymeric materials derived from biomass components, *Polym. Int.* **47**(3), 267-277 (1998).

2. Katalyse: Institut für angewandte Umweltforschung, *Leitfaden nachwachsende Rohstoff* (C. F. Müller Verlag, Heidelberg, 1998).
3. D. L. Kaplan, *Biopolymers from Renewable Resources* (Springer Verlag, Berlin, 1998).
4. D. R. Gilbert, *Cellulosic Polymers – Blends and Composites* (Hanser Verlag, Munich, 1994).
5. J. U. Otaigbe, H. Goel, T. Babcock, and J. Jane, Processability and properties of biodegradable plastics made from agricultural biopolymers, *J. Elastomers Plast.* **31**(1), 56-72 (1999).
6. R. Wagenführ, *Anatomie des Holzes – Strukturanalytik, Identifizierung, Nomenklatur, Mikrotechnologie* (DRW-Verlag, Leinfelden-Echterdingen, 1999).
7. J. Simon, H. P. Müller, R. Koch, and V. Müller, Thermoplastic and biodegradable polymers of cellulose, *Polym. Degrad. Stab.* **59**(1-3), 107-117 (1998).
8. E. Sjöström and R. Alen (Eds.), *Chemistry, Pulping and Papermaking* (Springer-Verlag, Berlin, 1999), pp. 8-9.
9. F. F. Nord and G. de Stevens, Lignins and lignification, *Encyclopedia of Plant Physiology* Vol. 10, edited by W. Ruhland (Springer-Verlag, Berlin, 1958), p. 389.
10. I. A. Pearl, *The Chemistry of Lignin* (Marcel Dekker, New York, 1967).
11. C. J. Biermann, *Pulping and Papermaking*, 2nd ed. (Academic Press, New York, 1996).
12. F. E. Brauns, *The Chemistry of Lignin* (Academic Press, New York, 1952).
13. F. E. Brauns, *The Chemistry of Lignin: Supplement Volume* (Academic Press, New York, 1960).
14. *Ullmann's Encyclopedia of Industrial Chemistry* Vol. A 15 (VCH-Verlag, 1992) p. 305.
15. *Bild der Wissenschaft*, Baumstark: Flüssiges Holz (Deutsche Verlags-Anstalt GmbH, Stuttgart, Feb. 2000), pp. 73-76.
16. *Holz und Kunststoffverarbeitung HK*, Flüssiges Holz (DRW-Verlag Weinbrenner GmbH & Co., Leinfelden-Echterdingen, Nov. 1999), pp. 44-47.
17. *Brand eins*, Die Holzverflüssiger (Brand 1 Verlagsgesellschaft mbH, Hamburg, May 2000), pp. 58-63.
18. *Das Magazin für Holz und Ausbau dds*, Ist mit Holz alles machbar? (Der Deutsche Schreiner Verlag GmbH, Stuttgart, Jan. 2001), pp.40-43.
19. *Kunststoff Information KI*, Tecnaro: Lignin-Werkstoff für Industrie-Zwecke, **30**(No. 1510), 7 (2000).
20. *Design-Report*, Geschmolzenes Holz (BLUE C. Verlag, Hamburg, Feb. 2001), p. 46.
21. *Konstruktionspraxis*, Aus nachwachsenden Rohstoffen, (Vogel Verlag und Druck GmbH & Co. KG, Würzburg, Dec. 2000), p. 38.
22. *EDM-Funkey*, Flüssiges Holz – phantastisch plastisch, (KEP Verlag, Stuttgart, Sept. 2001), p. 9.
23. J. Tröger and U. Heisel, Untersuchungen zur Zerspanbarkeit von ARBOFORM®, *Holzverarbeitung HOB*, 2001, in press.
24. N. Eisenreich, W. Eckl, E. R. Inone, H. Nägele, and J. Pfitzer, Arboform – a thermoplastic made of renewable resources, *Proceedings Electronic Goes Green 2000, Vol. 1, Technical Lectures* (Joint International Congress and Exhibition, Berlin, Germany, Sept. 2000).
25. H.-G. Elias, *An Introduction to Polymer Science* (VCH Verlagsgesselschaft mbH, Weinheim, 1997).
26. O. Schwarz, F.-W. Ebeling, and B. Furth, *Kuststoffverarbeitung* (Vogel Buchverlag Würzburg, 1999).
27. A. J. Bailey and O. W. Ward, Synthetic lignin resins and plastic, *Ind. Eng. Chem.* **37**, 12 (1945).
28. J. Murphy, *Reinforced Plastics Handbook* (Elsevier Advanced Technology, Oxford, 1994).
29. S. Thomas, Short natural fiber reinforced polymer composites, *Presentation at Unversität Stuttgart* (Stuttgart, 1999).
30. P. Eyerer, N. Eisenreich, E. Inone, H. Nägele, and J. Pfitzer; "Flüssiges Holz", HK Holz- und Kunststoffverarbeitung, Nov. 1999, p. 44.
31. H. Nägele and J. Pfitzer, Baumstark: Flüssiges Holz, Bild der Wissenschaft, Feb. 2000, p. 73.
32. O. Bobleter, Hydrothermal degradation of polymers derived from plants, *Prog. Polym. Sci.* **19**, 797-841 (1994).
33. O. Jedicke, N. Eisenreich, E. Steurenthaler, and H. Dümpert, Aquasolv® – Verfahren – Ein Konzept zur ganzheitlichen Verwendung von Biomasse, in: Verbindung mit "grüner Chemie", *Proceedings 2nd Scientific Meeting in Zero Emissions* (Frohnleiten, Technische Universität Graz, Austria, 1999), p. 107.
34. O. Jedicke and N. Eisenreich, Aquasolv® - Hydrothermolyse, The development of a process for completely use of biomass, *Proceedings 1st Biomass World Conference* (Sevilla, Spain, 2000).
35. O. Jedicke, N. Eisenreich, and H. Dümpert, Aqauasolv – Verfahren; Hydrothermales Aufschlußverfahren zur Gewinnung Nachwachsender Rohstoffe aus Biomasse, Achema 2000; 26. Ausstellungstagung Frankfurt am Main, *Proceedings Neue Chemische Prozeße und Reaktionstechnik* (Frankfurt, Germany, May 2000), p. 42.
36. O. Jedicke, I. Álvarez Ortega, and H. Nägele, Characterisation of polymeric composite materials based on natural components, lignin with wood and hemp fibers reinforcement, *Proceedings 4th Euromech Solids Mechanics Conference* (Metz, France, June 2000), p.34.

LIGNIN-BASED CARBON FIBERS

John F. Kadla*, Satoshi Kubo, Richard D. Gilbert and Richard A. Venditti

1. INTRODUCTION

1.1. Carbon Fibers

Carbon fibers are one of the most important engineering materials in advanced composites. They are lightweight, fatigue resistant materials that possess high strength and high stiffness. These unique properties result from their flawless structure and the development of highly anisotropic graphic crystallites orientated along the fiber axis during the production process.[1] Carbon fibers are manufactured by thermally treating fibers at 1000-2000 °C in an inert atmosphere while maintaining the fibrous structure. This is aided by a stabilization stage in which the precursor fibers are heated under tension at 200-300 °C in the presence of air. This causes crosslinking on the fiber surfaces, among other reactions, and prevents shrinking, melting and fusing.

The first commercially produced carbon filament was made from a cellulosic precursor for its application as incandescent lamp filament in 1879.[2,3] Its use as a reinforcing material in composite products began in the 1950's as the aircraft and aerospace industries led the search for new materials with properties superior to the then available metals. Cost was seldom a deciding factor. As the years passed, the advantages of advanced composite materials in other applications became apparent. Today carbon fiber composite products are routinely used in sports equipment, marine products, construction and the automotive industry to name a few. However, unlike in its infancy, the cost of carbon fiber production has limited its widespread use. The fundamental difficulty is reducing the cost of manufacturing the precursor, the so-called "white fiber." There are primarily three types of precursor materials of commercial significance; pitch (petroleum or coal), viscose rayon and polyacrylonitrile (PAN).[4] Of these three, PAN is the most important for structural applications.[5] It is an excellent precursor material that has been widely researched and is in wide commercial production. Almost eighty percent of commercially available carbon fibers are derived from PAN. However, current PAN technology is expensive, thereby limiting its utilization in lower cost general performance

* John F. Kadla, Department of Paper Science and Department of Chemical Engineering, North Carolina State University, Raleigh, North Carolina, 27695-8005.

applications, and research continues towards decreasing the PAN precursor costs, e.g. Amlon (BP Chemicals).[6]

The overwhelming success of PAN-based carbon fibers over rayon and pitch can be attributed to several key aspects.[4] Structurally, PAN has a faster rate of pyrolysis without much disturbance to its basic structure, and to the preferred orientation of the molecular chains along the fiber axis present in the original fiber. By contrast, carbon fibers from rayon suffer from extremely low carbon yield (20-25%) due to chain fragmentation, which eliminates the orientation of the precursor fiber. While improved properties can be achieved by stretch graphitization, this process is expensive and does not compensate for the low yields.

Pitch-based carbon fibers, on the other hand, provide properties not readily obtainable with PAN-based fibers. PAN-based fibers have excellent tensile strength at a modulus of 200 GPa, but the strength decreases as modulus is increased.[6] Pitch-based fibers have lower tensile strengths, but are capable of modulus levels up to the theoretical modulus of graphite, 1000 GPa, and have much better thermal and electrical conductivity properties than PAN-based fibers. Thus pitch and PAN-based fibers are really complementary to one another, each filling a different set of commercial needs.

In addition to rayon, pitch and PAN, many polymeric materials have been used to make carbon fibers.[7-16] However, only the big three, rayon, pitch and PAN have endured the high performance markets. Their price has dropped over the years, but remains high, accounting for over one-half of the production costs, too high for the general performance markets. The literature supports that recycled petrochemical polymers and fibers and renewable cellulosics and lignins, which are inexpensive and widely available, may be potential feedstocks for general performance carbon fibers.[7-16]

1.2. Lignin

As a raw material for carbon fiber production lignin (Scheme 1) offers some distinct advantages over PAN and pitch. Lignin, second only to cellulose in natural abundance, is readily available, relatively inexpensive and structurally appealing (*vide infra*). It is a high molecular weight aromatic biopolymer with a reported total worldwide production of approximately 50 million tons/year.[17] The byproduct of "wood-free" papermaking, technical or commercial lignins are predominately used as a fuel in the pulping chemical recovery process. However, as a fuel source it is very inefficient, producing less than ¼ as much energy as middle distillate fuels. Commercial lignins, such as those isolated from alkaline, acidic, or organic solvent-based processes, have undergone extensive fragmentation and degradation. As a result, a wide variety of lignins, in terms of chemical properties and structures can be obtained. Specifically, depending on the type and length of chemical processing, the lignins will vary in molecular weight, functional groups present, degree of condensation, types of intermonomeric linkages, and types and ratios of monomeric units. Due to this inherent chemical and molecular weight inhomogeneity, lignin and lignin derivatives have limited utility in applications demanding a constant well-defined feedstock. In fact, less than 2% of the total available lignin was reportedly being used in higher value products.[18] In 1996, total sales in lignin-based specialty products, such as animal bypass protein, agro chemicals, dispersants, adhesives and surfactants were reported at $600 million.[19]

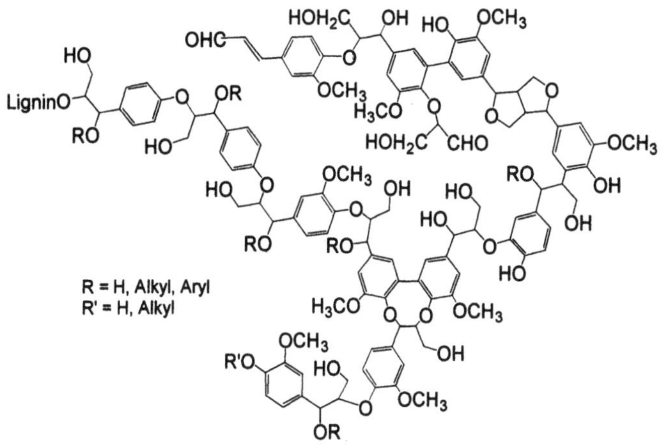

Scheme 1. Representative structure of lignin.

1.3. Lignin-based Carbon Fibers

Carbon fibers have been manufactured from lignin. One of the first reports of lignin-based carbon fibers was from Otani *et al.* in 1969.[20] Utilizing thiolignin, alkali lignin and lignosulfonates, carbon fibers were produced by both thermal and solvent spinning followed by thermal processing to 600-1000 °C. Lignin fibers were produced by thermal spinning of softwood and softwood/hardwood kraft lignin blends. A purification process was needed to remove impurities and facilitate fiber spinning. The fiber quality was poor, requiring slow winding rates (5-10 m/min) and producing low tensile strength (300 MPa) carbon fibers. Improvements in spinnability and tensile strength were obtained by using dry-spinning. Satisfactory lignin fibers were formed from kraft lignin, alkali lignin and lignosulfonate spun from water, alkaline water and glycerol. Blending with polyvinyl alcohol or polyacrylonitrile enhanced the spinnability and tensile strength of the fibers. Winding rates of 600 m/min were achieved when softwood kraft lignin was blended with polyvinyl alcohol (3/4, w/w). Fine fibers with excellent mechanical properties were produced. In the production of carbon fibers, an oxidation process, referred to as stabilization or "thermostabilization" is generally conducted to induce crosslinking to maintain fiber integrity during the subsequent carbonization process. It was expected that lignin-based carbon fibers, due to the high oxygen content in lignin, could be produced with shorter stabilization periods at lower temperature than other feedstocks. The various hydroxyl and ether groups, via oxygen-based radicals were expected to facilitate crosslinking within the lignin. In fact carbon fibers were prepared with very little thermostabilization, and tensile strengths as high as 800 MPa were obtained.[20]

Around the same time, Mansmann and co-workers reported the production of carbon fibers from a variety of dry-spun materials, including lignin (lignosulfonates) by simple addition of small amounts of polyethylene oxide or acrylic acid-acrylamide copolymers.[21] Although similar to the procedure of Ohtani, Mansmann employed acidic rather than neutral and/or alkaline spinning conditions. It was reported that the spinning under acidic conditions enhanced the mechanical properties of the lignin and carbon fibers. Furthermore, the amount of synthetic polymer required to enhance spinnability was much less in comparison to the method of Ohtani. The concentration of the polymer required for favorable spinning was dependent on its molecular mass. Only 0.3-0.6% (wt%) polyethylene oxide, DP = 100K, was needed for good spinning. In addition, dry-spinning from acidic conditions reduced the thermostabilization temperature to between 100-150 °C. Tensile strengths of ~ 100 MPa, and ~ 600-800 MPa and Young's modulus of 2-8 GPa, and 33.3 GPa were reported for the lignin and carbon fibers respectively.

Based on these encouraging results, pilot-scale production of lignosulfonate-based carbon fibers, Kayacarbon, were developed and made commercially available by Nippon Kayaku Co.[22] Carbon fiber was prepared by dry-spinning a lignosulfonate plasticized with polyvinyl alcohol (PVA). Pilot trials revealed that lignin-based carbon fibers could be produced at lower temperature with shorter stabilization periods than other feedstocks, such as pitch, because crosslinking reactions that produce the graphitic structure are facilitated by oxygen-based radicals derived from the various hydroxyl and ether groups within lignin. Close examination of the carbon fibers produced showed poorly developed fibrillar structures and lack of homogeneity between the crystalline planes. A large number of inclusions were found, arising from catalytic graphitization by sodium impurities, which contributed to a weakening of the fibers.[23,24] Microvoids of complicated size and shape were also observed.[25, 26] The sizes of microvoids, as well as of the solid phase around them, increased with increasing heat treatment temperature; stretching of the fiber promoted further increases. These flaws, combined with the lack of orientation resulted in low modulus and low tensile strength carbon fibers. As a result, these carbon fibers were abandoned because they did not meet the specifications of the high performance aerospace industry, which dominated the carbon fiber market.

Sudo and co-workers demonstrated that steam-exploded lignin could be converted into a pitch-like molten material using hydrogenolysis followed by solvent extraction and polycondensation under high temperature.[27, 28] The lignin-pitch had H/C ratios between 1.03-1.06, a softening point of 110 °C and melted at over 145 °C to form a viscous liquid with suitable properties for thermal spinning and thermostabilization. A continuous fine filament could be produced at a speed of over 100 m/min. The fibers were heated in air at 1-2 °C/min to 210 °C, converting the lignin fibers to an infusible material, enabling them to be carbonized at a heating rate of 5 °C/min up to 1000 °C in a stream of nitrogen. The typical properties of the lignin based carbon fiber were: Fiber diameter = 7.6 ± 2.7 μm; Elongation = 1.63 ± 0.29%; Tensile strength = 660 ± 230 MPa; Modulus of elasticity = 40.7 ± 6.3 GPa. Unfortunately, the over-all yield of the carbon fiber was only 20.7 %, well below that of other lignin based carbon fibers. To enhance the over-all yield hydrogenolysis was replaced by phenolation using creosote.[29, 30] This resulted in yields of 40.0-49.2%.

In addition to steam-exploded lignin, Alcell and acetic acid lignins were studied. The spinability of the phenolated Alcell was better than that of the other phenolated lignins. The phenolated Alcell lignin had a softening point of 150-170 °C and could be continuously spun for 500 seconds at a winding rate of 1,000 m/min. To improve the spinability of the phenolated acetic acid lignin, it was blended with pitch in the ratio of 1/1 (w/w). The over-all yield of the carbon fibers made by this method was substantially higher than that of the hydrogenolysis lignin. Over-all yields of 40.0-49.2% and 49.6-

57.3% were reported for the phenolated and pitch-added phenolated lignin-based carbon fibers respectively. Superior properties to that of the Kayacarbon were observed, with a tensile strength of 614 MPa, Young's modulus of 31.8 GPa and 1.94% elongation being reported for the phenolated acetic acid lignin carbon fiber.

More recently, Sano *et al.* produced lignin fibers suitable for carbon fibers by thermal spinning Organosolv lignin obtained from aqueous acetic acid pulping of birch.[31-33] Unlike the previously reported acetic acid lignin fibers, this system did not require any chemical modification (e.g., phenolysis). The spinnability of the Organosolv lignin was attributed to the polydispersity and partial acetylation of the lignin due to the pulping process. The raw acetic acid lignin showed poor spinnability due to the formation of pyrolyzed gaseous materials in the extruder nozzle. Thermal treatment at 160 °C under vacuum for 30 min removed this pyrolyzable material permitting excellent fiber spinning at more than 400 m/min. Tensile strength of 30.8 ± 3.5 MPa, Young's modulus of 3.59 ± 0.43 GPa and 0.81 ± 0.08% elongation was reported for this lignin fiber. This fiber was easily carbonized after a thermostabilization process (0.5 °C min^{-1} to 250 °C). Tensile strength of 355 ± 53 Mpa, Young's modulus of 39.1 ± 13.3 GPa and 0.98 ± 0.25% elongation was reported for this lignin fiber. The tensile strength increased with decreasing fiber diameter and was comparable to previously reported lignin-based carbon fibers if fine carbon fiber was made from this lignin. Over-all carbon fiber yield of 32.7% was obtained, comparable to that of isotropic petroleum pitch, ~30-35%, but lower than that reported for the phenolated acetic acid lignin of Sudo and co-workers.[29] Shortly there after, Sano *et al.* reported on the production of carbon fiber from softwood acetic acid lignin. The infusible softwood lignin was made fusible by removing the infusible high molecular mass fraction (~ 30% of the lignin) by successive extraction with various concentrations of acetic acid. Good fiber spinning was produced after thermal treatment of this lignin fraction. Moreover, this fiber was converted to carbon fiber without a thermostabilization process. Tensile strength of 26.4 ± 3.1 MPa, 147 ± 51 MPa, Young's modulus of 3.59 ± 0.43 GPa, 19.5 ± 5.5 GPa, and 0.71 ± 0.14%, 0.75 ± 0.27% of elongation was reported for this lignin fiber, carbon fiber respectively.

At about the same time, Itoh *et al.* reported that lignin prepared by the organosolv pulping using phenols and water could be thermally spun into lignin fibers for the production of carbon fiber.[34] Fiber spinning was achieved at a spinning temperature of between 150-250 °C without any lignin modification required. Tensile strength of 3-60 MPa was reported for the thermostabilized fiber, however, no information about the properties of the carbon fiber has been reported.

Although both petrochemical polymers and lignin were successfully used as carbon fiber precursors with good carbon fiber properties, these materials were gradually abandoned as the high-performance aerospace market, which relied increasingly on PAN and pitch, came to dominate carbon fiber production. Today however, carbon fibers are routinely being used in marine products, construction, the transportation industry and the like. Cost, not performance is now the driving force in their utilization. Thus, any technology that can produce a low cost fiber precursor, suitable for carbon fiber will have a great impact on these markets. As discussed, recycled petrochemical polymers and commercial lignins can. However, in each system discussed above, problems concerning production costs exist. Either plasticization or lignin modification was required, or noncommercially available lignins (e.g., steam-exploded, acetic acid) were used.

Therefore to overcome this we set out to utilize a commercially available technical lignin for the production of carbon fibers with properties suitable for the general purpose carbon fiber market.[35] Here we report the first lignin-based carbon fibers prepared from commercial kraft and Alcell lignins and blends with recyclable petrochemical polymers utilizing a thermal extrusion process.

2. EXPERIMENTAL

2.1. Materials

Three kinds of lignin namely Organosolv lignin (ALCELL - Repap), softwood kraft lignin (SWKL - Indulin ATTM – Westvaco) and hardwood kraft lignin (HWKL – Westvaco) were studied. The kraft lignin samples were desalted prior to fiber spinning/blending by repeated washing with distilled water while maintaining the pH below 5 with HCl. When the desired salt concentration, less than 1000 ppm was reached, as determined by ashing at 550 °C for one hour in air, the lignin was air-dried and ground to pass through a 60 mesh screen. Poly(ethylene oxide), (PEO) 100K, 300K and 600K molecular weight (Union Carbide Corporation), poly(propylene) (PP), poly(ethylene) (HDPE/LDPE) (Aldrich), and poly(ethylene terephthalate) (PET) (Goodyear) were used as received.

2.2. Lignin Analysis

Relative average molecular mass was measured by high performance size exclusion chromatography (Waters HPLC model 510 with UV and RI detectors) of acetylated lignin samples, calibrated against polystyrene standards.

Thermal analysis of the lignin was performed by differential scanning calorimetry (DSC) using a Perkin Elmer DSC-7. The analyses were conducted under a nitrogen atmosphere with heating and cooling rates of 20 °C/min. The glass transition temperatures, T_g's, of the samples were measured as onset temperatures of the step change in heat capacity in the DSC trace. Sample weights were 4.5 ± 0.5 mg. Fiber samples were cut into short pieces prior to analysis.

Scanning electron microscopy (SEM) was run on uncoated and gold-coated lignin and lignin-carbon fibers on a JEOL 6400F SEM (Accelerating Voltage: 10kV, Magnification: 500X - 160,000X).

2.3. Fiber Spinning

Thermal extrusion of the lignin fibers was performed using an Atlas Laboratory Mixing Extruder equipped with a 1/32 inch spinneret. The maximum winding rate was 74 m/min. Extrusion temperatures were varied between 130 °C and 240 °C, depending on the fiber blend.

2.4. Thermostabilization and Carbonization

The lignin-based fibers were first thermostabilized prior to carbonization. Specifically, twelve to eighteen inch long fibers were mounted on a stainless steel plate to maintain tension, and heated at various heating rates (12-180 °C/h) and held at 250 °C for 1 hour in an air atmosphere. The thermostabilized fibers were then carbonized under a nitrogen atmosphere at 1000 °C (heating rate 180 °C/h) using a modified muffle furnace.

2.5. Mechanical Strength of the Fibers

The mechanical strength of the fibers was measured according to the ASTM standard (D 3379-75) with an Instron model 4411 using a 0.5N load cell and is reported as the average of 20 fibers per sample.

3. RESULTS AND DISCUSSION

The goal of our research was to produce general performance grade carbon fibers from commercially available lignin (kraft and Alcell) and recyclable petrochemical polymers (PE, PP, PET, PEO), to ensure low raw material costs, as well as ample raw material for large-scale applications. Thermal spinning was used to minimize capital and operating costs by eliminating any spin solvent storage, recovery and purification.

3.1. Thermal Spinning of Lignin Fibers

Unlike Alcell lignin, kraft lignins contain inorganic impurities which hinder thermal motion and prevent thermal processing.[36] The kraft lignins were therefore washed with dilute HCl to remove the ash. Using water washing, it was possible to consistently produce lignins with low ash contents. Ash contents of 0.23% and 0.45% were obtained after washing for the SWKL and HWKL, respectively.

The potential for fiber formation of the Alcell and kraft lignins was first examined. Temperatures were identified at which the powders softened or could generate hand-drawn fibers. The Alcell lignin and HWKL could form fibers at temperatures above 140 and 200 °C respectively. The SWKL however, did not soften during heating at temperatures from 140 to 240 °C, instead charring occurred.

To enable fiber formation, a thermal pretreatment of the lignins at 145 °C under vacuum for one hour was required to remove volatile contaminents prior to mechanical blending.[31] The devolatilized lignin was then extruded using an Atlas Mixer/Extruder to produce thick rods or "sticks" that was used (discussed below) for subsequent re-extrusion into fine fibers (Figure 1).

The optimal conditions for fiber production were obtained by slowly increasing the temperature of the rotor and spinneret until fiber formation occurred, then held at temperature.

Figure 1. *Stick* formation (left) and resulting lignin fiber (right).

The spinning temperatures of Alcell and HWKL (Table 1) were lower than that of their thermal decomposition temperatures, 268 °C and 270 °C, respectively. Continuous spinning was achieved and satisfactory fibers were made from all samples. The spinning temperature is closely related to the molten viscosity of the sample, with a low spinning temperature indicative of a low molten viscosity.

Table 1. Spinning temperature for Alcell lignin and HWKL fibers

	Spinning temperature (°C)		
Lignin sample	Rotor	Nozzle	Take-up (m/min)
HWKL	180	195-228	74
Alcell lignin	155	145-165	74

It has been reported that during the course of thermal processing, dehydration and condensation/polymerization occur and contribute to the fluidity of the lignin.[32] Figure 2 shows the relative molecular mass of the Alcell lignin and HWKL after each thermal treatment step leading to lignin fiber production.

The effect of thermal processing on HWKL is more pronounced than on Alcell lignin. The HWKL increases linearly in relative molecular mass in going from the desalted lignin, $\overline{Mw} \sim 4200$ to the extruded HWKL fiber, $\overline{Mw} \sim 6300$. The Alcell lignin on the other hand increases from $\overline{Mw} \sim 3700$ to 4300 after heat treatment, but does not undergo any further substantial increase in $\overline{Mw}$ as a result of *stick* and lignin fiber formation. This can in part be explained based on the relative structures of the two lignins. Alcell lignin, produced from an acidic ethanolysis process, is well known to have a lower phenolic hydroxyl content at the same percent delignification relative to that of kraft lignin.[37] According to Kubo *et al.*[32] it is these phenolic hydroxyl groups that participate in the condensation reactions and contribute to the observed increase in $\overline{Mw}$.

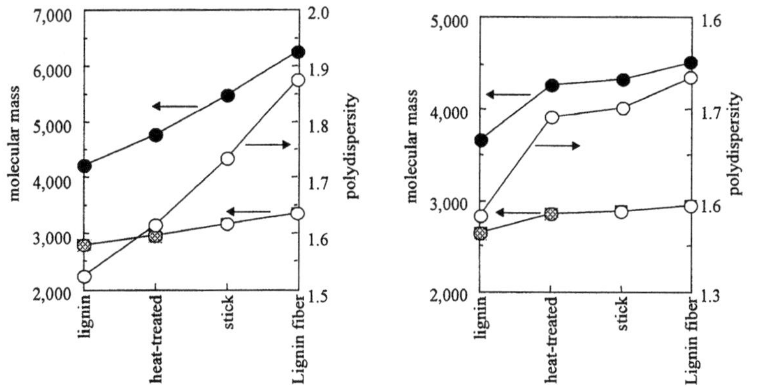

Figure 2. Effect of thermal processing on average molecular mass of HWKL (left) and Alcell (right) lignin (● $\overline{Mw}$; ● $\overline{Mn}$; ○ $\overline{Mw}/\overline{Mn}$).

3.2. Thermal Spinning of Lignin/Synthetic Polymer Blends

Polymer blending is a convenient method to develop products with desirable properties. The chemical and physical properties of the polymer blends are dependent on monomer type(s), molecular weight and distribution, composition of the respective polymers and miscibility. Most polymers, however, are immiscible due to the low entropy of mixing. Only through specific intermolecular interactions can miscible polymer blending occur and composite materials with desirable properties produced. We first set out to assess the effect of lignin/synthetic polymer blending on fiber spinning. Several different Alcell-, HWKL-based lignin fibers were produced with blend ratios of 0 to 25% PE, PP, PET or PEO respectively. The spinnability, or ease of fiber formation of the lignin/PEO series was better than that of the 100% lignin for both the Alcell lignin and HWKL. In particular, the spinnability of the Alcell/PEO (75/25) blend was excellent. Continuous spinning for more than 30 min was achieved with a spinning temperature of 120 – 157 °C. Figure 3 shows the various fibers produced from the HWKL/PEO blends. Interestingly, in the Alcell lignin blends, the 25% PEO blended fibers fused together on the take-up spools. This did not occur in the HWKL/PEO blends, and is likely due to the T_g of the Alcell/PEO fibers being below room temperature (Table 3).

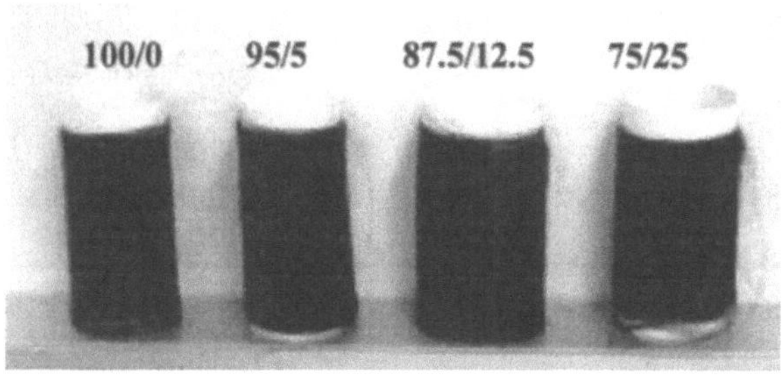

Figure 3. Melt extruded HWKL/PEO (w/w) fibers.

Figure 4 is an electron photomicrograph of a HWKL/PEO fiber at 500x and 160000x magnification. The fiber is continuous with a smooth surface when observed at low magnification, however at the higher magnification, a fine regular pattern is apparent.

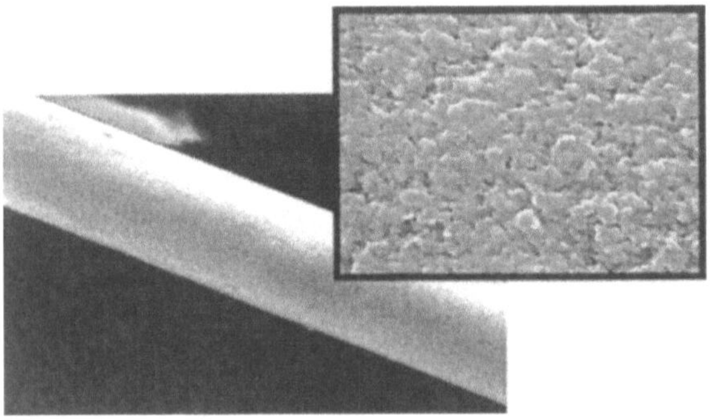

Figure 4. SEM micrograph of HWKL/PEO (99/1) blend fiber at 500X (insert at 160000X magnification).

In the case of PE, PP and PET lignin blends, spinnability decreased with increasing synthetic polymer incorporation. In fact fiber formation was not possible for the HWKL/PE blends, only short fiber fragments were obtained. For that reason, the focus will be on lignin/PEO blend fiber systems in the rest of this chapter. Table 2 lists the spinning parameters used for the various HWKL/polymer blend fibers.

Table 2. Spinning temperatures for HWKL blend fibers

Blend ratio (%)		Spinning temperature (°C)	
		Rotor	Nozzle
HWKL	PEO		
100	0	180	195-228
95	5	180	189-198
87.5	12.5	170	191-200
75	25	140	150-182
HWKL	PET		
95	5	180	208-228
87.5	12.5	190	213-224
75	25	195	223-231
HWKL	PP		
95	5	130	204-213
87.5	12.5	130	208-226
75	25	130	208-220

3.3. Thermal Analysis of Lignin-Based Fibers

Thermal analyses of the HWKL/PEO and Alcell/PEO blend fibers are listed in Table 3 and shown in Figure 5. The T_g of the HWKL (83.3 °C) is higher than that of Alcell lignin (68.2 °C), most likely due to the differences in chemical structure and molecular weight. It has been reported that Alcell lignin has a softening temperature[38] because its molecular mass is relatively low for an industrial lignin. The weight average molecular mass of the Alcell lignin used here is 3,600. This value is slightly lower than that of the HWKL (4,100). Alcell lignins also have lower phenolic hydroxyl group content, particularly condensed units, and a lower carboxylic group content than the corresponding HWKL produced from the same wood species (*vide supra*). As a result, thermal motion of the slightly higher molecular weight more condensed hydrophilic HWKL lignin is suppressed relative to that of the Alcell lignin. The result is a higher T_g and higher thermal energy required to induce flow for fiber formation.

Table 3. Glass transition temperature (T_g) of lignin/PEO blend fibers (°C)[a]

Lignin/PEO	HWKL	Alcell lignin
100/0	83.3	68.2
95/5	68.1	50.9
87.5/12.5	48.9	36.1
75/25	24.9	12.4
0/100	-63.0	-63.0

[a] Increasing the molar mass of the PEO from 100K to 600K required higher spinning temperature, but did not affect the observed T_g.

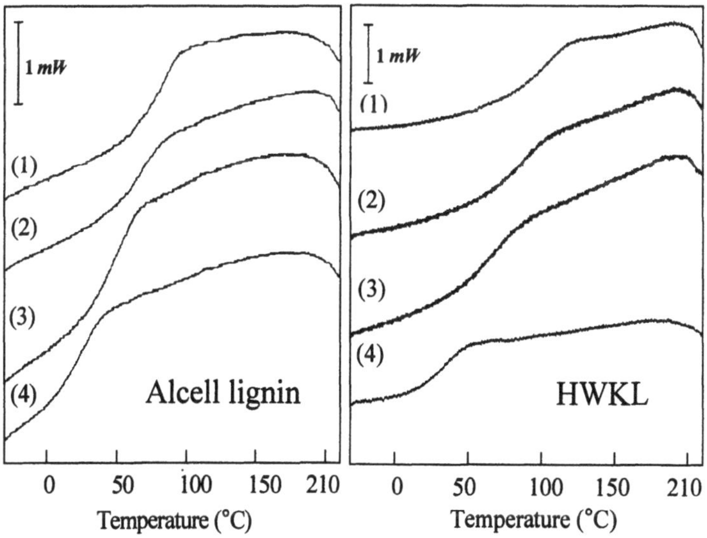

Figure 5. DSC thermograms of Alcell/PEO and HWKL/PEO bleand fibers (lignin/PEO (w/w); (1) 100/0; (2) 95/5; (3) 87.5/12.5; (4) 75/25).

It is apparent from the DSC traces (Figure 5), that both the Alcell and HWKL/PEO blend systems form miscible blends. In fact thermal and DRFT-IR analyses of the various Alcell/PEO and HWKL/PEO blend fibers reveal distinct differences in intermolecular interactions between Alcell/PEO and HWKL/PEO.[39] The melting peak of PEO (T_m = 69.4 °C), present in the mechanically blended powders was not in the thermograms of the melt spun blended fibers. Thus, polymer blend miscibility exists in the HWKL/PEO and Alcell/PEO blend fibers, where heat and shear assist the intimate mixing of these materials.

3.4. Carbonization of Lignin and Lignin/PEO Blend Fibers

In order to convert the lignin fibers produced from thermal spinning into carbon fibers a thermostabilization stage was necessary to prevent fiber fusing.[32] The thermoplastic character of the lignin is changed to thermosetting, enabling the lignin fibers to maintain fiber form during the subsequent carbonization. During the thermal stabilization stage the temperature was carefully increased at varying rates and the effect on fiber stability was observed. Fiber stability was dependent on thermostabilization conditions and PEO blend content. The 100% Alcell lignin fiber was very sensitive to the heating rate during thermostabilization. Increasing the heating rate above 12 °C/h resulted in the fibers in contact with one another fusing together. All the PEO blended Alcell lignin fibers could not be thermally stabilized under the conditions used. In contrast, the HWKL-based fibers were more thermally stable. The 100% HWKL fibers could be thermally stabilized at heating rates as high as 120 °C/h. The HWKL/PEO blend fibers however, as with the Alcell/PEO blend fibers, were not as thermally stable, fusing together at decreasing heating rates with increasing PEO content. However HWKL-PEO fibers with PEO blend contents of less than 5% (100K PEO) and 12.5% (600K PEO) could be stabilized. Figure 6 shows the effect of temperature on fiber stability.

Figure 6. Effect of increasing thermostabilization temperature rate on fiber stability. Shown are micrographs of 99% HWKL / 1% PEO (100K) fibers after thermostabilization at temperature rates of (i) 30 °C/h (10X magnification) and (ii) 90 °C/h (100X magnification).[35]

Reactions occurring during thermostabilization increase the T_g. As the temperature increases at the slow heating rate the T_g can increase faster than the temperature, maintaining the material in the glassy state (T_g>T): non-tacky. At higher heating rates the thermostabilization reactions are not able to maintain T_g>T and thus the material devitrifies entering the liquid/rubbery state, tacky, and thus fuses together. Gillham and co-workers have described such phenomena in continuous-heating-transformation (CHT) diagrams.[40, 41]

The carbon fibers produced from HWKL and its blends with PEO (Figure 7) do not have any micropores, and the cross-section reveals no microtextures generally found in pitch-based carbon fibers.[42]

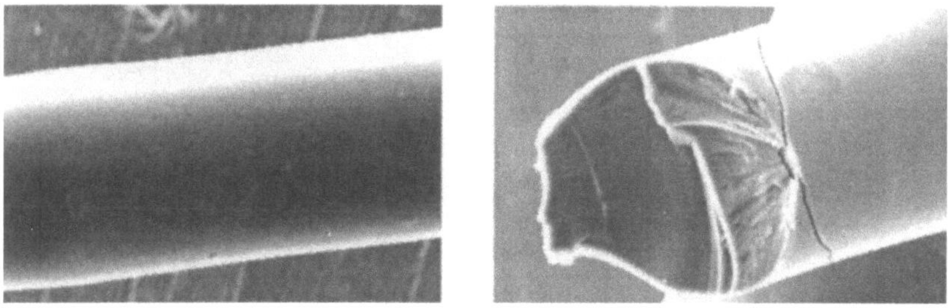

Figure 7. SEM of 99% HWKL / 1% PEO (100K) carbon fibers (1000X magnification).[39]

PEO blend content (≤ 5%) and molecular weight (100K – 600K) did not affect the mechanical properties of the carbon fibers (Table 4). The properties are slightly better than those previously reported for carbon fibers produced from acetic acid lignin (tensile strength 355 ± 53 MPa, modulus 39.1 ± 13.3 GPa)[33] and phenolated steam-exploded lignin (tensile strength 311-394 MPa).[29]

Table 4. Mechanical properties of HWKL, HWKL/PEO and Alcell lignin carbon fibers [35]

	Diameter (μm)	Tensile-strength (MPa)	Modulus (GPa)	Elongation (%)
HWKL	46 ± 8	422 ± 80	40 ± 11	1.12 ± 0.22
HWKL/PEO (100K) 97-3	34 ± 4	448 ± 70	51 ± 13	0.92 ± 0.21
HWKL/PEO (100K) 95-5	46 ± 3	396 ± 47	38 ± 5	1.06 ± 0.14
HWKL/PEO (300K) 95-5	44 ± 4	392 ± 89	44 ± 12	0.91 ± 0.16
HWKL/PEO (600K) 97-3	33 ± 2	458 ± 97	59 ± 8	0.79 ± 0.21
HWKL/PEO (600K) 95-5	63 ± 7	339 ± 53	33 ± 2	1.25 ± 0.26
Alcell lignin	31 ± 3	388 ± 123	40 ± 14	1.00 ± 0.23
Steam-exploded lignin[29]				
Hydrogenolysis	7.6	660	41	-
Phenolysis	-	400	-	-
Acetic Acid Lignin[33]	14	360	39	-
PAN[6]	7	3300	230	-
Isotropic Pitch[6]	14.5	720	32	-

It is noteworthy that in the acetic acid and phenolated steam-exploded lignin-based systems, the fiber diameters were between 7-14 μm, but the HWKL-based fibers and Alcell lignin fiber have significantly higher fiber diameter (30-63 μm). As expected both the tensile strength and modulus increased with decreasing fiber diameter (Figure 8). The solid lines are regressions from the actual data extrapolated to smaller fiber diameters to show expected results. If the HWKL-based carbon fibers had diameters less than 10 μm, the mechanical strength would be the same as that of isotropic pitch (tensile strength ~ 700MPa[29]) and suitable for general performance grades.[35]

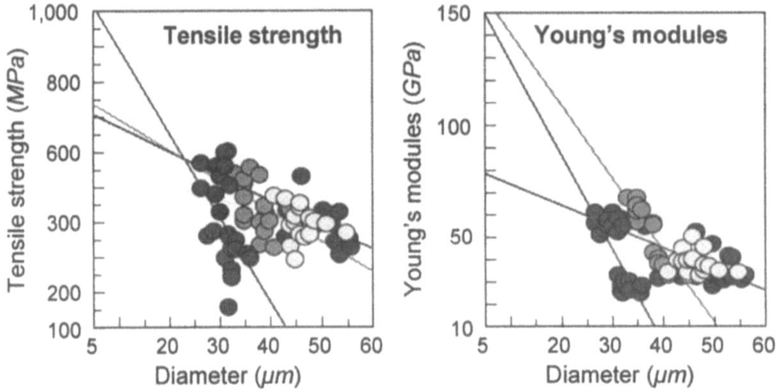

Figure 8. Relationship between mechanical properties and fiber diameter for HWKL- and Alcell lignin-based carbon fibers. The solid lines are regressions from the actual data extrapolated to smaller fiber diameters (HWKL ●; HWKL/PEO(100K) 97/3 ●; HWKL/PEO(100K) 95/5 ○; Alcell lignin ●).[35]

3.5. Carbon Fiber Yield

An important aspect in carbon fiber production is yield. Higher weight losses, or lower yields, result in higher production costs. Depending on the structure of the "white-fiber", yields from the various processing steps will vary. Table 5 summarizes the yields obtained for each process step, as well as the overall yield in the production of lignin-based carbon fibers (included are the reported yields for a commercial tar pitch[29] and acetic acid lignin[33]).

Table 5. Yields for each step in the production of carbon fibers from HWKL and Alcell lignin[35]

	Pretreatment (%)	Spinning (%)	Stabilization (%)	Carbonization (%)	Over-all (%)
HWKL	99.0	95.0	94.4	51.6	45.7
HWKL/PEO (97/3)	99.0	95.0	94.2	49.3	43.7
HWKL/PEO (95/5)	99.0	95.0	93.5	51.0	45.0
Alcell lignin	97.0	95.0	97.0	46.2	40.0
Acetic acid lignin[33]	95.2	96.7	88.0	40.1	32.7
Tar Pitch[29]	40.0	95.0	110.0	80.0	33.4

Unlike commercial tar pitch, the thermal pretreatment step had little effect on yield losses for the HWKL and Alcell lignin-based fibers. The thermal pretreatment step is utilized to remove the light fraction in commercial tar pitch, which hinder fiber formation. The large difference in pretreatment yield between lignin and pitch is due to the differences in the chemical components. Both are complex mixtures of aromatic compounds with alkyl side chains, but technical lignins contain far less components with low melting points.

In the lignin-based fiber systems, slight weight loss occurred during thermal spinning (ca. 5%) and stabilization (ca. 3-7%). These yield losses were accompanied by an increase in molecular weight (vide supra), the result of condensation and dehydration of the lignin macromolecules.[31] These results are opposite to those for the oxidative stabilization of pitch fibers,[29] in which weight/yield increases. In pitch, the formation of infused oxygen bridging and cross-linking occur analogous to those in lignin. However, these reactions require the alkyl components of pitch to be first oxidized prior to the dehydrogenation and cyclization reactions, thus the weight increases. In lignin, in which the side-chain is already highly oxidized, the condensation reactions release water and decrease weight. Although slightly lower yields are obtained from stabilization, the lignin-based carbon fibers have the advantage of needing a lower temperature and shorter residence time since cross-linking is facilitated by oxygen-based radical reactions within lignin.

In the lignin-based fibers, the majority of weight loss occurred during the carbonization step; 48-51% for HWKL-based fibers and ~54% for Alcell lignin fibers. This weight loss is substantially higher than that observed for pitch (20%). The purpose of carbonization (<1500 °C) is to eliminate all elements other than carbon in order to produce a graphite-like structure with a carbon content of greater than 96%.[6] Technical lignins, with elemental carbon contents of only 59-61%,[43] would be expected to have substantial weight loss, while pitch, >80% carbon,[6] should have higher yields. However, the yields obtained for both the HWKL and Alcell lignins are approximately 10% higher than those previously reported for the carbonization of organosolv lignins[32] and comparable to those from phenolated steam-exploded lignins.[29]

Overall yields were approximately 45% for the HWKL-based carbon fibers and 40% for the Alcell lignin-based carbon fibers. These results are significantly higher than those reported for tar pitch (33.4%),[29] and acetic acid pulped lignin (32.7%).[32] Thus, with respect to cost of production, HWKL and Alcell seem to be promising resources for carbon fibers.

4. ACKNOWLEDGEMENTS

Financial support was provide by Oak Ridge National Laboratories and is gratefully acknowledged. The scanning electron micrographs were courtesy of ORNL.

5. REFERENCES

1. J. Riggs, in: *Encyclopedia of Polymer Science and Engineering*, edited by H. F. Mark and J. I. Kroschwitz (Wiley, New York, 1985), p. 17 v.
2. J. W. Swan, British Patent: 4933 (1880).

3. T. A. Edison, U.S. Patent: 223,398 (1880).
4. J. B. Donnet and O. P. Bahl, in: *Encyclopedia of Physical Science and Technology* (Academic Press, New York, 1987), p. 515.
5. Developed in the 1940's by DuPont and Union Carbide, PAN began to be used as a carbon fiber precursor material in the 1960's through work done in Japan (Toray) and Great Britain (Courtaulds).
6. J. B. Donnet, *Carbon Fibers*, 3rd Rev. and Expand Ed. (Marcel Dekker, New York, 1998), viii, 573.
7. H. Hatori, Y. Yamada, and M. Shiraishi, Inplane orientation and graphitizability of polyimide films. 2. Film thickness dependence, *Carbon* **31**(8), 1307-1312 (1993).
8. D. E. Sliva and W. G. Selley, Continuous method for making spinnable polyacetylene solutions convertable to high strength carbon fibers, U.S. Patent: 3928516 (1975).
9. C. M. Krutchen, Melt extrudable polyacetylene copolymer blends, U.S. Patent: 3852235 (1974)
10. R. W. Kobayashi and R. D. Zaldivar, Carborane catalyzed graphitization of polyarylacetylene (PAA), U.S. Patent: 5288438 (1994).
11. A. Nagasaki, H. Ashitaka, Y. Kusuki, D. Oda, and T.Yoshinaga, Process for producing carbon fiber, U.S. Patent: 4131644 (1978).
12. S. Horikiri, J. Iseki, and M. Minobe, Process for producing carbon fiber, U.S. Patent: 4070446 (1978).
13. T. Araki, H. Takita, and K. Asano, Verfahren zum herstellen von kolenstoffasern, German Patent: 2024063 (1970).
14. I. Seo, T. Oono, and Y. Murakami, Process for producing raw material pitch for carbon materials, European Patent: 381493 (1990)
15. M. Shiokawa and T. Matsumoto, Production of pitch-based carbon fiber, Japanese Patent:1282349 (1989).
16. H. Ashitaka, Y. Kusuki, S. Yamamoto, Y. Ogata, and A. Nagasaka, Preparation of carbon-fibers from syndiotactic 1,2-polybutadiene, *J. Appl. Polym. Sci.* **29**(9), 2763-2776 (1984).
17. W. G. Glasser and S. S. Kelley, in: *Concise Encylopedia of Polymer Science and Engineering*, edited by J. I. Kroschwitz (Wiley, New York, 1990), p. 544.
18. D. Fengel and G. Wegener, *Wood: Chemistry, Ultrastructure, Reactions.* (W. de Gruyter, Berlin, 1984), xiii, 613.
19. W. G. Glasser, R. A. Northey, and T. P. Schultz, *Lignin: Historical, Biological, and Materials Perspectives* (ACS Symposium Series 742, American Chemical Society, Washington, D.C., 2000), xv, 559.
20. S. Otani, Y. Fukuoka, B. Igarashi, and K. Sasaki, Method for producing carbonized lignin fiber, U.S. Patent: 3,461,082 (1969).
21. M. Mansmann, G. Winter, P. Pampus, H. Schnoring, and N. Schon, Stable lignin fibers, U.S. Patent: 3,723,609 (1973).
22. "Kayacarbon" Manufacturer's brochure, N.K.C.L., *"Kayacarbon"Manufacturer's brochure, Nippon Kayaku Co. Ltd.*
23. D. J. Johnson, I. Tomizuka, and O. Watanabe, Fine-structure of lignin-based carbon-fibers, *Carbon* **13**(4), 321-325 (1975).
24. D. J. Johnson, I. Tomizuka, and O. Watanabe, Fine-structure of pitch-based carbon-fibers, *Carbon* **13**(6), 529-534 (1975).
25. I. Tomizuka and D. J. Johnson, Microvoids in pitch-based and lignin-based carbon fibres as observed by x-ray small angle scattering, *Yogyo-Kyokai-Shi* **86**(4),186 (1978).
26. I. Tomizuka, T. Kurita, Y. Tanaka, and O. Watanabe, Voids in the carbon fibers produced from lignin and PVA, *Yogyo-Kyokai-Shi* **79**(12), 460 (1971).
27. K. Sudo and K. Shimizu, A new carbon-fiber from lignin, *J. Appl. Polym. Sci.* **44**(1),127-134 (1992).
28. K. Sudo, M. Okoshi, and K. Shimizu, Carbon-fiber from lignin--improvement of conversion process of lignin, Abstracts of Papers of the American Chemical Society, **195**, 107-CELL (1988).
29. K. Sudo, K. Shimizu, N. Nakashima, and A. Yokoyama, A new modification method of exploded lignin for the preparation of a carbon-fiber precursor, *J. Appl. Polym. Sci.* **48**(8),1485-1491 (1993).
30. K. Sudo and K. Shimizu, Method for manufacturing lignin for carbon fiber spinning, U.S Patent: 5,344,921 (1994).
31. S. Kubo, N. Ishikawa, Y. Uraki, and Y. Sano, Preparation of lignin fibers from softwood acetic acid lignin - Relationship between fusibility and the chemical structure of lignin, *Mokuzai Gakkaishi***43**(8), 655-662 (1997).
32. S. Kubo, Y. Uraki, and Y. Sano, Preparation of carbon fibers from softwood lignin by atmospheric acetic acid pulping, *Carbon* **36**(7-8), 1119-1124 (1998).
33. Y. Uraki, S. Kubo, N. Nigo, Y. Sano, and T. Sasaya, Preparation of carbon-fibers from Organosolv lignin obtained by aqueous acetic-acid pulping, *Holzforschung* **49**(4), 343-350 (1995).
34. K. Itoh, Prepareation of lignin fiber, Japanese Patent: H1239114 (1989)
35. J. F. Kadla, S. Kubo, R. D. Gilbert, R. A. Venditti, W. Griffith, and A. L. Compere, Lignin-based carbon fibers for composite fiber applications, submitted to *Carbon* (2001).

36. E. Corradini, E. A. G. Pineda, and A. A. W. Hechenleitner, Lignin-poly (vinyl alcohol) blends studied by thermal analysis, *Polym. Degrad. Stab.* **66**(2), 199-208 (1999).
37. Y. Lui, S. Carriero, K. Pye, and D. Argyropoulos, A comparison of the structural changes occurring in lignin during alcell and kraft pulping af hardwoods and softwoods, in: *Lignin: Historical, Biological, and Materials Perspectives*, edited by R. A. Northey, W. G. Glasser, and T. P. Schultz (ACS Symposium Series 742, American Chemical Society, Washington, D.C., 2000), p. xv, 559.
38. E. K. Pye and J. H. Lora, The Alcell Process - a proven alternative to kraft pulping, *Tappi J.* **74**(3), 113-118 (1991).
39. J. F. Kadla, S. Kubo, R. A. Venditti, and R. D. Gilbert, Lignin-based thermoplastics: Effect of intermolecular interactions on polymer blends, *Polymer Preprints - PMSE* (2001).
40. J. K. Gillham and J. B. Enns, On the cure properties of thermosetting polymers using torsional braid analysis, *Trends in Polymer Science* **2**(12), 406-419 (1994).
41. G. Wisanrakkit and J. K. Gillham, Continuous heating transformation (CHT) cure diagram of an aromatic amine epoxy system at constant heating rates, *J. Appl. Polym. Sci.* **42**(9), 2453-2463 (1991).
42. J.-B. Donnet and R. C. Bansal, *Carbon Fibers* (International fiber science and technology series, Marcel Dekker, New York, 1984), v. 3. p. 291.
43. S. Y. Lin and C. W. Dence, *Methods in Lignin Chemistry* (Springer series in wood science, Springer-Verlag, Berlin, 1992).

THE USE OF LIGNOSULFONATES AS WATER REDUCING AGENTS IN THE MANUFACTURE OF GYPSUM WALLBOARD

Robert A. Northey*

1. INTRODUCTION

The chemical treatment of woody materials to produce fibers (pulps) also produces a nearly equal amount of soluble chemicals consisting mainly of modified lignin and carbohydrate degradation products. In most chemical pulp mills, this material is burned to create energy for the recycling of pulping chemicals. A small number of pulp mills, mostly acid sulfite mills, purify the lignin streams and market this material commercially. In 1998, close to one million tons of lignosulfonates were used worldwide in a variety of industrial applications.[1] The largest use of this material is as a dispersant of particle slurries such as concrete, clays, gypsum, and dyes.[2] In these applications, lignosulfonates must compete with synthetic dispersants such as condensation polymers of formaldehyde with sulfonated naphthalene. Much effort, therefore, has been directed towards increasing the dispersing ability of lignosulfonates in these systems through chemical modifications and physical separations. These processes, however, can also affect other properties of the particle slurries such as setting time, air entrainment, and color. It is very important, therefore, to look at the entire system when considering a lignin modification. In this chapter, the effect that lignosulfonate components and chemical modification reactions have on the function of lignosulfonate dispersants in the manufacture of gypsum wallboard will be discussed.

1.1. Manufacture of Gypsum Wallboard

In the manufacture of gypsum wallboard, industrial stucco is mixed with water and a variety of additives to form a slurry, which is deposited between paper liners and allowed to harden. Stucco, which is referred to as plaster outside North America, is produced through calcination of either mined gypsum or synthetic gypsum from flue gas

* Robert A. Northey, College of Forest Resources, University of Washington, Seattle, WA 98195.

Edited by Thomas Q. Hu, Kluwer Academic/Plenum Publishers, 2002

desulfurization. Stucco from mined gypsum typically contains over 80% calcium sulfate hemihydrate ($CaSO_4$ ½H_2O) along with small amounts of clays, carbonates, and other minerals. Stucco from synthetic gypsum contains mostly calcium sulfate hemihydrate. Calcination of these materials is accomplished either through kettle calcination or through air suspension calcination. These two processes impart significantly different setting and water requirement characteristics to the stuccos.

Enough water is added to stucco in a mixer to form a slurry, which will flow across the paper. Only a small portion of the added water is needed to hydrate the calcium sulfate hemihydrate to gypsum ($CaSO_4$ 2H_2O). The excess water must be removed in a drying stage after the board has hydrated. Dispersants, such as lignosulfonates, are added during the mixing stage at levels in the range of 0.1 to 0.6% on wallboard solids to reduce the amount of water needed to obtain a slurry of the desired viscosity. With plants using kettle-calcined stucco, this can significantly reduce the amount of water that needs to be removed and thus lower the energy costs. With air suspension calcination, the amount of water that can be removed using a dispersant is much less but other benefits related to the overall flowability of the slurry are obtained.

The process of hardening (setting) of gypsum wallboard occurs through the hydration of the calcium hemihydrate to the dihydrate (gypsum). The rate of this process is very important in industrial applications. The mixture must stay fluid long enough to exit the mixer and flow across the paper. Past this point, it is desirable for the rate of hydration to proceed rapidly. The rate of set determines the speed at which the production line runs.

1.2. The Use of Lignosulfonates in Gypsum Wallboard

Lignosulfonates have a long history of use as dispersants in the gypsum wallboard industry. Currently, a variety of different types of lignosulfonates are used as dispersants. The simplest and cheapest of these products normally have been desugared through fermentation, separation, or chemical oxidation processes. Typical addition rates for these types of products are approximately 1-3 pounds of lignosulfonate solids per thousand feet of one-half inch wallboard. The competition for lignosulfonates in this market is synthetic condensation polymers of sulfonated naphthalene with formaldehyde.[3] These polymers not only are highly efficient dispersants, but they also do not retard the setting of the wallboard as do the lignosulfonate products.

Improved dispersing lignosulfonate products have been developed to compete with the synthetic dispersants through a variety of chemical treatments and membrane separation techniques. While the dispersing abilities of these products are increased, so too are the production costs. Unfortunately, many of the treatments, which increase dispersion, also increase the retarding nature of the product.

In this chapter, the effect that each individual lignosulfonate component has on the dispersion and set retardation of industrial stucco slurries will be discussed. Much of this information comes from previously published studies.[4, 5] New information, however, on the effect of chemical and mechanical treatments on lignosulfonates will also be presented.

2. THE EFFECT OF INDIVIDUAL LIGNOSULFONATE COMPONENTS

Lignosulfonates are made up of many components all of which can affect how this material functions in gypsum wallboard. The major component in lignosulfonates from the acid sulfite pulping of hardwoods or softwoods is sulfonated lignin (60-80%). This three-dimensional network polymer is primarily responsible for the dispersion of the slurry particles. This is thought to occur through a combination of electrostatic and steric stabilization. Other components found in commercial lignosulfonates include sugars, sugar acids, wood extractives, and inorganic compounds. The amount and types of these compounds are dependent upon the wood species and the pulping procedures utilized.

2.1. Non-Lignin Chemicals

2.1.1. Sugars

Sugars typically are the second most abundant chemicals found in lignosulfonate products after lignin. The majority of commercial lignosulfonate products are manufactured using acid sulfite pulping technology and can contain up to 30% free sugars. Typically, however, the sugar level is reduced through fermentation, ion exchange, ultrafiltration, or chemical oxidation techniques. This typically reduces the sugar content to between 1-5%. Commercial products from neutral sulfite pulping will contain some free sugars but mainly large percentages of oligosacharides.

In a previous investigation, it was shown that sugars have no effect on the dispersion or retardation of stucco slurries[4] (Table 1). This was proven through the addition of large amounts of xylose to a purified lignosulfonate dispersant while keeping the total charge of sulfonated lignin on stucco constant. Similar results were obtained with other sugars and with larger sugar polymers. Therefore, sugars present in lignosulfonate dispersants actually dilute the dispersing ability of the material. During the manufacture of gypsum wallboard, sugars are in fact added to reduce the over drying of the ends of the board.

2.1.2. Sugar Acids

During acid sulfite pulping, sugars are oxidized by bisulfite ions to aldonic acids (sugar acids). These compounds typically make up approximately 4-7% of the total lignosulfonate solids. At neutral to alkaline conditions, an additional number of mono- and dicarboxylic acids are formed. Previous studies have shown that although aldonic acids function as dispersants in concrete slurries, they have no effect on the dispersion of gypsum (Table 1). This is also true for the various salts of aldonic acids. Gluconic acid, however, was found to severely retard the set of stucco slurries. This was also true for the calcium, sodium, and potassium salts of gluconic acid. Other salts, such as ferrous, ferric, aluminum, and manganese were much less retarding. This discovery led to the development of a less retarding dispersant.[6]

Table 1. Effect of calcium lignosulfonate and non-lignin constituents on the dispersion and retardation of stucco slurries

Component	Water reduction[a] (%)	Set time[b] (minutes)
Control (water)	N/A	19
Calcium lignosulfonate	13	26.3
Xylose	0	18.8
Gluconic acid	0	32.6
Calcium gluconate	0	26.6
Ferrous gluconate	0	19

[a] 0.2% use level.
[b] 0.6% use level.

2.1.3. Wood Extractives

The wood species used in sulfite pulping typically contain very low levels of wood extractives or they have been subjected to a treatment to remove extractives before pulping. Therefore, there are typically only very limited amounts of these types of compounds in commercial lignosulfonate products. Even the low levels of compounds such triterpenoids or fatty acids found in most commercial lignosulfonates will cause air entrainment problems in concrete. In gypsum studies, it has been found that strongly chelating extractives can cause retardation.[7] The levels of these compounds are typically so low in lignosulfonate samples, however, that their contribution to set retardation is negligible.

A class of wood extractives rarely discussed in industrial applications is the proteins. These compounds are found in wood at levels of around 1%. Analysis of a typical softwood fermented calcium lignosulfonate indicated nitrogen levels of approximately 0.15-0.2%. Further analysis of a highly purified sodium lignosulfonate which had been fractionated into different molecular weight fractions through ultrafiltration indicated that the nitrogen was evenly spread throughout the entire molecular weight range. The exact nature of these nitrogen compounds after acid sulfite pulping is not known.

A series of experiments were conducted to test the effect of protein, peptides, and amino acids on the dispersion and set retardation of stucco slurries. Casein and gelatin were used to model protein. Peptone, an acid degraded protein was used to model peptides. A series of 14 amino acids were also tested. None of these classes of nitrogen compounds functioned to any extent as dispersants in stucco slurries.

In set time tests, the nitrogen compounds retarded the set of the stucco slurry to different extents (Table 2). The proteins, casein and gelatin, did not retard the set while the peptide (peptone) was very set retarding. All of the amino acids retarded the setting time of stucco slurries to some extent. Most were negligible to somewhat retarding but three compounds, aspartic acid, glutamic acid, and cystine were very set retarding. It is important to note that these compounds were added to stucco slurries at levels significantly higher than what could be found in lignosulfonate samples. When cystine was added alone at levels equivalent to 1% of lignosulfonate solids, the set retardation was negligible. However, when this level of cystine was added with another set retarding agent, the level of set retardation due to cystine was measurable.

Table 2. Effects of nitrogen compounds on stucco setting time

Nitrogen compounds	Negligible retardation	Somewhat retarding	Very retarding
Amino acids	Proline Valine Lysine Phenylalanine Histidine Methionine Threonine	Leucine Arginine Glycine Asparagine	Aspartic acid Glutamic acid Cystine
Peptides			Peptone
Protein	Casein Gelatin		H_2O_2 oxidized Casein
Control		Ca Lignosulfonate	

Additional tests were run in which casein was treated with an alkaline solution of hydrogen peroxide. This treatment resulted in the formation of a very set retarding material. No attempt was made to determine what changes had been made to the protein. As will be discussed later, many of the treatments used to improve the water reducing ability of lignosulfonate samples involve alkaline oxidation. Although the dispersing ability of these products is improved, they also typically become highly set retarding. It is possible that a good portion of this retardation is due to the degradation of proteins. It is also important to note that alkaline degraded protein products are marketed commercially as gypsum retardants to keep stucco slurries from setting in the mixer.

2.2. Sulfonated Lignin

In most commercial lignosulfonate products, sulfonated lignin is the major constituent making up between 60 to 100% of the product. In acid sulfite pulping, lignin is solubilized through sulfonation of the benzylic carbon of the phenyl propane unit. In commercial products, the lignin is sulfonated anywhere from 0.3 to 0.6 per phenyl propane unit. Lignosulfonates are also produced from kraft lignin using methods such as sulfomethylation to sulfonate the aromatic ring. Both methods of sulfonation produce excellent dispersants.

There are many variables related to the nature of the lignosulfonate polymer that could possibly affect both dispersion and set retardation properties. Included in this list are fiber raw material, the pulping method, the base of the lignosulfonates, the molecular weight distribution,[8] the amounts and types of functional groups present,[9] and any structural changes caused through modification reactions. Many of these variables have been investigated in the hopes of producing improved dispersants.

2.2.1. *Effect of Fiber Source and Pulping Base on Dispersion*

Commercial lignosulfonate products are currently manufactured exclusively from softwoods and hardwoods. Typically, these products are available as calcium, ammonium, or sodium salts although other salts can be obtained. An experiment was undertaken to test the effect of fiber source and base on the dispersion and set retardation of stucco slurries. Hardwood and softwood fermented lignosulfonates were obtained from the same pulp mill. Additionally, the softwood lignosulfonates were base exchanged from calcium to sodium and ammonium salts. Because the hardwood lignosulfonates contained a higher level of unfermentable five carbon sugars, the lignosulfonate concentrations were lower than the softwood product. Additionally, a softwood lignosulfonate sample was amine salt purified to obtain a sodium sample of nearly pure sulfonated lignin.

Testing of all these materials in a stucco slurry demonstrated that there was not much of a significant difference in dispersing ability between the hardwood and the softwood samples (Table 3). Most of the small difference can be attributed to the lower level of sulfonated lignin in the hardwood material. Similarly, the same trend can be seen between the unfermented, the fermented, and the amine salt purified lignosulfonates. The difference in dispersing ability correlates to the concentration of sulfonated lignin. It is important to note, that this correlation does not work between lignosulfonates from different processes or between products that have been modified. Finally, it can be seen that similar dispersing abilities were found for the calcium, sodium, and ammonium lignosulfonate salts.

Table 3. Effect of fiber source, fermentation and cooking base on the dispersion of stucco slurries

Lignosulfonate products	Sulfonated lignin (%)	Water reduction[a] (%)
Softwoood, unfermented, calcium	60	8
Softwood, fermented, calcium	77	10.2
Softwood, fermented, sodium	77	10.1
Softwood, fermented, ammonium	77	10.3
Softwood, amine salt purified, calcium	99	12.5
Hardwood, fermented, calcium	72	9.4

[a] Use level 0.2% of stucco solids.

2.2.2. *Effect of Lignin Molecular Weight on Dispersion*

As demonstrated in the previous section, increasing the concentration of sulfonated lignin in a lignosulfonate product improves the stucco dispersing ability. One method of accomplishing this is by ultrafiltration. This technique is also used to increase molecular weight. Previous studies have shown that both the very high and low range molecular weight lignosulfonates are less efficient dispersants in stucco slurries than the mid range lignosulfonate fractions.[5] This was proven through testing a series of highly characterized narrow molecular weight lignosulfonate fractions prepared through ultrafiltration (Figure 1). Dispersing ability of the lignosulfonates improved with increasing molecular weight until the highest molecular weight fraction. As was the case with the correlation between percent lignosulfonates and water reduction, this molecular weight versus dispersing ability curve is specific to only this lignosulfonate material.

The ability of lignosulfonates to disperse stucco particles is directly related to the amount of sulfonated lignin adsorbed onto the surface of the stucco particles. It has been determined that as molecular weight increases, so does the percentage of sulfonated lignin adsorbed. The curve shown in Figure 2 is for an unmodified, fractionated softwood lignosulfonate. It can be seen that at low molecular weights, only a very small percentage of the sulfonated lignin is adsorbed. As the molecular weight increases, much higher percentages of lignin are adsorbed until a point is reached where only small increases are obtained with large increases in molecular weight. At this point, adsorption becomes less important than the number of molecules in solution so the higher molecular weight fractions are less dispersing. The curve in Figure 2 is only valid for the specific lignosulfonate. Modifications of this lignosulfonate will dramatically change the shape of this curve.

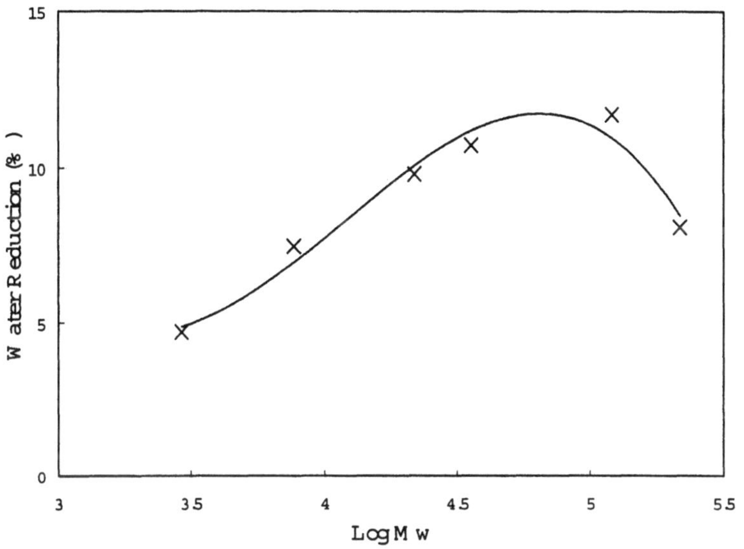

Figure 1. Effect of the molecular weight of lignosulfonate on its ability to disperse stucco.

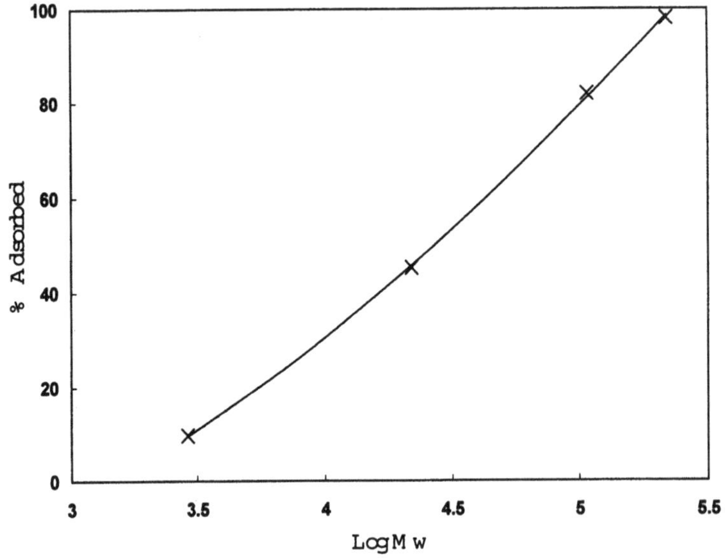

Figure 2. Effect of sulfonated lignin molecular weight on percentage adsorbed on stucco particles.

2.2.3. Lignosulfonate Modification Reactions

Over the last several decades, a large number of modified lignosulfonate products have been developed for use as dispersants in the manufacture of gypsum wallboard. Generally, the modifications used in the production of these products involve the use of some type of oxidation reaction. The simplest of these reactions involve boiling a lignosulfonate solution at an alkaline pH over a long period of up to twenty-four hours. Harsher reaction conditions include ozone oxidation and high temperature alkaline oxidation as used in the formation of vanillin. These reactions all produce improved dispersing products that are also quite set retarding. Ultrafiltration of these products produces improved dispersing and lower retarding products.

There are other products that work quite well as gypsum dispersants that do not fit the above description of modified lignosulfonate products. One of these is kraft lignin that has been sulfomethylated. These are highly dispersing and low set retarding products. A unique lignosulfonate product with improved dispersing and reduced set retarding nature is produced from lignosulfonates through a mild alkaline hydrolysis followed by a base exchange to the ferrous salt.[6]

In all of these modified products, improved dispersion is obtained through increased adsorption of sulfonated lignin on the stucco particles. Table 4 lists several modified lignosulfonate gypsum dispersants along with the percent adsorbed and water reduction. Exactly why adsorption is increased in these products is not known. Chemical analysis of the functional groups does not explain the differences. The degree of sulfonation is not reduced in most of these products, the carboxyl content is increased to only a small extent, and the molecular weight is hardly changed.

As can be seen in Table 4, sulfomethylated kraft lignin is highly adsorbed and therefore an excellent stucco dispersant. An important observation about this is that these products have lower molecular weights than do the lignosulfonates in this table.

Table 4. Effects of lignosulfonate modifications on adsorption and water reduction

Product	Manufacturing procedure	Adsorbed (%)	Water reduction (%)
LIGNOSITE® 458 Sodium Lignosulfonate	Fermented sodium lignosulfonate	25	8.4
LIGNOSITE® 260 Sodium Lignosulfonate	Alkaline air oxidized lignosulfonate	37	11.3
LIGNOSITE®NR Sodium Lignosulfonate	Alkaline treated ferrous lignosulfonate	44	11.2
STARFLO™ G	Alkaline air oxidized, ultrafiltered lignosulfonate	55	13.0
Polyfon T	Sulfonated kraft lignin	61	13.3
Polyfon F	Sulfonated kraft lignin	68	13.5

2.2.4. *Set Retardation*

Lignosulfonate products that have had nearly all of the non-lignin chemicals removed are nearly non-set retarding in stucco slurries. Compared to synthetic dispersants, there is a very small amount of retardation with these lignosulfonate products. It is possible to quantify the amount of retardation only at excessively high levels of addition. It is not possible to determine whether this limited amount of retardation is due to the sulfonated lignin polymers or due to trace amounts of impurities still associated with the lignin. When the sulfonated lignin products are purified after modification reactions, the resulting lignin products are also nearly non-retarding even though the modified products are highly retarding.

3. CONCLUSIONS

The choice of dispersants in the gypsum wallboard industry will continue to be based on cost and performance. In many cases, the low cost of lignosulfonate products will continue to allow these materials to compete in this market even though they are not as dispersing as synthetic products and they retard the set of the board. Research into these problems has now shown that the majority of set retardation is due to non-lignin impurities that are good chelating agents such as sugar acids and protein degradation products. Removal of these products through ultrafiltration eliminates most of the set retardation. In order to achieve greater dispersing efficiency, several options are available. The first involves increasing the content of sulfonated lignin in the product through removal of non-lignin compounds using ultrafiltration or fermentation. The second option involves increasing the molecular weight of the product through ultrafiltration that will cause the sulfonated lignin to be more highly adsorbed on the surface of the stucco particles and thus acts as a better dispersant. The third option involves improving the adsorption of the sulfonated lignin through a variety of chemical modification reactions. Although these modification procedures are known to increase adsorption, the exact mechanism for these improvements has yet to be determined.

4. EXPERIMENTAL

4.1. Materials

The stucco used in these tests included a kettle calcined stucco from Georgia-Pacific Gypsum in Quanah, Texas, a kettle calcined stucco from the Georgia-Pacific facility in Wilmington, Delaware and a kettle calcined stucco from the Domtar Gypsum facility in Tacoma, Washington. These industrial stuccos roughly contained calcium sulfate hemihydrate (~80%), anhydrous calcium sulfate (~2%), clays (~5%), carbonates (~5%), and various other impurities. In some of these samples, there were also various amounts of fibers depending on the amount of recycled material utilized.

The lignosulfonate products were obtained from the Georgia-Pacific calcium based acid sulfite pulp mill in Bellingham, Washington. The softwood lignosulfonates came from a mixture of approximately 90% Western Hemlock (Tsuga heterophylla) and 10% other wood species. The hardwood lignosulfonates came mainly from Red Alder (Alnus rubra). The hexose sugars were removed from the lignosulfonate at the pulp mill through fermentation by yeast (Saccharomyces cerevisiae). The basic calcium lignosulfonate used in these studies was LIGNOSITE® CX Calcium Lignosulfonate. Modifications to this calcium material provided the following other products used in this study.

1. LIGNOSITE® 458 Sodium Lignosulfonate: A standard commercial sodium lignosulfonate product.
2. LIGNOSITE® 17 Ammonium Lignosulfonate: A standard commercial ammonium lignosulfonate product.
3. LIGNOSITE® NR: This material was produced in the laboratory through the method described in reference 6. The product is a neutralized iron (III), calcium, potassium lignosulfonate.
4. LIGNOSITE® 260 Sodium Lignosulfonate: A standard commercial alkaline oxidized sodium lignosulfonate product.
5. STARFLO™ G: A commercial ultrafiltered higher molecular weight chemically modified lignosulfonate product with a Mw of approximately 53,000.
6. Molecular weight fractionated lignosulfonates: Molecular weight fractions produced through ultrafiltration as described in reference 10.
7. Amine salt purified lignosulfonate: This material was produced in the laboratory using methods describe by Lin.[11] The lignosulfonates were extracted in the tripentylamine salt form using butanol as the solvent. This method produces a sulfonated lignin, which is nearly free of non-sulfonated materials.
8. Polyfon T and Polyfon F were obtained from WESTVACO Inc.
9. Xylose, gluconic acid, calcium gluconate, sodium gluconate, and ferrous gluconate were commercially available from Aldrich as were all of the amino acids, proteins, and peptide.

4.2. Stucco Testing

The absorption of the lignosulfonate onto industrial stucco was determined by measuring sulfonated lignin concentration of a solution before and after contact with the stucco using UV spectrometry. Fifty grams of stucco was weighed into a 6-ounce sealable plastic bag. A solution of water and lignosulfonate was added to the bag and allowed to stand for 30 seconds. The contents of the bag were then thoroughly mixed by gently massaging the bag for 3 ½ minutes. The bag was centrifuged at 2500 r.p.m. for 2 minutes. The supernatant liquid was filtered through a 0.2 mm nylon filter to remove any particulate matter and pH of the filtrate was adjusted to 4-5. The amount of lignosulfonate in the filtrate was determined specrophotomerically by measurement of the UV absorbance at 232.5 nm using a Perkin Elmer Lambda 3 UV-VIS spectrophotometer. The amount of lignosulfonate absorbed onto the stucco was calculated as the difference in lignosulfonate concentration before and after stucco contact.

The dispersing ability of the lignosulfonates was measured using a water reduction test in which the amount of water required to produce a stucco slurry of a specific viscosity was determined. For each test, 200 grams (g) of stucco were added to a mixture of water, 0.04 g of sodium citrate, and 0.4 g of a lignosulfonate dispersant. The sodium citrate was added as a retarder to prevent the stucco slurry from hardening before the test could be performed. After the stucco was added, the mixture was allowed to soak for 30 seconds and then stirred for 30 seconds. The slurry was then immediately poured to fill a brass cylinder having a 1.615 inch internal diameter and a height of 4.5 inches. The bottom of the cylinder was positioned two inches above a glass plate. The stucco slurry was allowed to flow through the orifice onto the plate to form a stucco patty. The amount of water required to form a six-inch diameter test patty under this configuration with the lignosulfonate dispersant in the mixture was compared to the amount of water required to make a six-inch patty under the identical configuration without the lignosulfonate in the mixture (i.e., with water and sodium citrate alone). This difference in water is reported as percent water reduction. Higher water reduction percentages are indicative of improved dispersing ability. The test for assessing set retardation of a lignosulfonate composition involves measurement of the time required for the stucco slurry containing the lignosulfonate to harden as determined by following the rise in temperature of the setting slurry because of the heat of hydration.

One hundred grams of stucco was dried at 105 °C for 2 hours, and then cooled in a desiccator. Before use, 0.2 g of land plaster (finely ground gypsum) was added to the cooled stucco. The stucco was added to a 150-ml paper cup containing a mixture of water, 0.2 g of potassium sulfate, and the lignosulfonate or other compounds of interest. The land plaster and potassium sulfate are commonly added set accelerators for commercial stucco. The amount of water used for each lignosulfonate tested was equal to 50% of the amount of water required (with each respective dispersant) to make a six-inch patty with 200 g of stucco as determined above in the water reduction test.

After addition of the stucco to the water mixture, the slurry was allowed to soak for 30 seconds and then stirred for 30 seconds. A thermocouple was then placed in the center of the mixture and the temperature rise versus time was recorded until the temperature began to fall. The set time was recorded as the time required for the measured temperature to achieve 95% of the total measured rise between the starting temperature and the maximum temperature.

5. REFERENCES

1. J. D. Gargulak and S. E. Lebo, Commercial use of lignin-based materials, in: *Lignin: Historical, Biological, and Materials Perspectives, ACS Symposium Series 742*, edited by W. G. Glasser, R. A. Northey, and T. P. Schultz (American Chemical Society, Washington D.C., 2000), pp. 304-320.
2. R. A. Northey, Low cost uses of lignin, in: *Emerging Technologies for Materials and Chemicals from Biomass, ACS Symposium Series 476*, edited by R. M. Rowell, (American Chemical Society, Washington D.C., 1992), pp. 146-175.
3. K. C. French and P. A. Leeming, Reducing the water demand of gypsum plaster, Great Britain Patent: 1,086,869a (1989).
4. R. A. Northey and K. D. Hefton, Water-reducing agent advancements for gypsum wallboard slurry thinning, in: *Inorganic-Bonded Wood And Fiber Composite Materials Volume 4*, edited by A. A. Moslemi (Forest Products Society, 1995), pp. 69-74.
5. R. A. Northey, C. Lee, C. R. Matz, and S. D. Krammer, The relationship between lignosulfonate physical properties and the ability to reduce inorganic slurry viscosity, *Proc. 10th Int. Symp. Wood Pulping Chem. Oral Presentations* (Yokohama, Japan, 1999), pp. 496-499.
6. R. A. Northey and S. K. Madsen, Modified lignosulfonate dispersant for gypsum, U.S. Patent: 5,286,412 (1994).
7. M. H. Simatupang and X. X. Lu, Influence of wood extractives on hardening o plaster of Pairs and on the manufacture of gypsum-bonded particle board, *Holz als Roh-und Werksoff* **43**, 325-331 (1985).
8. J. C. Le Bell, The relation between the structure of lignosulfonates and their effect as stabilizers for latex particulate dispersions, *Colloids Surf.* **9**(3), 237-251 (1984).
9. S. Y. Lin and W. J. Detroit, Chemical heterogeneity of technical lignins – significance in lignin utilization, *Proc. Intl. Symp. Wood Pulping Chem. Vol. 4* (Stockholm, Sweden, 1981), pp. 44-52.
10. R. F. Buchholz, J. A. Neal, and J. L. McCarthy, Some properties of paucidisperse gymnosperm lignin sulfonates of different molecular weights, *J. Wood Chem. Technol.* **12**(4), 447-469, (1992).
11. S. Y. Lin, Commercial spent pulping liquors, in: *Methods in Lignin Chemistry, Springer Series in Wood Science,* edited by S. Y. Lin and C. W. Dence (Springer-Verlag, New York, 1992), pp. 75-80.

MODIFIED KRAFT LIGNIN AND ITS USE FOR SOIL PRESERVATION

Kyoko Katsumata and Gyosuke Meshitsuka*

1. INTRODUCTION

The rapid decrease of forest land in the world, particularly that of the tropical rain forest, is a matter of most concern to us. According to a recent report, 1,690 x 10^4 hectares of tropical rain forest are disappearing every year due to increase in forest burning and timber production. Although tropical rain forest has extremely high biomass productivity when it is in sound conditions, its ecosystems are not necessarily very tolerant to drastic changes to their conditions. Once a forest is lost for some reason(s), reforestation may be difficult due to acidic or other poor soil conditions.

Inhibited growth of trees in acidic soils has been attributed primarily to the low pH condition, but also to the high aluminum (Al) ion concentration in the soil water.[1, 2] At a low pH, an increased amount of Al will dissolve into the soil water[3] and eventually inhibit the growth of plant roots.[4-7] Negative effect of Al is observed not only in the forest land, but also in the field of agricultural cultivation. Although the degree of growth inhibition by Al is dependent on plant species, remarkable effect has been found for wheat, barley and lettuce even at 1 ppm of Al in the soil water.[4] In some swamp areas of South East Asia, most of the trees cannot survive because of the low pH and high Al concentration in the soil water. For example, in a peat swamp area of Narathiwat Province in southern Thailand, only Melaleuca trees (*Melaleuca cajuputi*) can grow after clear cutting of the tropical forest. This may be attributed to the loss of surface soil containing humic substances critical for the protection of plants from Al toxicity.

We have been studying the chemical modification of industrial kraft lignin to mimic the chemical structures of humic substances and evaluating them as new soil conditioning agents.[8, 9] In this paper, we will discuss the interaction between the chemically modified lignins and Al with an emphasis on their structural characteristics.

* Kyoko Katsumata, Conservation Science Laboratory, Graduate School of Fine Arts, Tokyo National University of Fine Arts and Music, Tokyo, Japan. Gyosuke Meshitsuka, Laboratory of Wood Chemistry, Department of Biomaterial Sciences, Graduate School of Agricultural and Life Sciences, The University of Tokyo, Tokyo, Japan.

2. MATERIALS AND METHODS

2.1. Alkaline-Oxygen Treatment of Industrial Kraft Lignin

A commercial softwood kraft lignin (6 g) purified by extraction with aqueous dioxane/diethyl ether was dissolved in 0.1 M NaOH (300 mL) and treated in a 500-mL glass autoclave under 0.3 MPa pressure of oxygen at 70 °C for 4 and 8 h, respectively (Sample OL 4 and OL 8) (Figure 1 and Table 1). For the preparation of the control sample (KL), an alkaline solution of the kraft lignin was kept in the air at room temperature (~ 20 °C) for 4 h. Reaction mixtures were purified by dialysis and evaporated to dryness *in vacuo.*

2.2. Sulfonation of Kraft Lignin

The same purified, commercial kraft lignin (6 g) was dissolved in 0.1 M NaOH (300 mL) containing 15 or 5 g of sodium sulfite (Na_2SO_3) and treated in a 500-mL glass autoclave at 70 °C for 4 h under a constant oxygen pressure of 0.3 MPa (Table 1). The reaction mixtures were purified by dialysis (Figure 1) and evaporated to dryness in *vacuo.* Lignosulfonate (LSA) was selected as a reference sample.

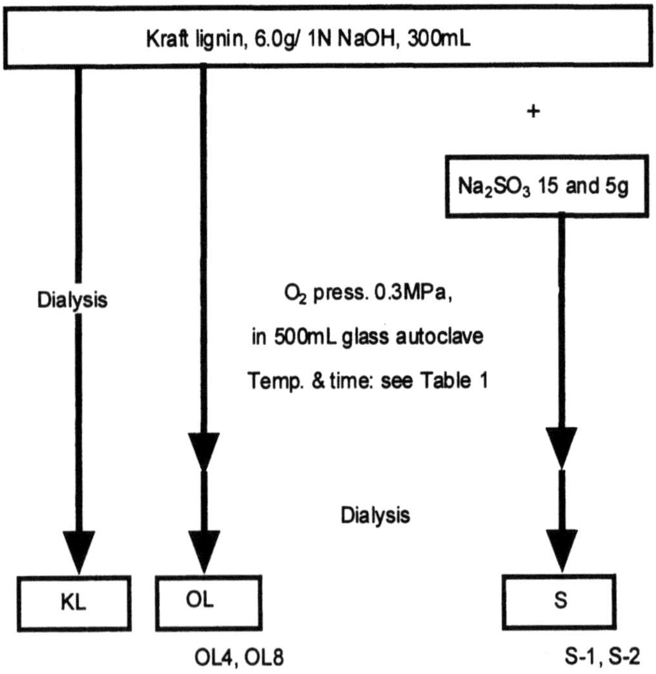

Figure 1. Scheme of chemical modification of kraft lignin by radical sulfonation and alkaline-oxygen treatment.

Table 1. Sample preparation conditions

Sample	Na_2SO_3 (g)	O_2 press. (MPa)	Temp. (°C)	Time (h)	Treatment
OL 4	0	0.3	70	4	Alkali-oxygen
OL 8	0	0.3	70	8	Alkali-oxygen
S-1	15	0.3	70	4	Sulfonation
S-2	5	0.3	70	4	Sulfonation
KL	0	Atm. press.	~ 20	4	Control
LSA	-	-	-	-	Lignosulfonate

2.3. Determination of Functional Group in Modified Lignin

A modified lignin (20 mg) was dissolved in 10 ml of methyl cellosolve/water (1/9) and eluted through a cation exchange resin (Amberlite IR-120) column. The eluted solution was diluted to 50 ml with a fresh methyl cellosolve/water mixture (1/9). Amount of acidic groups including sulfonic acid, carboxylic acid and phenolic hydroxyl groups were estimated by conductometric titration with 0.01 M NaOH.

2.4. ^{27}Al NMR

Various amounts of a modified lignin were dissolved in water with a known amount of aluminum nitrate. The solutions were adjusted to pH ~ 4.5 with KOH or HNO_3. ^{27}Al NMR spectra were recorded on a Bruker AC300 spectrometer at ambient temperature. The measurements were run in a 10-mm tube and the signal of D_2O in capillary was used for locking the signal.

2.5. Plant Growth Test

2.5.1. Soil Conditions

A modified lignin was mixed with Toyoura standard sand (silt type with particle size of 197 to 203 μm) at a concentration of ~ 0.1% lignin over sand (778 mg lignin/1000 g sand). The sand contained 16% of water, in which the plant nutrient in the form of the liquid type HYPONEX (Murakami Bussan Ltd.) and various amounts of Al, in the form of aluminum sulfate, $Al_2(SO_4)_3 \cdot 14\text{-}18\ H_2O$, were added. The final Al concentration was adjusted to 0 - 50 ppm.

2.5.2. Culture Solutions

Malkanthi's method[10] was modified and used to prepare culture solutions in a 2-L plastic container. Concentrations of the macro nutrients in the solutions were 0.6 KNO_3, 0.04 KH_2PO_4, 0.2 $NaNO_3$, 0.4 $CaCl_2 \cdot 2H_2O$ and 0.2 $MgSO_4 \cdot 7H_2O$ mM. Those of the micro nutrients in the solutions were 4.0 Fe-citrate, 6.0 H_3BO_3, 1.0 $MnCl_2 \cdot 4H_2O$, 0.1 $ZnSO_4 \cdot 7H_2O$, 0.06 $CuSO_4 \cdot 5H_2O$ and 0.02 Na_2MoO_4 μM. $KAl(SO_4)_2 \cdot 24H_2O$ was used as the Al source, and its concentration was varied between 0 to 50 ppm. A calculated amount of a modified lignin as an aqueous solution was added to the culture solution. The pH of the culture solution was adjusted to 4.5 with diluted aqueous HNO_3 or KOH, and readjusted twice a day to maintain such a pH throughout the experiment.

2.5.3. Growth Test

2.5.3a. Soil experiment. Radish (*Raphanus sativa var. radicula Pers.*) seeds were treated with a gibberellin (10 ppm) solution at 4 °C overnight, and then germinated on a filter paper at 25-30 °C for 3-4 days under natural light. A young seedling with a root of about 4 cm length was transplanted to a Teflon sheet-coated aluminum container (20 x 16 x 0.5 cm) (Figure 2) containing a Toyoura sand soil described in *2.5.1.* The container was kept at 25-30 °C for 10 days under natural light.

2.5.3b. Culture solution experiment. Young seedlings were transferred into the culture solution in a plastic container. Each plant was kept on a nylon net that was floating on the culture solution (Figure 3) through which air was bubbling. The plants were grown in a phytotron at 23.5 °C with irradiation of light of over 5000 lux for 14 h a day.

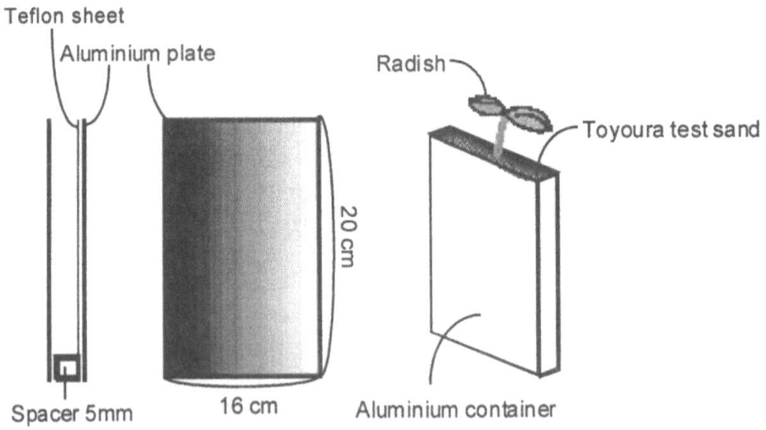

Figure 2. Aluminum container used for the root growth experiment of radish in soil culture.

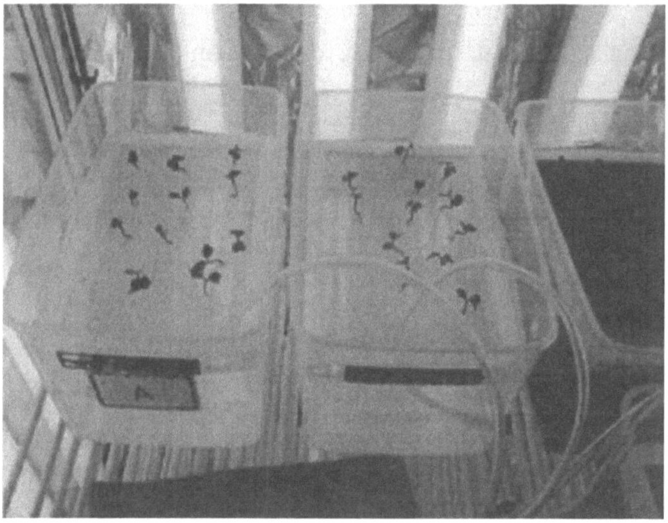

Figure 3. System used for the root growth experiment of radish in solution culture. Radish seedlings were put on nylon net floating on the aerated solution.

2.5.4. Determination of Plant Growth Rate

As a measure of the growth rate of radish, elongation of its root was determined by two methods, namely, the neutron radiography and the line intersection methods. The former method, initially introduced to observe water in a plant body by Nakanishi,[11-14] is based on the intensity of a neutron beam passed through the sample. When the neutron beam is irradiated, the number of neutrons penetrating the material decreases with respect to the hydrogen content, or the water content in the sample because more than 80% of hydrogen in a living plant is present in the form of water. Neutrons transmitted through the sample are converted by a Gd converter to γ-rays, to which X-ray film adjacent to the converter is exposed and an image of the sample obtained. The darkness in the image corresponds to the water content in the plant sample.

Neutron radiography was performed using an atomic reactor, JRR-3M, at Japan Atomic Energy Research Institute. Each sample was fixed on an aluminum cassette in which X-ray film (Kodak-SR) and a Gd converter were enclosed *in vacuo*. The aluminum cassette with the sample was set in a thermal neutron chamber, and the neutron beam was irradiated on them for 19 s. The thermal neutron flux was 2.6×10^8 n / $cm^2 \cdot s$. After development of the X-ray film, the image on the film was viewed by a scanner (Sharp JX325M) and processed on a Macintosh computer, emphasized, skeletonized, and evaluated by Adobe photoshop 3.0J and NIH image 1.59 software. The line intersection method was conducted according to the procedure reported by Böhm.[15] In this method, numbers of intersections between the radish root and the regular pattern of lines (0.5 cm x 0.5 cm) were counted. The root length was then calculated according to the following formula:

L (Length of root, cm) = 0.39 x N (Number of intersections).

2.6. Hematoxylin Staining of the Root

Hematoxylin staining[16, 17] of the root was used to confirm the adsorption of Al on its surface. Four days radish seedlings were kept in solutions containing 0.2 mM $CaCl_2$, 5 ppm of Al and 0 – 700 ppm of a modified lignin for 6 h, and were washed with deionized water for 30 min with air bubbling. These seedlings were then stained with a solution containing 0.2% of hematoxylin and 0.02% of KIO_3 for 30 min. Excess hematoxylin was washed with deionized water and the root color was observed under a stereomicroscope.

3. RESULTS AND DISCUSSION

3.1. Chemical Characteristics of the Modified Lignins

Conductometric titration indicates the existence of one or two types of acidic groups, namely the strongly acidic (A) and the weakly acidic groups in the modified lignins. The weakly acidic group can be further classified into moderately acidic (B) + (C) and very weakly acidic (D) ones. As shown in Table 2, no strongly acidic group was found in OL 4 or KL. An increased amount of moderately acidic groups (B) and (C) for OL 4 and S-2 can be attributed to the oxidative demethylation and/or cleavage of the lignin aromatic structures.[18,19] Phenolic hydroxyl groups in catechol structures formed by oxidative demethylation may be an important part of the acidic group (C) while muconic acid structures are the likely candidates for the acidic group (B). The strongly acidic groups found in and only in the sulfonated lignins can be attributed to the sulfonic acid groups, although it is not clear whether these acid groups are located on the side chains, the aromatic rings or the degraded aromatic structures of the modified lignins.[20-24]

Table 2. Acidic group (eq./200 g) in various lignin samples

Sample	Strongly acidic group (A) $-SO_3H$	Weakly acidic groups (B) + (C) (-COOH + Phenolic OH)	(D) Phenolic OH	Toal
S-1	0.153	0.241 + 0.097	0.123	0.614
S-2	0.135	0.270 + 0.086	0.135	0.626
OL 4	-	0.368 + 0.147	0.163	0.678
KL	-	0.259 + 0.099	0.188	0.546

All values are shown on the basis of 200 g of lignin samples. Conductometric titration was done on 20 mg of lignin smaple in 50 mL methyl cellosolve/H_2O (1/9) using 0.01 M NaOH.

3.2. Effect of Modified Lignin on Al Toxicity

The toxic effect of Al in the soil water on the growth of plants generally appears as the reduced elongation of the root of the plants.[1, 25-29] However, the toxic effect of Al is dependent on plant species. In the case of rye, different toxicity has also been found within the same species.[30] At the present time, there is not a single explanation for the different toxicity for various plant species. Secretion of organic acids from plants as a response to Al can be one of the possible explanations. Al-tolerant maize[31] and Al-tolerant leguminous trees[32] secrete citrate. Taro[33] and Buchwheat[34-36] secrete oxalate, and wheat secretes malate.[37] Secretion of both citrate and malate have been found for rye[37] that has a particularly high tolerance to Al. These organic acids can combine with Al ions and thereby protect plants from the toxic effect of Al.[25,37-42] Similar effect may be expected for the soil organic materials[43,44] that can catch not only Al but also many other metal ions. As for the function mechanism of soil organic materials, contribution of carboxylic acids and phenolic hydroxyl groups has been confirmed by FT-IR spectrometry.[45-50] In this context, an alkaline oxygen treated lignin (OL 4) containing an increased amount of carboxylic acids (Table 2) may be expected to also have a protecting function for plants against Al toxicity.

Young radish seedlings with root length of about 4 cm was transplanted to Toyoura standard sand containing liquid type HYPONEX as nutrients, Al and different types of the modified lignins. After 10 days in the phytotron the length of the root was measured using the neutron radiography[11-14] and line intersection[15] methods. Figure 4 shows images of the radish roots in soil obtained using the neutron radiography method. The left side is an image of the root with Al addition while the right side is an image of the root with Al and OL 4 addition. As can be seen, the growth of the root shown on the right side is remarkably improved, indicating a positive effect of the modified lignin. Root length determined from these images was comparable with that obtained from the line intersection method.

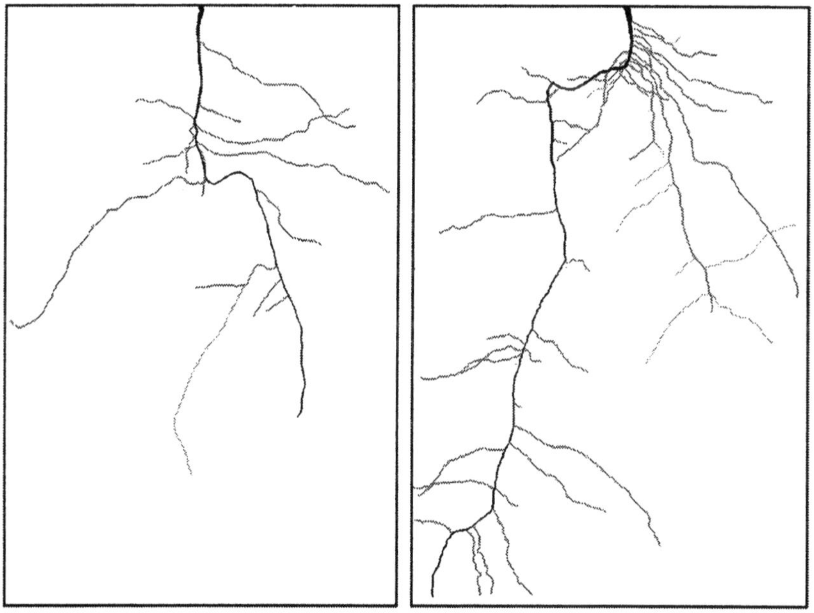

Figure 4. Effect of a modified lignin (OL 4) (875 ppm) on root growth recovery in soil containing Al (28.8 ppm). Left: Al only, Right: Al and OL 4. Images were obtained using neutron radiography.

The length of the radish root under different soil culture conditions determined by the line intersection method is shown in Figure 5. Addition of 28.8 ppm of Al as $Al_2(SO_4)_3 \cdot 14\text{-}18H_2O$ to the soil led to a decrease in the root length at an initial pH of 4.4 - 4.8. When a modified lignin (OL 8) at a dosage of 875 ppm was added to the soil, the toxic effect of Al disappeared and a much larger root elongation was observed at an initial pH ≤ 4.8. This indicates that the modified lignin not only can remove the toxic effect of Al, but it can also have a positive effect on the root elongation. When the initial pH was ~ 5.0, neither a toxic effect of Al nor a positive effect of the modified lignin could be observed, presumably due to the precipitation of the lignin by Al in the soil solution. It is important to note that the pH of soil water is not stable and that some Al may be adsorbed to or released from the soil into the soil water at a low pH. Therefore, it is difficult to illustrate the effect of the modified lignin at an exact pH and Al concentration in the soil experiment.

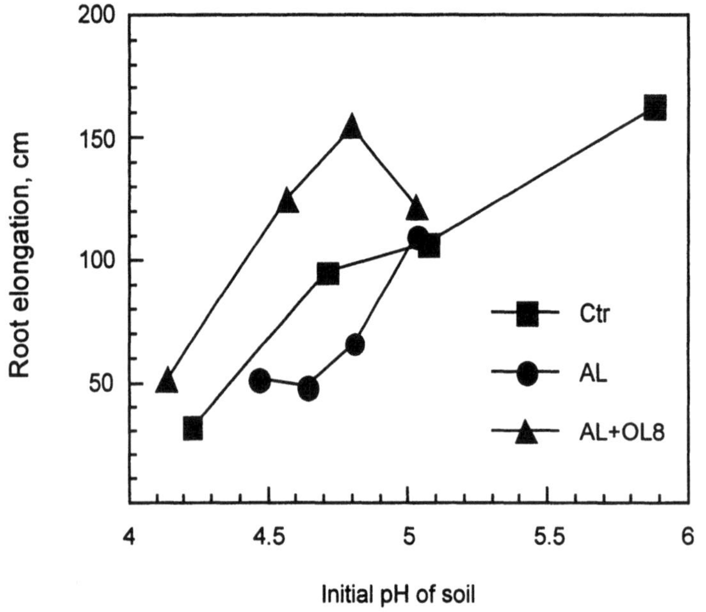

Figure 5. Effect of modified lignin (OL 8) (875 ppm) on the growth recovery of Al-treated (28.8 ppm Al) radish roots with the initial pH of the soil being adjusted by H_2SO_4 or KOH.

The radish growth test was conducted on culture solution at an exact pH and Al concentration. It is interesting to note that the growth of the root was practically prohibited in the absence of a lignin and in the presence of only 5 ppm of Al (Figure 6). However, with an addition of KL, root elongation slowly increased at the beginning and finally reached to the same level as that without Al at a concentration of ~ 350 ppm of KL. It is also interesting to note that KL itself had a positive effect on the growth of the radish root. When 70 ppm of KL was added to the culture solution, root elongation was about twice as much as that of the control condition.

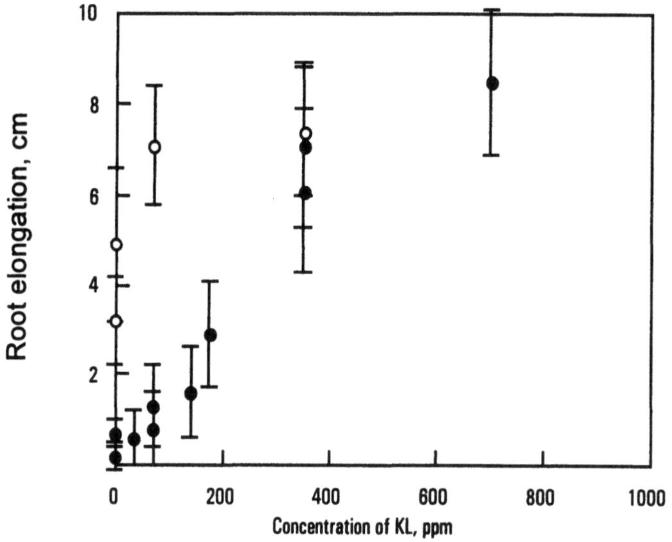

Figure 6. Effect of kraft lignin (KL) on root elongation in solution culture containing Al (5 ppm), pH 4.5, 3 days. Open circle: no Al, close circle: with Al.

When a modified lignin (OL 4) at a dosage of 140 ppm was added to the Al containing soil, the toxic effect of Al was clearly removed (Figure 7). The amount of the modified lignin was less than half of the unmodified KL needed to cancel the toxic effect of Al. At ≥ 350 ppm of OL 4, root elongation was larger with Al than without Al. In other words, the modified lignin appears to have a synergistic effect in the presence of Al to accelerate the growth of radish root.

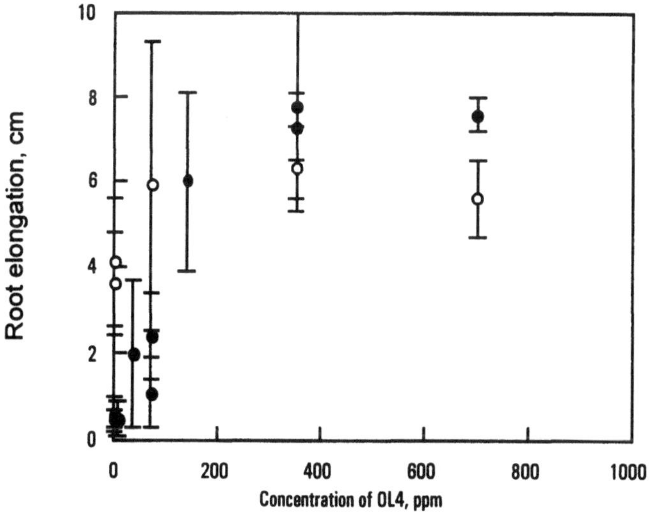

Figure 7. Effect of a modified lignin (OL 4) on root elongation in solution culture containing Al (5 ppm), pH 4.5, 3 days. Open circle: no Al, close circle: with Al.

3.3. Interaction between Al and the Modified Lignin

It has been reported that cell formation at the root tip is seriously affected by the presence of Al in the culture solution to the extent that root elongation will eventually be prohibited.[26,51,52] In order to confirm the effect of the modified lignin on the root elongation, presence of Al on the surface of the root tip was studied using the hematoxylin staining technique.[16, 17] When treated with hematoxylin, root tip grown in a solution culture without Al was not stained at all, but the one grown in the presence of Al was colored purple indicating the formation of hematoxylin-Al complex (Figure 8). In other words, Al is present on the surface of the root tips. Root tips grown in a solution containing 5 ppm of Al and 350 ppm of OL 4, however, were only slightly colored purple. This indicates that adsorption of Al to the root tips was effectively prohibited by the presence of a modified lignin, leading to the reduction of the toxic effect of Al.

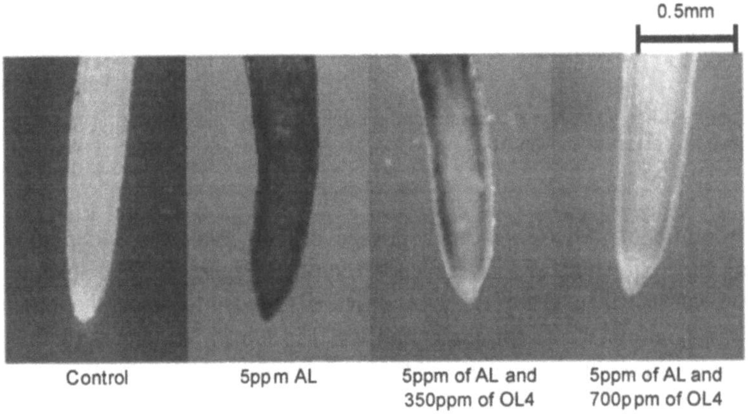

Figure 8. Difference of Al absorption on the root tip of radish kept in solution containing Al with or without addition of a modified lignin (OL 4). Al was detected by hematoxilin staining.

In order to understand the function mechanism of the modified lignin, its interaction with Al was studied. If 350 ppm of KL or 140 ppm of OL 4 is necessary to trap 5 ppm of Al (Figures 6 and 7), the number of functional sites in OL 4 should be at lease twice as many as that in KL. If a functional site in lignin forms a complex with a molecule of Al, then the number of functional site in 200 g of lignin samples can be roughly estimated to be 0.11 for KL and 0.28 for OL 4. Thus, the chemical modification of KL by alkaline-oxygen treatment is very effective in introducing the functional sites to lignin and eventually leading to the removal of the Al toxicity.

A considerable amount of proton was released when a modified lignin was mixed with Al in an aqueous solution as shown in the titration curve of the OL 4 and Al mixture (Figure 9). Much lower pH values were observed for the OL 4 and Al mixture than OL 4 or Al solution alone. This is an indication of the formation of a complex between the modified lignin and Al. It is interesting to note that a release of proton had also been found when soil organic materials was in contact with metal ions.[43] In order to understand the nature of a complex, determination of the stability constant for the complex will be very important.[38,53-56] The stability constants for the Al-maleate[55] and Al-citrate[38] complexes have been reported.

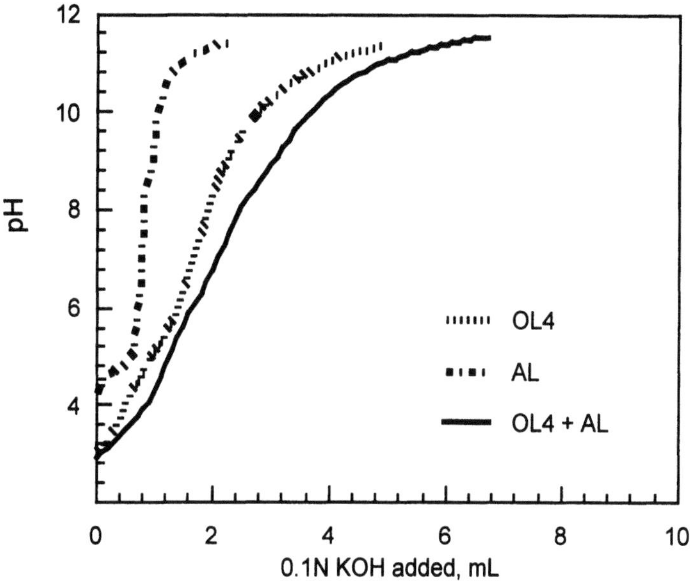

Figure 9. Titration curves showing proton release by complex formation between a modified lignin (OL 4) and Al. Al: 25.5 μmol, OL 4: 92.1 mg.

Another evidence for the strong interaction between lignin and Al was obtained by ^{27}Al NMR spectrometry. Formation of Al-maleate and Al-citrate discussed earlier has been confirmed by ^{27}Al NMR.[41] Al-oxalate complex in buckwheat root was found to be formed by coodination of three oxalate molecules to the Al ion.[36] An aqueous solution of Al (pH ~ 4.5) gives clear and sharp peaks at 63 and 0 ppm in the ^{27}Al NMR spectrum (Figure 10), ascribable to polymeric and monomeric aluminium ion, respectively.[57-60] When a small amount of a modified lignin (~ 1-2 mg) (OL 8) was added to 1 ml of 5 μmol Al solution, those peaks started to decrease (Table 3). When ~ 5 mg of the modified lignin was used, those peaks were not obvious any more. When 20 mg of the lignin was used, the peaks completely disappeared without any noticeable amount of Al–precipitation (Table 3). Disappearance of Al-signals in the presence of a modified lignin may be due to the asymmetrical structures of modified lignin-Al complexes, although at this moment, real structures of the complexes are not known. More studies are needed to fully understand ^{27}Al NMR spectrum of the lignin-Al complex.

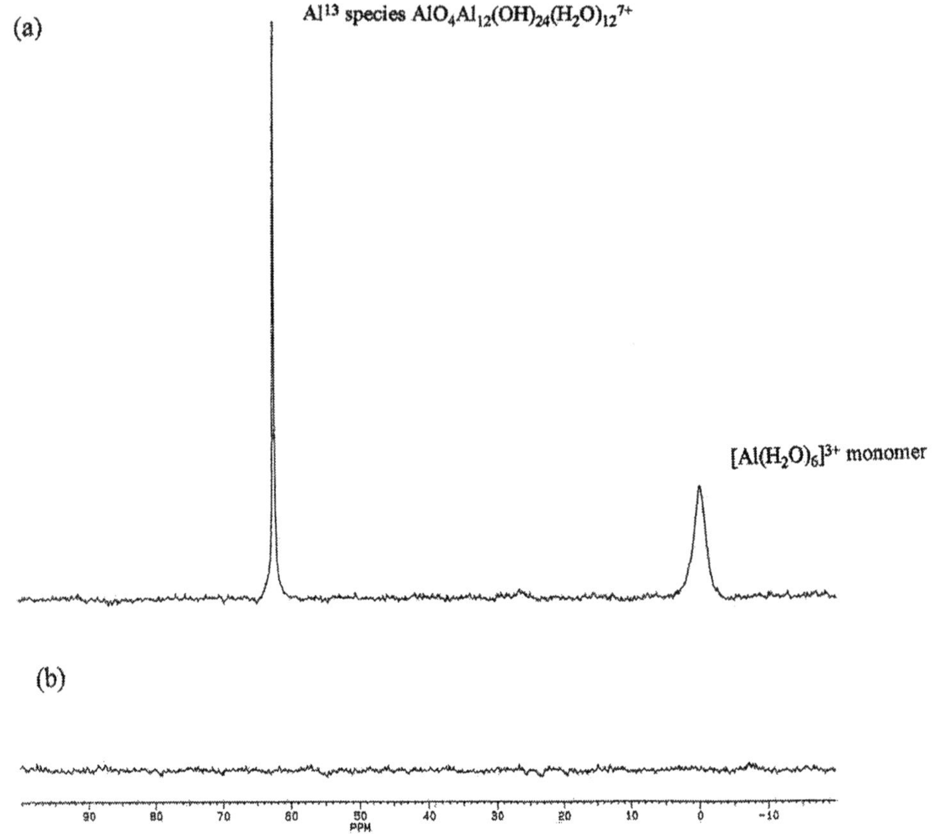

Figure 10. ^{27}Al NMR spectrum of 1 mL of 5 μmol $Al(NO_3)_3$ solution without (a) or with (b) 20 mg of OL 8 in acetone-d_6.

Table 3. Detection of polymeric and monomeric Al by ^{27}Al NMR

$Al(NO_3)_3$ μmol/ml	OL 8 (mg/ml)	pH	Solubility	^{27}Al NMR Area 63 ppm	0 ppm
5	0	4.45	Yes	1.8	10.5
5	1.1	4.45	No	0.1	6.8
5	2.1	4.43	No	-	3.1
5	5.1	4.59	No	-	-
5	10.0	4.70	No	-	-
5	19.9	4.70	Yes	-	-

4. CONCLUSIONS

1) Modified lignin prepared by alkaline-oxygen treatments of industrial kraft lignin contained an increased amount of carboxylic acid and phenolic hydroxyl groups while those prepared by radical sulfonation introduced a strong acid group, presumably sulfonic acid group.
2) The modified lignins could effectively remove Al toxicity to plant root growth at pH ~ 4.5, as shown by the root length determination using the neutron radiography and line intersection methods.
3) Formation of complexes between the modified lignins and Al was confirmed by the release of proton in the modified lignin-Al system and the disappearance of Al signals in ^{27}Al NMR spectrum in the presence of modified lignin.

5. REFERENCES

1. J. Olesyn, P. Karolewski, M. J. Giertych, A. Werner, M. G. Tjoelker, and P. B. Reich, Altered root growth and plant chemistry of pinus sylvestris seedlings subjected to aluminum in nutrient solution, *Trees* **10**, 135-144 (1996).
2. D. L. Godbold and G. Jentschke, Aluminium accumulation in root cell walls coincides with inhibition of root growth but not with inhibition of magnesium uptake in norway spruce, *Physiol. Plant* **102**, 553-560 (1998).
3. D. L. Sparks, in: *Environmental Soil Chemistry* (Academic Press, San Diego, 1995), pp. 203-217.
4. R. Aimi and M. Taka, Cell-physiolosical studies on the effect of aluminum on the growth of crop plant, *Bull. Nat. Inst. Agr. Sci., Series D* **12**, 331-396 (1964).
5. F. Adams and Z. F. Lund, Effect of chemical activity of soil solution aluminum on cotton root penetration of acid subsoils, *Soil Sci.* **101(3)**, 193-198 (1966).
6. A. Tanaka and S. A. Navasero, Aluminum toxicity of the rice plant under water culture conditions, *Soil Sci. Plant Nutr.* **12(2)**, 55-60 (1966).
7. O. C. Magistad, The aluminum content of the soil solution and its relation to soil reaction and plant growth, *Soil Sci.* **20**, 181-225 (1925).
8. K. Saito, T. M. Nakanishi,M. Matsubayashi, and G. Meshitsuka, Development of new lignin derivatives as soil conditioning agents by radical sulfonation and alkaline-oxygen treatment, *Mokuzai Gakkaishi* **43**, 669-677 (1997).
9. K. S. Katsumata, M. Maruyama, and G. Meshitsuka, Reduction of aluminum toxicity to radish by alkaline oxygen treated kraft lignin, *J. Wood Sci.* **47**, 129-134 (2001).
10. D. R. R. Malkanthi, K. Yokoyama, T. Yoshida, M. Moritsugu, and K. Matsushita, Effects of low pH and Al on growth and nutrient uptake of several plants, *Soil Sci. Plant Nutr.* **41(1)**, 161-165 (1995).
11. T. M. Nakanishi, S. Inanaga, and H. Kobayashi, Non-destructive analysis of rape plant pod by neutron radiography, *Radioisotopes* **40**, 126-128 (1991).
12. T. M. Nakanishi, S. Matsumoto, H. Kobayashi, Morphological change of plant root revealed by neutron radiography, *Radioisotopes* **41**, 638-641 (1992).
13. T. M. Nakanishi, S. Matsumoto, H. Kobayashi, Water hydrology by neutron radiography when water absorbin polymer was added to the soil, *Radioisotopes* **42**, 26-29 (1993).
14. T. M. Nakanishi, S. Matsumoto, and A. Tsuruno, water movement near the soybean root by neutron radiography, *Radioisotopes* **43**, 451-456 (1994).
15. W. Böhm, in: *Method of Studying Root Systems* (Springer-Verlag, Berlin, Heidelgberg, New York, 1979), pp. 132-135.
16. K. Ono, Y. Yamamoto, A. Hachiya, and H. Matsumoto, Synergistic inhibition of growth by aluminum and iron of tobacco (*nicotiana tobacum* l.) cell in suspension culture, *Plant Cell Physiol.* **36(1)**, 115-125 (1995).
17. T. Kataoka, M. Mori, T. M. Nakanishi, S. Matsumoto, and A. Uchiumi, Highly sensitive analytical method for aluminum movement in soybean root through lumogallion staining, *J. Plant Res.* **110**, 305-309 (1997).

18. J. Gierer and F. Imsgard, in: *The Reactions of Lignin with Oxygen and Hydrogen Peroxide in Alkaline Media, Chemistry of Delignification with Oxygen, Ozone, and Peroxide* (University Publishers, Tokyo, 1980), pp.137-150.
19. R. C. Eckert, H. Chang, and W. P. Tucker, Oxidative degradation of phenolic lignin model compounds with oxygen and alkali, *Tappi* **56**, 134-138 (1973).
20. G. Meshitsuka and J. Nakano, Studies on water solubilization of lignin (1), *Japan Tappi* **34**, 349-355 (1980).
21. G. Meshitsuka and J. Nakano, Studies on water Solubilization of lignin (2), *Japan Tappi*, **34**, 743-749 (1980).
22. M. Watanabe, M. Sakumoto, G. Meshitsuka, A. Ishizu, and J. Nakano, Radical sulfonation of lignin. water solubilization of acid-hydrolysis lignin, *Mokuzai Gakkaishi* **34(5)**, 428-435 (1988).
23. M. Watanabe, G. Metshitsuka, and A. Ishizu, Radical sulfonation of lignin II. Water solubilization of acid hydrolysis lignin, *Mokuzai Gakkaishi* **36(10)**, 876-882 (1990).
24. M. Watanabe, G. Meshitsuka, and A. Ishizu, Radical sulfonation of lignin III. Application to various lignins, *Mokuzai Gakkaishi* **38(2)**, 173-179 (1992).
25. C. D. Foy, R. L. Chaney, and M. C. White, The physiology of metal toxicity in plants, *Ann. Rev. Plant Physiol.* **29**, 511-566 (1978).
26. C. D. Foy, in: *Soil Acidity and Liming – Agronomy Monograph No. 12* (2nd ed.) (ASA-CSSA-SSSA, Madison, 1984), pp. 57-97.
27. M. A. Pavan, F. T. Bingham, and P. F. Pratt, Toxicity of aluminum to coffee in ultisols and oxisols amended with $CaCO_3$, $MgCO_3$, and $CaSO_4 \cdot 2H_2O$, *Soil Sci. Soc. Am. J.* **46**, 1201-1207 (1982).
28. T. Wagatsuma, M. Kaneko, and Y. Hayasaka, Destruction process of plant root cells by aluminum, *Soil Sci. Plant Nutr.* **33 (2)**, 161-175 (1987).
29. H. L. Van, S. Kuraishi, and N. Sakurai, Aluminum-induced rapid root inhibition and changes in cell-wall components of squash seedlings, *Plant Physiol.* **106**, 971-976 (1994).
30. O. Pinto-Cárinide and H. Guedes-Pinto, Differential aluminum tolerance of Portuguese rye populations and North European rye cultivars, *Agronomie* **20**, 93-99 (2000).
31. D. M. Pellet, D. L. Grunes, and L. V. Kochian, Organic acid exudation as an aluminum tolerance mechanism in maize (*Zea mays* L.), *Planta* **196**, 788-795 (1995).
32. H. Osawa, K. Kojima, and S. Sasaki, in: *Excretion of Citrate As an Aluminium-Tolerance Mechanism in Tropical Leguminous Trees, Plant Nutrition – For Sustainable Food Production and Environment*, edited by T. Ando *et al.* (Kluwer Academic Publishers, Japan, 1997), pp. 455-456.
33. Z. Ma and U. C. Miyasaka, Oxalate exudation by taro in responce to Al, *Plant Physiol.* **118**, 861-865 (1998).
34. J. F. Ma, S. J. Zheng, and H. Matsumoto, Detoxyfying aluminum with buchwheat, *Nature* **390**, 569-570 (1997).
35. S. J. Zheng, J. F. Ma, and H. Matsumoto, High aluminum resistance in buchwheat I. Al-induced specific secretion of oxalic acid from root tips, *Plant Physiol.* **117**, 745-751 (1998).
36. J. F. Ma, S. Hiradate, and H. Matsumoto, High aluminu resistance in buchwheat II. Oxalic acid detoxifies aluminum internally, *Plant Physiol.* **117**, 753-759 (1998).
37. X. F. Li, J. F. Ma, and H. Matsumoto, Pattern of aluminum-induced secretion of organic acids differs between rye and wheat. *Plant Physiol.* **123**, 1537-1543 (2000).
38. J. E. Gregor and H. K. J. Powell, Aluminium(III) – citrate complexes: A potentiometric and ^{13}C NMR study, *Aust. J. Chem.* **39**, 1851-1864 (1986).
39. N. V. Hue, G. R. Graddock, and F. Adams, Effect of organic acids on aluminum toxicity in subsoils, *Soil Sci. Soc. Am. J.* **50**, 28-34 (1986).
40. L. M. Shuman, D. O. Wilson, and E. L. Ramseur, Ameriaoration of aluminum toxicity to sorghum seedlings by chelating agents, *J. Plant Nutr.* **14(2)**, 119-128 (1991).
41. G. L. Kerven, P. L. Larsen, L. C. Bell, and D. G. Dewards, Quantitative ^{27}Al NMR spectroscopic studies of Al(III) complexes with organic acid ligands and their comparison with GEOCHEM predicted values, *Plant & Soil* **171**, 35-39 (1995).
42. J. F. Ma, Role of organic acids in detoxification of aluminum in higher plants, *Plant Cell Physiol.* **41**, 383-390 (2000).
43. S. S. Khanna and F. J. Stevenson, Metallo-organic complexes in soil: I. Potentiometric titration of some soil matter isolates in the presence of transition metals, *Soil Sci.* **93**, 298-305 (1962).
44. M. Schenitzer, Reaction between fulvic acid, a soil humic compound and inorganic soil constituents, *Soil Sci. Soc. Amer. Proc.* **33**, 75-81 (1969).
45. M. Schenitzer and S. I. M. Skinner, Organo-metallic interactions in soils: 3. Properties of iron- and aluminum-organic matter complexes, prepared in the laboratory and extracted from a soil, *Soil Sci.* **98**, 197-203 (1964).

46. M. Schnitzer and S. I. M. Skinner, Organo-metallic interaction in soils: 4. Carboxyl and hydroxyl groups in organic matter and metal retention, *Soil Sci.* **99(4)**, 278-284 (1965).
47. M. Schnitzer and I. Hoffman, Thermogravimetric analysis of the salts and metal complexes of a soil fulvic acid, *Geochimica et Cosmochimica Acta* **31**, 7-15 (1967).
48. K. H. Tan, Formation of metal-humic acid complexes by titration and their characterization by differential thermal analysis and infrared spectroscopy, *Soil Biol. Biochem.* **10**, 123-129 (1978).
49. A. Shinagawa, N. Miyauchi, and T. Higashi, Preparation of Al-humates and their aluminum content and cation-exchange capacity, *Soil Sci. Plant Nutr.* **28(1)**, 1-7 (1982).
50. R. F. Howe, X. Lu, F. Hook, and W. D. Johnson, Reaction of aquatic humic substances with aluminium: a ^{27}Al NMR study, *Mar. Freshwater Res.* **48**, 377-383 (1997).
51. L. V. Kochian, Cellular mechanisms of aluminum toxicity and resistance in plants, *Annu. Rev. Plant Physiol. Plant Mol. Biol.* **46**, 237-260 (1995).
52. M. Sivaguru, F. Baluska, D. Volkmann, Hubert H. Felle, and J. Horst, Impacts of aluminum on the cytoskeleton of the maize root apex. Short-term effects on the distal part of the transition zone, *Plant Phisiol.* **119**, 1073-1082 (1999).
53. Von G. Schwarzenbach and H. Ackermann, Kompexone V. Die athylendiamin-tetraessigsäure, *Helv. Chim. Acta* **30**, 1798-1804 (1947).
54. Von G. Sahwarzenbach and W. Biedermann, Kompezone IX. Titration von metallen mit athylendiamin-tetraessigsäure H4Y. Endpunktsindikation durch pH-effekte, *Helv. Chim. Acta* **31**, 459-465 (1948).
55. L. Mu, V. Y. Young, and N. B. Comerford, Formation constants of aluminum(III)-maleate complexes determined by potentiometric titration, *J. Chem. Eng. Data* **38**, 481-485 (1993).
56. N. T. Coleman, A. C. McClung, and D. P. Moore, Formation constants for Cu(II)-peat complexes, *Science* **24**, 330-331 (1956).
57. J. W. Akitt, N. N. Greenwood, and G. D. Lester, Hydrolysis and dimerization of aqueous aluminium salt solutions, *Chem. Comm.* 988-999 (1969).
58. J. W. Akitt, N. N. Greenwoodm B. L. Khandelwal, and G. D. Lester, ^{27}Al nuclear magnetic resonance studies of the hydrolysis and polymerization of the hexa-aquo-aluminium(III) cation, *J. Chem. Soc. Dalton Trans.* 604-610 (1972).
59. J. W. Akitt and A. Farthing, Aluminum-27 nuclear magnetic resonance studies of the hydrolysis of aluminium(III). Part 2. Gel-permiation chromatography, *J. Chem. Soc. Dalton Trans.* 1606-1608 (1981).
60. J. W. Akitt and A. Farthing, Aluminium-27 nuclear magnetic resonance studies of heteropolyanions containing aluminium as heteroatom, *J. Chem. Soc. Dalton Trans.* 1615-1616 (1981).

NITROGENOUS FERTILIZERS FROM LIGNINS - A REVIEW

Klaus Fischer and Rainer Schiene*

1. INTRODUCTION

The desire to apply and utilize lignin as a plant nutrient is prompted by two aspects. The first one is the chemical relationship between lignin and humus. The lignin component of plants is a basic chemical substance for the formation of soil organic matter (SOM). Humic substances are required for the soils to become fertile. At the same time they essentially influence the structure and the biological activity of the soils. The second aspect is the availability of great quantities of technical lignin, a by-product in the chemical processes of wood utilization. Nowadays, this happens almost exclusively in chemical pulp manufacture. However, the hydrolysis lignin from the processes of wood hydrolysis is also noteworthy from the historical point of view. Over the past few decades, chemical pulp manufacture has increased on a global scale, and so has the amount of by-product lignin. In the modern industrial methods of chemical pulping technical lignins are used as a valuable fuel and they are also incorporated into the pulping chemical recovery system. Efforts made for their more effective material utilization, however, continue to be significant, in particular because of the fact that lignin is a sustainable, renewable material.

If one intends to utilize lignin as a plant nutrient, the following has to be taken into account:

1) Due to the chemical pulping processes, technical lignin differs greatly from the natural lignin (protolignin). This refers to the kind and number of C-C and C-O-C bonds, functional groups, as well as the molecular weight and molecular weight distribution.
2) Technical lignin forms the main component of the spent liquors of chemical pulp manufacture. However, the spent liquors comprise yet a number of other components, thus, being a mixture of substances of a complicated and variable composition. The organic components partly consist of low-molecular weight

* Klaus Fischer and Rainer Schiene, Dresden University of Technology, Institute of Plant and Wood Chemistry, Pienner Str. 19, D 01737 Tharandt, Germany.

substances from degradation reactions of the entire wood substance, or they are special derivatives from the polysaccharides and extractives of the wood species used. The inorganic components comprise pulping chemicals and their conversion products as well as the ash components of the wood species. If the spent liquors are directly applied to the soils, undesirable side effects may occur. It appears to be more favorable to isolate/purify the lignin from the spent liquor. This is carried out in alkaline pulping by precipitation of alkaline lignin from the black liquor by pH lowering. The isolation of the lignosulphonates directly from spent sulfite liquors, however, requires a greater effort and results in an increasing cost.

3) Technical lignins and those in the form of spent liquors contain insufficient amounts of macronutrients, especially nitrogen and phosphorus.

Historically, sulphite spent liquor was the first to be investigated (Melms, Schwenzon 1967). It was almost a century ago that neutralized sulphite spent liquor was reported to have a positive effect on soils deficient in humus (Stutzer 1913, in Rautavaara 1967).

The fact that the sulphite spent liquors contain very little nitrogen (C/N > 100) as well as phosphorus is disadvantageous. An exception to this is the ammonium bisulphite spent liquor.

In order to utilize the organic matter in spent sulfite liquors despite its low N content, various ways were pursued (Rautavaara 1967). One way was to add spent sulphite liquors to the compost of sewage water, sewage sludge and waste materials (garbage). Another way was to mix spent sulphite liquors with conventional fertilizers. Others included the manufacture of soil conditioners using microbiological methods. Although a number of positive findings were obtained, the results did not meet the expectations. The positive effect on plant growth was too low and the profitability was called into question. Subsequently, chemical reactions suitable for the conversion of technical lignins into products containing greater quantities of nitrogen, as characterized by a low C/N ratio (C/N < 6), and applicable as organo-mineral N fertilizers started to be investigated.

Recently, humus has come more and more into the focus. According to ecological findings, the humus situation has aggravated. Humus plays an important role in soil development and plant growth such as the supply of nutrients, the soil microbiology and structure, and the soil regime of water and air. The increase of world population and industrial development in a broader sense has caused the devastation of soils including the humus layer. Hence, the improvement of the humus situation is one of the most significant goals of rehabilitation measures. The demand for high-grade humus cannot be met in the long run since the natural humus formation is a very long-term process. Therefore, efforts are being taken on a global scale to make available humus substitutes or to accelerate the natural process of humus formation (Fischer, K. *et al.*, 2001).

2. PATHWAYS OF MANUFACTURING N-CONTAINING LIGNIN PRODUCTS

The reactivity of lignins opens up a number of possibilities for their conversion into N-containing products according to Allan (1971). This refers to their reactions:

- with basic nitrogen compounds such as ammonia and amines to form ammonium salts of their acidic groups, or Schiff bases and desulfonated lignin sulfonates
- with aniline hydrochloride to form N-containing diphenyl methane structures
- with isocyanates to form urethanes
- with propylene oxide and subsequent with diisocyanates to form urethanes
- with ethylene imine (aminoethylation)
- with hydroxylamine and phenylhydrazine to form oximes and phenylhydrazones, respectively
- with formaldehyde and dimethylamine or piperidine (Mannich reaction)
- with hexamethylene tetramine
- with diazonium compounds (coupling)
- with cyanuric chloride
- with formaldehyde and urea (condensation)
- with HNO_3 and NO_x, respectively.

The targets for a practical application differed greatly in this connection. An application as N fertilizer requires specific product properties. Naturally, the N-containing products from lignins have to be plant-compatible, and the nitrogen has to be available to the plants. Furthermore, they should be industrially producible in great quantities by feasible technologies and at a low cost.

Among the above-mentioned possible reactions only the condensation with formaldehyde and urea, the reaction with ammonia, and the reaction with HNO_3 and NO_x, respectively, have been applied.

The reaction of lignin with HNO_3 comprises bond cleavages, substitutions and oxidations (Dence, 1971). N-containing products having quinone structures (quinone nitropolycarboxylic acids) could be produced from hydrolysis lignins. The ammonium salts of these N-containing products were found to stimulate plant growth when mixed with hydrolysis lignins (Chudakov *et al.*, 1968 a, b). This process was also applicable to spent sulfite liquors (Sapotnitskii, 1981). Using the two-step HNO_3-oxidation and ammonisation, N-humic fertilizers were produced from Kraft black liquor as well (Gonzalez *et al.*, 1992).

However, oxidative ammonolysis, i.e. the oxidation in the presence of basic nitrogen compounds, in particular ammonia, has to be regarded as the most important and versatile method for manufacturing N-containing products.

3. OXIDATIVE AMMONOLYSIS IN ORGANIC CHEMISTRY

Since the 1950s the oxidation of organic compounds by oxygen/air in the presence of ammonia (oxidative ammonolysis) has been systematically investigated and applied as a technological process for producing nitriles. Hydrocarbons (aromatic hydrocarbons and aromatic heterocycles, cycloalkanes and cycloalkenes, alkenes) can be converted to nitriles. The reaction takes place in the vapor phase in the presence of catalysts at temperatures > 300 °C. Regarding the overall reaction, a complex of radical mechanisms and intermediate stages have been suggested. Oxidation by-products such as aldehydes, ketones, acids, anhydrides and lactones can also be formed during oxidative ammonolysis. Their yields depend on reaction parameters such as temperature, ammonia concentration and reaction time. At elevated temperatures they may react with NH_3 in a catalytic vapor phase ammonolysis, and amines, imines as well as amides and nitriles of the acids are formed (Suvorov *et al.*, 1965; Scheme 1, details according to Suvorov *et al.*, 1965).

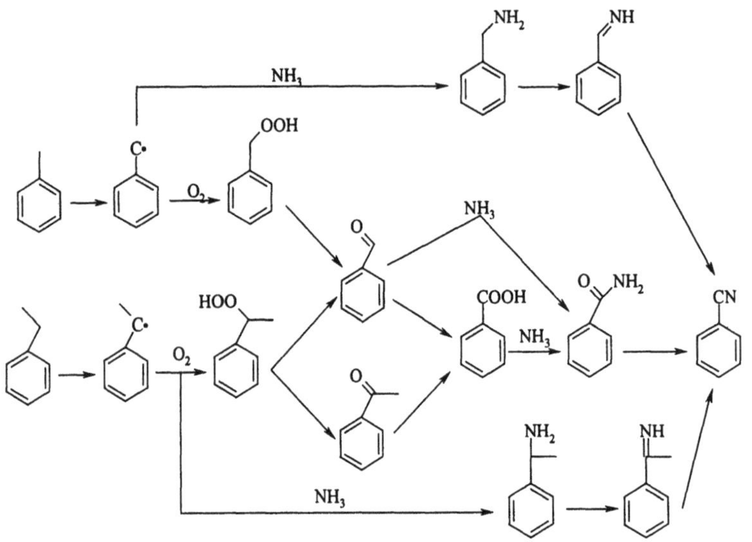

Scheme 1. Proposed reactions in the oxidative ammonolysis of alkylbenzene.

4. MANUFACTURE OF N–CONTAINING PRODUCTS BY OXIDATIVE AMMONOLYSIS OF LIGNINS

4.1. Processes for Manufacturing Organomineral N-Fertilizers

The experimental possibilities for oxidative ammonolysis of lignins are limited by their polymeric properties. The general method applied is the oxidation of the lignins in an aqueous ammonia solution, suspension or mixture. Either air or oxygen may serve as an oxidant. An alternative method is the reaction of the solid lignins with a gaseous mixture of ammonia and oxygen or air, of which, however, only a few experimental examples are known.

4.1.1. Conventional Processes in Aqueous Media under Pressure

These processes aimed at producing organomineral N fertilizers, i.e. products characterized by a relatively high N content and a low C/N ratio, e.g. < 6. The oxidative ammonolysis of lignins is strongly influenced by the reaction parameters such as the concentration of the oxidant and ammonia as well as temperature, with the concentration and dispersion of the oxidant in the reaction mixture being particularly important.

The initial work in this area dates back a long time ago. Recent and fossil plant stock (Ehrenberg, 1929; Ehrenberg and Heimann, 1929; Caro *et al.* 1929; and Caro and Frank, 1930), spent liquors from pulping process (Franz and Palm, 1930), hydrolysis lignin (Duiker, 1959), fossil fuels (Grosskinski and Klempt, 1949, 1950) as well as humic substances (peat, brown coal/lignite) (Nettesheim, 1958) had been used as parent materials. These materials reacted with NH_3 (up to 12%) and O_2-containing gases such as air at 50 - 300 °C under a pressure of > 8 atm for a reaction time of up to 12 h to give products with a total nitrogen content (N_{tot}) of up to 21%.

Marked advance was achieved by Flaig (Flaig *et. al*, 1959; Flaig and Hingst, 1960) and subsequently by several other German research teams (Bratzler and Aalrust, 1962; Hingst *et al.*, 1962; Bratzler and Thormann, 1962; Hingst *et al.*, 1963). In their investigations exclusively technical lignins, both spent sulfite liquors and alkali lignin were converted to products with an N_{tot} of up to 24%. In this connection, favorable ranges of the reaction parameters (temperature: up to 130 °C, pressure: 20 – 150 atm, reaction time: up to 10.5 h, NH_3: 10 - 25%) were ascertained.

Flaig and Söchtig were the pioneers in assessing the oxidative ammonolysis of lignins for artificial or forced humification (Flaig and Söchtig, 1973; Söchtig and Flaig, 1973), as well as in emphasizing the distinct properties of the resulting, so-called N-lignins where the nitrogen is found in different chemical bonds. The N-lignins have been defined as slow-releasing N-fertilizers. They inhibited nitrification and had a favorable impact on phytophysiology (Flaig and Söchtig, 1974). A process for making these lignins was implemented in Germany by LURGI-Gesellschaften in 1969, and tested in an experimental plant (Wiesner, 1971).

At Tharandt Institute we initially investigated the oxidative ammonolysis of technical lignins using air as an oxidant at high pressures in a bubble reactor. This will be described in Session 4.2 in connection with the development of a novel method.

The work of the research teams in the former Soviet Union also focused on the oxidative ammonolysis of spent sulfite liquor but with a thorough examination of the influences of the technological parameters and, for the first time, also of the kinetics of the process (Kaganskii *et al.*, 1968; Mazhara *et al.*, 1969 a, b). Products with an N_{tot} of up to 20.5% were obtained at 40 - 180 °C under a pressure of up to 60 atm (pNH_3 up to 7 atm) using a reaction time of up to 4 h.

Hydrolysis of wood was performed on an industrial scale in the former Soviet Union. The oxidative ammonolysis of the hydrolysis lignin was comprehensively dealt with (Chudakov *et al.*, 1969; Kazarnovskii and Chudakov, 1972, 1973; Mekler *et al.*, 1974). Products with an N_{tot} of up to 22.5% were obtained at 180 - 250 °C under a pressure of 10 - 50 atm with a reaction time of 3 - 6 h and NH_3 of 6%.

Even the natural lignin in the lignocellulose material could be subjected to oxidative ammonolysis and thereby dissolved. This principle was investigated in depth and a pulping process for wood was derived. The dissolved and isolated N-containing lignin was a useful by-product and could be utilized as a fertilizer (Ioffe and Sergeeva, 1972; Ioffe *et al.*, 1975; Kopnin and Ioffe, 1975).

Other investigations concentrated exclusively on the conversion of spent sulfite liquor (Sulan *et al.*, 1975, 1978; Ishibashi *et al.*, 1978). Products with an N_{tot} of up to 19% were obtained by Sulan *et al.* at 125 - 129 °C under a pressure of 30 - 32 atm using a reaction time of 1 - 25 h and an excess of NH_3; by Ishibashi *et al.* at 90 - 120 °C under a pressure of 4 -10 atm using a reaction time of 4 h and NH_3 of 15 – 30 mol/kg (dry matter).

In the more recent work the objectives for oxidative ammonolysis of technical lignins differed greatly. They included the improvement of the process itself, and studies of the properties and application of the products, as well as the kinetics and the mechanism of the reaction (Meier *et al.*, 1994; Lapierre *et al.*, 1994; Varela *et al.*, 1999; Capanema *et al.*, 1998, 1999). Products with an N_{tot} of up to 18.8% were obtained by these authors at a temperature range of 100 - 175 °C under a pressure of 6 - 18 atm using a reaction time of 50 - 283 min and NH_3/lignin (w/w) of 0.30 – 0.71, and in one case 2.7% NH_4OH.

4.1.2. Conversion of Solid Lignins Using Gaseous Mixture of NH_3 and Air

This interesting and experimentally exacting variant has been reported only in a few papers hitherto. The principle refers to normal pressure at elevated temperatures (> 200 °C). N-lignins of maximally 18% N_{tot} were produced from the hydrolysis lignins at 200 - 250 °C (Zakis *et al.*, 1976). In a fluidized bed-reactor N-lignins containing more than 20% N_{tot} were obtained from Kraft lignin at 210 °C (Zuniga *et al.*, 1992; Martinez *et al.*, 1999). Even from wood in the form of sawdust solid products containing a maximum of 27% N_{tot} could be manufactured at a temperature of 320 °C. More than 90% of N_{tot} were found to be organically bonded (Kim *et al.*, 1981).

4.1.3. Conversion of Lignins in Aqueous Media Using Alternative Oxidants

Research teams in the former Soviet Union also applied alternative oxidants in the oxidative ammonolysis of lignins and studied model compounds to elucidate the structures and mechanisms. The oxidative ammonolysis using H_2O_2 was performed (Kazarnovskii, 1974; Kazarnovskii and Chudakov, 1976). Products with an N_{tot} of up to 23.1% were obtained from various lignins (Dioxane-, Klason-, Willstätter-, Methanol- and Alkali-lignin) at 200 °C under elevated pressures using a reaction time of 3 h and NH_3/H_2O_2 (6% NH_3).

Zakis *et al.* studied the conversion of Alkali and Klason-lignins using the system of $NH_3/S_2O_8^{2-}$ in an aqueous solution at room temperature (20 °C) and normal pressure (Zakis *et al.*, 1973 a, b; Zakis, *et al.*, 1974; Zakis and Neiberte, 1978 a, b) using a reaction time of up to 100 h and 20% NH_3. Products with an N_{tot} of up to 20% were obtained.

4.1.4. Related Processes

In this context, multistage processes involving oxidation and subsequent treatment with ammonia (ammoniation) have been studied. By definition and strictly speaking, these processes do not belong to oxidative ammonolysis. In the second step ammonia, however, reacts with an oxidized material without the exclusion of oxygen. Conversions of the oxidized material by oxidative ammonolysis are still conceivable, although at a lower extent and dependent on the reaction conditions (Caro *et al.*, 1929; Ehrenberg, 1930). The same applies to the processes described in the literature, in which the lignin material is converted in one step exclusively by means of ammonia. The involvement of oxygen, however, cannot be excluded (Erasmus, 1928; Davis and Scholl, 1933; Kleinert 1952; Murashkevich *et al.*, 1970; Forostyan and Kovalchuk, 1972).

4.2. Development of a Novel Method for Manufacturing Artificial Humus

4.2.1. Investigations of Spent Sulfite Liquors

During the first phase of our work the oxidative ammonolysis of spent sulfite liquors was investigated. Preferentially, fermented spent ammonium bisulfite liquor from the pulping of beech was used as the starting material. A method was developed according to the principle of a bubble reactor with an additional mechanical stirring and using an air stream as the oxidizing gas. This method enabled the conversion of technical lignins into N-modified products, the so-called AO lignins, in aqueous ammoniacal solution at elevated pressures and temperatures (Tables 1 and 2, Figure 1). The AO-lignins contained up to 19% N_{tot}, the chemical bondings of which differ. Because of its diverse hydrolysability N_{tot} was classified according to Flaig into NH_4^+-N, amide-N and not readily hydrolyzable nitrogen (sob N) (Alber *et al.*, 1977; Pruzina, 1979; Schiene *et al.*, 1979; Fischer, F. *et al.*, 1983; Fischer, F. *et al.*, 1985; Fischer, K. and Schiene, 1990).

Table 1. Reaction conditions for manufacturing AO lignins

Pressure (MPa)	Temp. (°C)	Reaction time	Rate of air (L/h)	NH_3/spent liquor, dry matter	NH_3/mixture
2 - 12	105 - 200	Up to 90 min	200 - 1200	0.133 – 0.616	4.3 – 15.5%

Table 2. Composition of the AO lignins

C %	N_{tot} %	NH_4^+-N[a]	Amid-N[a]	sob N[a]	C/N	OCH_3 %
32 - 37	15 - 19	33 - 44%	8 - 16%	47 - 56%	<2.5	3 - 6

[a] % of N_{tot}.

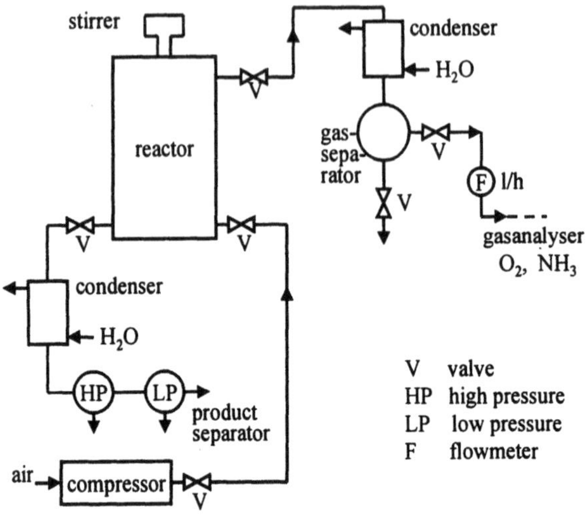

Figure 1. Scheme of the high-pressure experimental plant for the production of AO lignin.

The influence of the reaction parameters was examined in great details at constant concentrations of lignin and NH_3 (Miletzky 1985; Miletzky and Schiene, 1985). The process is greatly controlled by the reaction temperature and the oxygen supply. Within the range of 105 - 180 °C, N_{tot} of the AO lignins increases proportionally with the temperature from 11.5 - 18.7%. The relative N distribution in various types of bonds is modified in a characteristic way. The content of NH_4^+-N decreased, while that of sob-N increased (Figure 2). In the bubble reactor the oxygen supply could be adjusted by controlling the air-flow rate; an increase from 200 - 1200 L/h led to an increase of N_{tot} by 30% (from 10.3 to 13.4% N_{tot}).

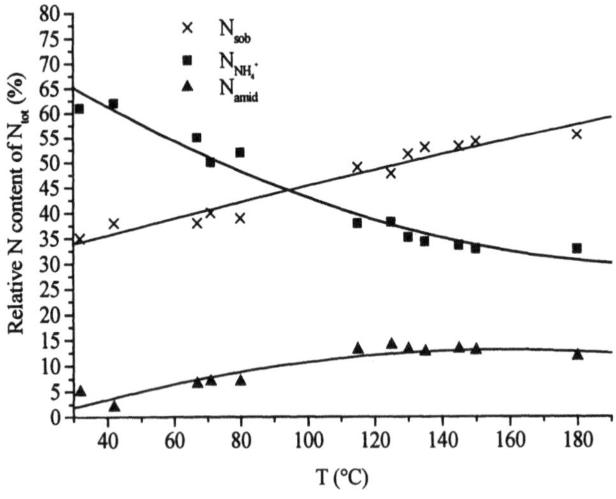

Figure 2. Nitrogen distribution (relative of N_{tot}) vs. reaction temperature (NAO and AO process).

Investigations into the kinetics of the process revealed a direct relationship between the incorporation of N_{tot} into the AO lignins as well as the decrease of OCH_3 groups and oxygen consumption (Miletzky, 1985; Miletzky and Schiene, 1985). The results pointed to zero-order reactions for both the incorporation of N_{tot} and the decrease of OCH_3 groups with the supply of oxygen to the reaction site appearing to be the rate-controlling step. Nitrogen incorporation and N distribution at a constant temperature depended on the reaction time with N_{tot} increasing with time (Figure 3). The total amount of both NH_4^+-N and sob N also increased. The relative content of NH_4^+-N in N_{tot} decreased but that of sob N increased. After an initial increase the amide-N content remained fairly constant but started to decrease slightly after 40 min or so (Figure 4). This suggested that amides were formed as consequential reactions of ring cleavages.

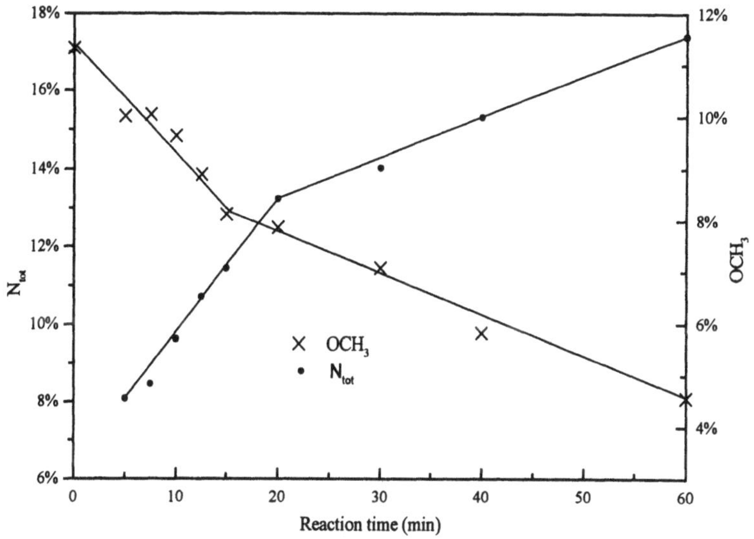

Figure 3. N_{tot} and OCH_3 content of N lignins vs. reaction time (AO process).

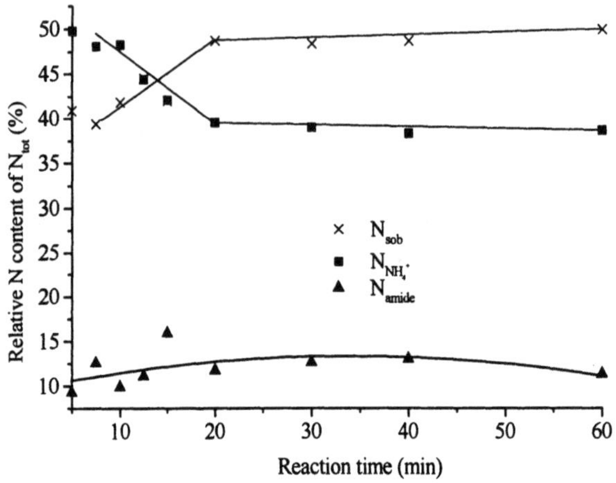

Figure 4. Nitrogen distribution vs. reaction time (AO process).

Overall, the results revealed the advantages of the AO method using air as oxidant. Products (AO lignins) of up to 19% N_{tot} can be readily produced. The properties of the products such as the N distribution that is important as a fertilizer can be controlled by varying the reaction temperature and/or time. This method can be allocated to the class of high-pressure processes described in section 4.1.1. Using this method conventional technical lignins have been converted into N-containing products with up to 24% N_{tot}. The final products are characterized by a C/N ratio of < 6 and can be treated primarily as long-term, organomineral fertilizers. The starting lignins are greatly modified (degraded as well as condensed) in their structures.

In the next stage of our investigations into oxidative ammonolysis of spent sulfite liquor, a novel method/process based on the principle of effective oxygen feeding utilizing the surface properties of spent sulfite liquor was developed (Miletzky 1985; Miletzky *et al.*, 1987, 1990; Schiene *et al.*, 1985; Fischer, K. *et al.*, 1990). In this method/process, the oxidation gas, air or pure oxygen is fed into the reaction mixture. An effective gas-liquid interface allowing for the exchange of the reactants is formed. The process runs under normal pressure and high oxygen input at < 100 °C for up to 4 h with up to 10% spent liquor (dry matter/mixture) and 1.4 – 8% NH_3/mixture. When comparing the novel process with the AO process, both the common nature and the particular specifics become obvious. In the novel method/process, i.e. the normal pressure process (NAO process) (Figure 5) the concentrations of lignin and ammonia in the reaction mixture are distinctly lower than in the high-pressure process (AO process). The conversion takes longer at this lower reaction temperature (<100 °C). A lesser amount of N_{tot} is incorporated into the NAO lignins than into the AO lignins; NAO lignins comprise up to ca.11% N_{tot} in various kinds of N bonds (Table 3). In addition, The NH_4^+-N content is higher and the sob N content is lower in the NAO lignins than those in the AO lignins (Tables 2 and 3).

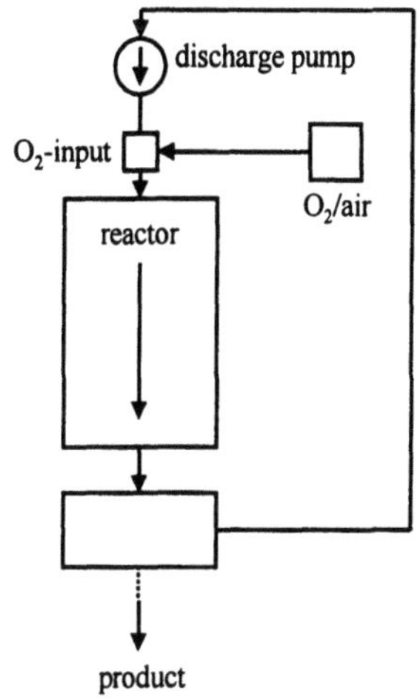

Figure 5. Scheme of the normal pressure process (NAO).

Table 3. Composition of the NAO lignins

C %	N_{tot} %	NH_4^+-N[a]	Amid-N[a]	sob N[a]	C/N	OCH_3 %
39.5 - 40.9	up to 10.72	50.9 - 54.6	7.9 - 8.3	37.0 - 40.9	4.3 - 8.8	up to 6.16

[a] % of N_{tot}; starting sulfite spent liquor N_{tot} 3.6 %.

A good supply of oxygen at the reaction site forms the foundation of the NAO process. In this process the dominating reaction parameters are also the reaction temperature and time. A longer reaction time and an increase of temperature cause the increase of N_{tot}. The N distribution in both methods (AO and NAO processes) mainly depends on the reaction temperature. This holds true for the examined temperature range of 35 – 180 °C. An increase of temperature leads to an increase of the relative content of sob N and a decrease of the NH_4^+-N content. The relative content of amide-N increases with temperature up to 120 °C, thereafter remains fairly constant, and starts to drop from 150 °C (Figure 2).

Characteristic steps of oxidative ammonolysis of spent sulfite liquors are the degradation (demethoxylation and/or demethylation) of OCH_3 groups (Table 4) as well as desulfonation with the formation of sulfate (Table 5). In general, the degradation of OCH_3 and the desulfonation increase along with intensifying N modification. With comparable N_{tot} (~ 11%), the degradation of OCH_3 groups in the NAO process is, however, more pronounced, suggesting a more effective oxidative attack on the aromatic units.

Table 4. Degradation of the OCH_3 groups in the AO and NAO lignins

	N_{tot} %	[a]OCH_3%
AO lignin	17.4	4.56
AO lignin	10.7	8.89
NAO lignin	10.72	6.16

[a] Starting material OCH_3 = 11.3%.

Table 5. Desulfonation in the AO and NAO lignins

	N_{tot} %	%S[a] as SO_4^{2-}
AO lignins	> 14	up to 3.0
AO lignins	5 - 14	0.8 - 1.8
NAO lignins	up to 10.7	up to 1.4

[a] Starting material %S as SO_4^{2-} = 0.4 - 0.8.

The oxidative ammonolysis causes both degradation and condensation in lignin. Investigations of N lignins and their fractions by means of gel filtration on Sephadex G 100/ G 50 and nitrobenzene oxidation have shown that the incorporation of nitrogen is associated with condensation reactions in both the AO and the NAO processes. More products with high M_w (M_w > 100,000) are formed in the AO process than the NAO process.

The nitrobenzene oxidation shows that the typical lignin structures, of which the characteristic products of oxidation may be formed, are still available in the N lignins (see Section 4.2.2). They are altered and decreased by oxidative ammonolysis. The decline is most apparent in the low-molecular weight degradation products and quite pronounced in the high-molecular weight range, a fact attributable to the preceding condensation reactions.

A higher incorporation of N with N_{tot} > 12% corresponds to a higher degree of modification for the AO-lignins than for the NAO lignins. The case is vice versa in N lignins with N_{tot} < 12%. The results reflect the stronger oxidative attack on the aromatic units in the NAO process, a fact being in conformity with the degradation of the OCH_3 groups.

N lignins may be produced from different spent sulfite liquors using both processes. The N_{tot} in the products is influenced by the base cation of the spent liquor and the type of lignin. Ammonium bisulfite spent liquor from beech gives a higher incorporation of N than Ca- or Mg-bisulfite spent liquor from spruce. This is due both to the NH_4^+ cation and thus the N-starting level and to the less condensed hardwood lignin having a higher OCH_3 content.

The properties of the N lignins are also affected by possible reactions of carbohydrates present in the reaction mixture. N_{tot} and the content of sob N of the lignins produced in a high-pressure process were shown to increase with increasing content of monosaccharides (Sulan *et al.*, 1978). As a matter of fact, sugar was shown to react with aqueous ammonia and at increased temperatures heterocyclic products, mainly imidazoles and pyrazines, were formed (Kort 1970). When oxidative ammonolysis of glucose was carried out in model experiments using the AO (Pruzina, 1979) and the NAO processes, N-containing products with a high sob N content were obtained (Table 6).

Table 6. Products from oxidative ammonolysis of glucose

Product	N_{tot} %	NH_4^+-N[a]	Amide-N[a]	sob N[a]
AO glucose	25.0	18.8	10.4	70.8
NAO glucose	14.1	29.8	14.2	56.0

[a] % of N_{tot}.

To further clarify the role of sugars, oxidative ammonolyses of spent sulfite liquor, substances with basic lignin structures and monosaccharides were conducted, respectively, in an aqueous system of $^{15}NH_3/H_2O_2$. Subsequently, the ^{15}N-containing products were investigated by means of ^{15}N NMR. The spectra of all the products showed typical signals for NH_4^+, amide and amine nitrogens. Only the products from monosaccharides showed signals at -205.2 to -208.7 ppm interpreted for azole, and signals at -57.3 to -64.5 ppm interpreted for azole or azine and also for imine structures (Schiene *et al.*, 1996).

In pot experiments of forest plants, N-lignins from conventional spent sulfite liquors and prepared using the AO process were found to contain physiologically active substances which temporarily seemed to have a negative effect on plant growth (Fiedler and Schmidt, 1979, 1981, 1983). These influences are attributable to substances in the lower molecular weight range. The effects are obviously connected to products that may be formed from the residual carbohydrates in the spent sulfite liquors (Fischer, K. *et al.*, 1990). The process of oxidative ammonolysis is to be controlled in a way that the formed heterocyclic structures are humus-like incorporated in the macromolecular organic matrix.

4.2.2. *Investigations of Sulfur-free Lignin*

In our further work sulfur-free technical lignins obtained from Organosolv pulping processes were selected as the starting material. Owing to the 1991 state-of-the-art technical lignin produced in the ORGANOCELL process, pulping with NaOH/methanol was predestined for this purpose. The objectives were to illustrate the effects of various pulping parameters and conditions of lignin isolation (precipitation with CO_2 at pH 9 or with H_2SO_4 at pH 4) on parent lignin structures and properties, on oxidative ammonolysis and on the quality of the N lignins.

Lignins (from spruce) manufactured and isolated under defined conditions were made available by ORGANOCELL GmbH. From these lignins which were characterized in details (Zier 1996, Zier *et al.*, 1997), N modified products (NAO and AO lignins) were produced according to the normal pressure process and the high-pressure process, respectively (Fischer, K., 1991; Fischer, K. *et al.*, 1992,1993, 1995; Schiene *et al.*, 1992) (Table 7).

The AO lignins and NAO lignins contain up to ~20% and ~5% N_{tot}, respectively (Table 8). AO lignins have a much lower C/N ratio and OCH_3 content than the NAO lignins.

Table 7. Reaction parameters for manufacturing AO and NAO lignins from ORGANOCELL lignins

Principle	Pressure (MPa)	Temperature (°C)	Time (h)	Lignin/mixture (%)	NH_3/mixture (%)
AO	5-15	110 - 225	1	5 - 15	17.4
NAO	normal	70 - 80	5	up to 10	up to 7

Table 8. Composition of the ORGANOCELL AO and NAO lignins

Principle	C %	N_{tot} %	NH_4^+-N[a]	Amid - N[a]	sob N[a]	C/N	OCH_3 %
AO	44 - 53	7.8 - 20.4	13.1 - 28.9	10.0 - 19.2	55.0 - 74	3.1 - 6.9	2.8 - 9.5
NAO	49 - 59	up to 4.7	9.8 - 41.4	13.6 - 20.5	43.6 - 72.6	11.7 - 18.9	10.9 - 12.1

[a] % of N_{tot}.

N incorporation was shown to correlate with the reaction pressure in the high-pressure process. In the two processes, it is, however, the reaction temperature that exerts the greatest influence on N incorporation and the distribution of the types of bonds. For example, in the AO process N_{tot} continuously increases with temperature up to a maximum of ~170 °C, thereafter it starts to drop slightly. A similar tendency is also found for NH_4^+-N, amide-N and sob N (Figure 6).

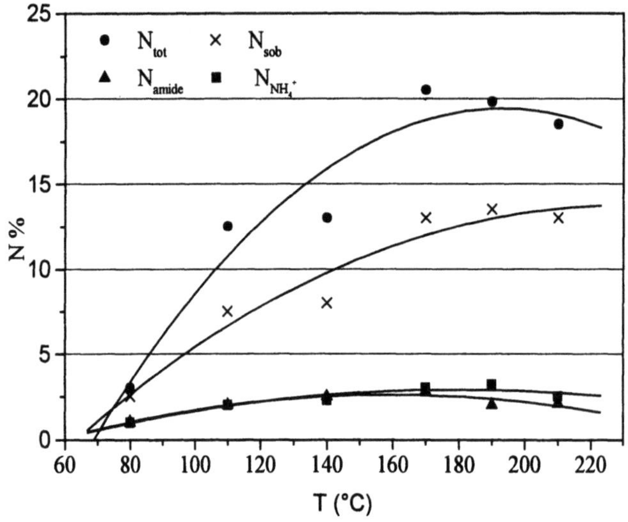

Figure 6. N incorporation and its distribution vs. reaction temperature in the oxidative ammonolysis of ORGANOCELL lignin using the AO method.

The N_{tot} of the N lignins and its distribution to various types of bonds also depend on the structure/properties of the parent lignins. These are essentially controlled by the conditions of their precipitation from the spent liquor (Zier 1996, Zier *et al.*, 1997). Precipitation at pH 9 causes a fractionation of the higher molecular weight component from the lower molecular weight one with the latter remaining in solution. Differences in the parent lignin, attributable to the cooking parameters are, however, slightly offset by oxidative ammonolysis and thus do not affect the N lignins to a greater extent. The properties of N lignins are essentially influenced by the differences in parent lignin structures due to the pH value of lignin precipitation (pH 4 or 9) (Table 9). The NAO lignins were produced from a number of parent lignins under identical reaction conditions. There are only small differences in total nitrogen incorporation between the various parent lignins within one type of precipitation. The differences are more pronounced for the lignins precipitated at different pH's. The AO lignins were produced from various parent lignins by oxidative ammonolysis using diverse reaction parameters. In general, a greater amount of N_{tot} is incorporated into the lignins precipitated at pH 4. In comparison with the NAO lignins the differences are not so pronounced. This is due to the markedly stronger reaction conditions of the AO-process, in particular the reaction temperature and the NH_3 concentration.

Table 9. N products from ORGANOCELL lignins precipitated at different pH's

Sample	Precipitation pH	N_{tot}%	NH_4^+-N[a]	Amide-N[a]	sob N[a]	OCH_3%
NAO	4	4.24 - 4.74	32.6 - 41.5	14.9 - 19.0	43.6 - 50.6	11.2 - 12.1
	9	2.9 - 3.80	9.8 - 13.2	16.8 - 20.5	66.6 - 72.6	10.9 - 12.1
AO	4	12.19 - 16.21	19.6 - 28.9	13.7 - 16.3	55.0 - 66.7	3.23 - 6.11
	9	10.41 - 14.5	14.6 - 20.6	14.3 - 18.6	62.2 - 70.5	2.79 - 5.43

[a] % of N_{tot}.

The conditions used for the precipitation of the parent lignins essentially control the course of reaction in the NAO process, leading to marked differences between the corresponding N lignins (Figures 7 and 8). pH 9 lignins are less reactive, the entire conversion taking place more slowly. The incorporation of N_{tot} is distinctly lower, with the relative N-distribution being completely different. The relative NH_4^+-N content is very low due to the Na^+ content of the parent lignins while the relative sob N content is very high. During the course of the reaction the already very low absolute values of NH_4^+-N and amide-N increase only slightly; whereas, the absolute value of sob-N increases distinctly. The relative N distribution undergoes only minor changes.

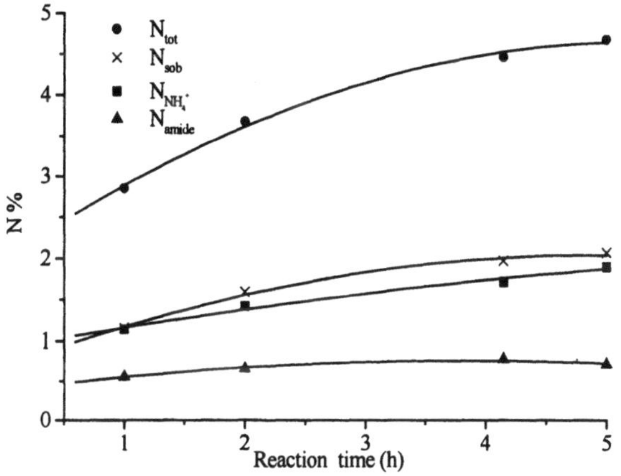

Figure 7. N incorporation and its distribution vs. the reaction time in oxydative ammonolysis of ORGANOCELL lignin (pH 4 precipitated) using the NAO method.

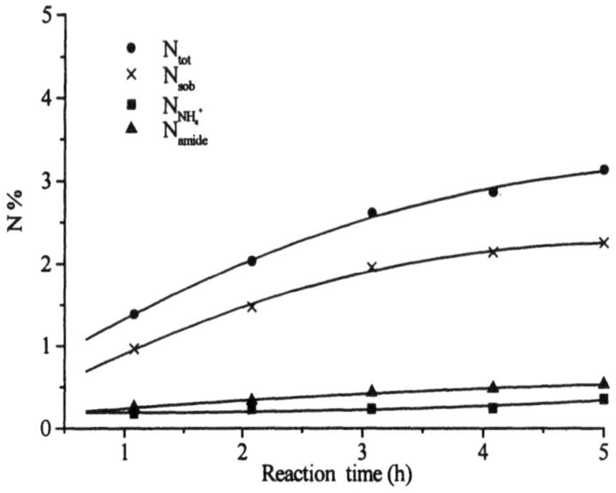

Figure 8. N incorporation vs. the reaction time in oxydative ammonolysis of ORGANOCELL lignin (pH 9 precipitated) using the NAO method.

The two methods are similar in regards to the degradation of OCH_3 groups typical of oxidative ammonolysis, this being a linear relationship to the incorporation of N_{tot}.

The spectra of the N lignins reveal the typical features of lignins including their expected changes. The UV spectra of the NAO and the AO lignins show the characteristic absorptions of the lignin spectra with a maximum at 280-290 nm and a minimum at 260-270 nm. With increasing N modification the characteristic absorption maximum loses its intensity at 280 nm. The modifications in the UV spectra of the N lignins are caused by the formation of new chromophores (e.g. α–carbonyl groups) from auxochromic groups (e.g. hydroxyl and amino groups) due to oxidation and condensation reactions during oxidative ammonolysis.

Almost all the specific bands typical of guaiacyl lignin are found in the FTIR spectra of AO lignin. However, the bands between 1000-1400 cm^{-1} are greatly reduced, especially at 1512 cm^{-1} (aromatic skeletal vibrations). In contrast, a band between 1600-1700 cm^{-1} (aromatic C=O vibrations) is broadened and enlarged. Additionally, a number of new bands can be found, which are characteristics of the following N-functional groups: ammonium-N at 3200 and 1400 cm^{-1}; amides at 3200, 1630, 1600 and 1400 cm^{-1}.

A strong broadening of the lines is observed in the ^{13}C-NMR spectra (in DMSO-d_6) of NAO lignins from various precipitated lignins. In NAO lignins from the pH 9-precipitated lignins this is most pronounced. Changes are discernible both in the aromatic (110-160 ppm) and in the aliphatic structures (0-50 ppm). Obviously, new aliphatic structures are formed in the NAO lignin, which can be attributed to ring fission. Within the aromatic range, it is the signals for C-1 and C-4 that are subjected to the changes. These two carbon atoms are points liable to attack by oxidation. As expected for the oxidation processes, a distinct increase in intensity as well as the occurrence of additional peaks within the carbonyl range (> 160 ppm) are also observed.

Comparative investigations by means of relevant degradation of selected N lignins as well as their parent lignins have been conducted to determine to which extent the typical lignin structures are changed in oxidative ammonolysis (Krusche *et. al.*, 1994). In the nitrobenzene oxidation, the total yield of vanillin/vanillic acid decreases with increasing N modification. The drastic decrease in highly modified N lignins points to the fact that there are few degradable non-condensed structures left. The lignin appears to be condensed to a large extent. The various precipitations of the parent lignins cause only minor changes. Likewise, the degradation products guaiacol and catechol, which are characteristics of the nucleus exchange, decrease with increasing N modification, with guaiacol to a higher extent than catechol. In the oxidative $KMnO_4/NaIO_4$ degradation of N-lignins the most important, typical lignin carboxyl esters from non-condensed and condensed structural units could be identified. This is demonstrated in Table 10. In general, it becomes apparent that the total yield of degradation products decreases in N lignins with increasing N modification. In this regard, the decrease of veratric acid, resulting from non-condensed units, is more pronounced than those of isohemipinic acid, metahemipinic acid, 2,2',3-trimethoxydiphenylether-4',5'-dicarboxylic acid and 5,5'-dehydro-diveratric acid from C- and O-condensed structure, respectively. The total yields of degradation products are generally higher in N lignins from parent lignins precipitated at pH 4 than from those precipitated at pH 9.

Table 10. Percentages of degradation products as compared to the parent lignin

Degradation/Degradation products	N-Lignins / N_{tot}			
	3-5 % (NAO process)		11-16 % (AO process)	
Precipitation of the parent lignin	pH 9	pH 4	pH 9	pH 4
	yield [% of parent lignin]			
$KMnO_4$-/$NaIO_4$ – degradation:				
Veratric acid	14.2	43.2	0 - 15.3	
Isohemipinic acid	23.7	67.4	6.4 - 59.9	n. d.
m-Hemipinic acid	49.6		18.0 - 77.2	
Trimethoxydiphenylether-dicarboxylic acid	-	50.4	0 - 10.1	
5.5-Dehydrodiveratric acid (all as carboxylic acid methyl esters)	17.6	63.1	0 - 32.8	
$\Sigma_{degradation\ products}$	17.4	51.1	2.9 - 24.4	
Nitrobenzene oxidation:				
Vanillin	64.9	56.8	15.1	13.8
Vanillic acid	86.3	86.6	51.1	58.3
$\Sigma_{degradation\ products}$	69.2	61.7	22.4	21.1
Nucleus exchange:				
Guaiacol	39.6	72.4	3.6	4.4
Catechol	42.3	22.4	31.2	2.5
$\Sigma_{Guaiacol\ /\ Catechol}$	39.8	53.2	5.6	3.7

The pyrograms of pyrolysis-gas chromatography/mass spectrometry (Py-GC/MS) (Krusche *et al.*, 1994) reveal both typical components of lignin (guaiacol, substituted guaiacols, phenoles and methyl substituted phenols) and N-containing substances. Because of the relatively low content of organically bonded nitrogen, the peaks of the N-containing pyrolysis products are small. The typical lignin components decrease with increasing N modification of the lignin, in agreement with previous findings. At the same time the N compounds increase and become detectable in N lignins with N_{tot} > 8% with the main product being vanillic acid nitrile. With N_{tot} > 14%, phthalimides and N-heterocycles such as pyrroles and pyridines can also be identified. In the pyrograms of strongly N-modified lignins (N_{tot} ca. 19-20%) produced at high reaction temperatures, only phenol, guaiacol as well as methyl-substituted phenols and guaiacols and other N compounds such as 3-methyl-1H-pyrrole and 4-cyanopyridine can be detected. This points to a strong degradation of the side chain and N incorporation in the aromatic ring in connection with intensified condensation reactions at high reaction temperatures. However, it is not entirely clear whether the N-containing compounds found are original structural elements of the N lignins or those formed during pyrolysis. The assumption that nitrile structures are present in the original N lignins is confirmed by other investigations.

4.3. Structures of N-Lignins, Types of Nitrogen Bonds and Mechanisms of Conversion

The N-lignins are to be used as humus-like fertilizers. It is important to know the types of nitrogen bonds and the reactions taking place during the oxidative ammonolysis of lignins, in particular in comparison with the formation of humic substances in nature. Oxidative ammonolysis is characterized by the reaction of reactive oxygen species with lignin as a polymer of a particular molecular weight distribution in an alkaline medium in the presence of the nucleophile ammonia. Regarding this constellation, a complex array of reactions has to be assumed.

It was already during an early stage of research in characterizing the N-modified products that a distinction was made between the various types of nitrogen bonds (Ehrenberg and Heimann, 1929; Caro and Frank, 1931; Kleinert 1952; Grosskinski and Klempt, 1949).

In the aspect of plant nutrition, Flaig and Söchtig (1973) have particularly underlined that the nitrogen incorporated is hydrolyzable and plant-available to various extents. Being organically bonded it exists both in the lower- and in the higher-molecular (> 10,000) proportions of N lignin. Low-molecular weight phenolic acids may be formed in analogy with the natural lignin degradation. Phytophysiological effects such as the inhibition of nitrification suggest the existence of chinoid structures (Flaig *et al.*, 1959; Flaig and Söchtig, 1973, 1974; Flaig 1976).

Preliminary kinetic investigations (Mazhara and Sapotnitskii *et al.*, 1969 a,b) suggested a first-order reaction for the conversion of spent sulfite liquor using an excess quantity of ammonia, oxygen and water. In the starting phase desulfonation takes place. The OCH_3 groups are also degraded during the first two hours. The optimal reaction temperature is 80 - 120 °C with the polymerizations increasing at T > 120 °C.

Zakis and Neiberte (1978a) gave a survey of the 1978 state-of-the-art of oxidative ammonolysis including the various goals such as the manufacturing of fertilizers and pulping. The types of nitrogen bonds in the order of decreasing degree of hydrolyzability were classified as mineral nitrogen, amide, amine nitrogen and non-hydrolysable heterocyclic nitrogen (Kazarnovskii *et al.*, 1976; Kazarnovskii and Chudakov, 1976). The investigations into the conversion of lignins by means of ammonia/peroxydisulfate revealed an analogy with the NH_3/O_2 system. The N lignins so formed contained various nitrogens equally as NH_4^+, amide and in a type of bond that was more difficult to hydrolyze. Amine structures were assumed and a number of NO structural types such as =N-OH, -N=O, $-NO_2$, -N-NO etc. excluded (Zakis and Neiberte, 1973b, 1978b; Zakis *et al.*, 1973a; Zakis *et al.*, 1974).

One of the potential uses of oxidative ammonolysis is for delignification. This potential use gave rise to investigations into the course and control of wood pulping using oxidative ammonolysis. The results prove that lignin is degraded according to a radical-chain mechanism (Deineko, 1989). The most important centers of reaction in lignin are the phenolic OH group and the α-C atom. These two centers take a similar share in the overall process (Deineko and Kolotov, 1989).

Recently, research has been focused again more on the conversion of technical lignins to long-term fertilizers. The investigations have shown that technical lignins from various pulping processes differ as to their reactivity (Meier *et al.*, 1994); concerning N-incorporation, the reactivity decreases in the following order: lignosulfonate, Kraft lignin, Organocell lignin, Kraft lignin (CO_2 precipitation), alkali/soda lignin, ASAM lignin.

FTIR spectra of various N-lignins point to drastic changes during oxidative ammonolysis. Aromatic structures and ether as well as alcohol functional groups are distinctly decreased. The conspicuous increase of the intensities of a broad band between 1550 and 1750 cm^{-1} with a maximum at 1660 cm^{-1} as well as a growing band at 1400 cm^{-1} corresponds to C=O, C-N, and C-N-H of the amides.

New insights into the reactivity of the side-chain and the condensation reactions were obtained by means of thioacidolysis (Lapierre *et al.*, 1994). The degree of condensation of an already heavily-condensed Kraft lignin can still be increased by oxidative ammonolysis. During the reaction β-O-4 ethers are cleaved, enol ether structures are converted and new C-C bonds are formed. Along with the decrease of β-O-4 ethers the formation of vanillic acid amide is enhanced. The original catechol structures decrease. The newly formed vanillic acid structures in the N lignins exist as ammonium salts. Both the amide and the nitrile of vanillic acid could be determined by means of GCMS. An original occurrence of the nitrile can be ruled out, since the FTIR spectrum does not comprise a corresponding band. Fatty acids and resinous acids contribute to the N_{tot} of N lignins in the form of ammonium salts.

In order to further elucidate the structures of N-lignins prepared from Organosolv lignins (Session 4.2.2), in particular the types of N bond and their formation, ^{14}N- and ^{15}N-lignins, relevant model compounds and DHP were subjected to oxidative ammonolysis. The products were then examined by a number of analytical methods such as ^1H-, ^{13}C- und ^{15}N-NMR spectroscopy, GCMS and IR-spectroscopy (Potthast 1994; Potthast *et al.*, 1996, 1997). The model compounds used were 1-(3,4-dimethoxyphenyl)-2-(2-methoxyphenoxy)propane-1-one (**1**), 1-(4-hydroxy-3-methoxyphenyl)-2-(2-methoxyphenoxy)propane-1-one (**2**) and 1-(4-methoxyphenyl)-2-(2-methoxyphenoxy)ethane-1-one (**3**), respectively (Scheme 2).

The reactions (maximal N incorporation N_{tot} = 16.8 – 22.7%) were done in a high-pressure autoclave at 130 °C under 10 bar O_2 with 17.5% NH_3 and $^{15}NH_3$, respectively, for 150 min.

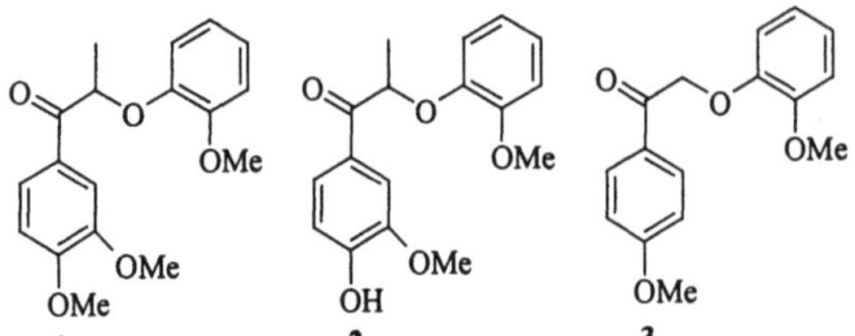

1 2 3

Scheme 2. Structures of the model compounds **1-3**.

The results of ^{15}N-NMR (Table 11) show that there is a good conformity among the N types of bonds in the ^{15}N products. This holds true for aliphatic and aromatic amides, urea and ammonium ions. By application of the two-dimensional sequence ^{15}N-^{1}H-COSY the formation of urea could be unambiguously verified. However, some differences do exist, particularly in the nitrile structures, and the structures of propionamide and formamide. By means of FTIR, nitrile structures were detected in the N-products derived from lignin (maximum around 2200 cm^{-1}) and identified by GCMS as substituted benzonitriles from lignin and the models **1** and **3** without free 4-OH groups (4-methoxybenzonitrile from model **3**; 3,4-dimethoxybenzonitrile from model **1** and lignin; 4-hydroxy-3-methoxybenzonitrile from lignin). Vanillic acid nitrile was already found during previous investigations of N-lignin by means of Py-GC/MS (Krusche *et al.*, 1994).

Table 11. ^{15}N-NMR data of the N-products from the reaction of lignin, model compounds and DHP

Bond type	δ in ppm	$^{1}J_{N\text{-}H}$ in Hz	Found in
Not assigned	-3.61		all
Subst. benzonitrile	-126.83		model **1**
Not assigned	-226.7		model **1**, **3**, ^{15}N-DHP
Formamide	-266.3		^{15}N-DHP
Acetamide	-270.63	89.12	all
Propionamide	-271.44		model **2**, lignin
Subst. benzamide	-279.66	89.11	all
Urea	-304.00	87.89	all
Ammonium ions	-358.62		all

Nitrogen bonded to the aromatic rings could be identified by a reductive method with subsequent GCMS analysis, however, it could not be unequivocally assigned. The ^{1}H-NMR spectra of N-lignin reveal distinctly reduced signals for OCH_3 and the aromatic protons due to the heavy degradation of the lignin skeleton. In the N-products of the models this is even more pronounced.

The findings of these investigations can be discussed on the basis of the following mechanism: The formation of nitriles begins with a nucleophilic attack of ammonia on the α-carbonyl groups of the lignin structures to give imines. Hydrogen is abstracted by hydroxyl radicals resulting in the cleavage of the α, β-C-C bond and formation of nitrile. Nitrile is not formed, if there is a free phenolic OH-group *para* to the α-carbonyl group. The electron density at the α-carbonyl group is higher by resonance-stabilization in this case as shown by quantum mechanical calculation on relevant model compounds.

Nitrogen may be bonded to the aromatic rings via 1,2-addition of ammonia to o-quinone structures. Quinone imines are formed, and consecutive reactions are conceivable (Scheme 3 in congruency with the proposal of Zakis and Neiberte, 1978a).

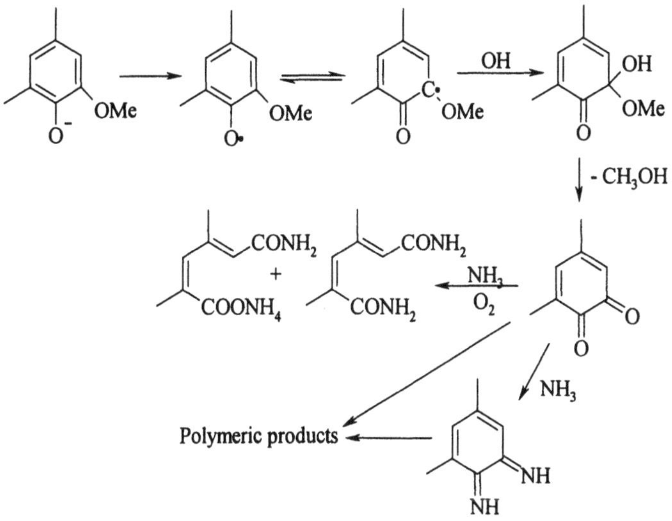

Scheme 3. Possible pathway of the oxidative ammonolysis of structures with free phenolic group (model **2**)

The formation of formamide in N-DHP can be explained by a β, γ-C-C cleavage with formation of formaldehyde from the CH_2OH-group followed by oxidation to formic acid. Acetamide results from cleavage of the side chain along with the formation of acetic acid.

The degradation of non-phenolic lignin structures can be explained as follows: β-aryl ether bonds and side chains are cleaved. Aromatic compounds continue to exist in the form of 3,4-substituted acids or derivatives such as amide and nitrile. Aliphatic acids result from the cleavage of the side chain. Aromatic rings are degraded to yield aliphatic dicarboxylic acids. The acids are converted into ammonium salts, and amides may be formed. Oxidative degradation to CO_2 leads to the formation of urea (Scheme 4).

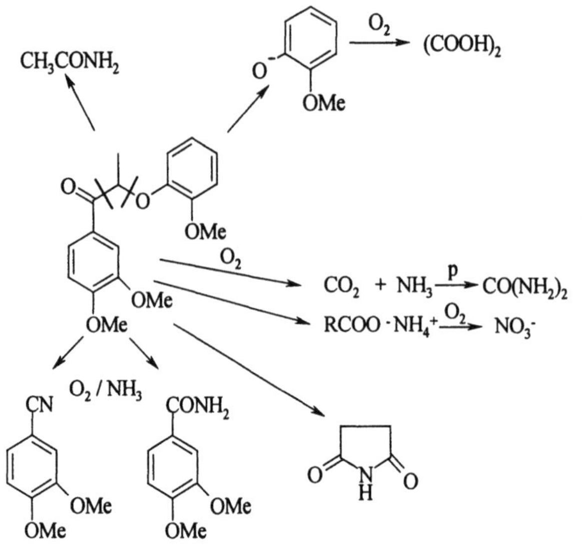

Scheme 4. Reaction products of the oxidative ammonolysis of model **1** (Potthast *et al*, 1997).

By combination of ^{15}N functionalisation with FTIR, Py-GC/MS and NMR (Varela *et al.*, 1999) previous findings that in oxidative ammonolysis of Kraft lignin (150 °C, 15 bar O_2, up to 283 min.) N incorporation occurs through oxidation of ether and alcohol groups and degradation of aromatic hydrocarbons could be corroborated. The CPMAS spectra show increasing ammonium and amide/amine signals during progressing conversion. By means of Py-GC/MS and depending upon the reaction time, up to 27 N species have been found. Of these, however, only benzonitrile and vanillic acid nitrile could be identified.

The following conclusions were drawn based on the investigations into the mechanism of oxidative ammonolysis of Organosolv lignin obtained from hardwood species (Repap, 100 and 130 ° C, 8 and 12 bar O_2, respectively, 150 min) (Capanema *et al.*, 1998, 1999):

- N incorporation increases with increasing reaction temperature, the reaction pathway, however, remains unaffected
- Both decarboxylation and dehydration take place
- Formation of amide and related structures takes place via ring cleavage and muconic acid intermediate, aliphatic or aromatic amines, however, cannot be ascertained.

4.4. Natural and Artificial Humification of Lignins

As a polyphenol and a source of phenolic compounds, lignin plays a key role in the natural process of humus formation. Models have been derived explaining humus formation, of which a summarizing introduction is given in the literature (Schnitzer and Khan 1972; Hayes *et al.*, 1989; Haider, 1996).

A model devised by Flaig (1988) shows that high-molecular weight lignins are degraded into small fractions by reactions with reactive oxygen species in an aerobic process due to microbial attack according to radical mechanisms. In this connection, the C and OCH_3 contents are reduced while the N and O contents as well as the contents of OH, C=O and

COOH groups increase. Reactive o-diphenol structures and o-quinoid structures are formed. They may react with proteins and their degradation products to yield N-containing polymers, i.e. precursors of humic acids or directly humic acids. On very simplified terms, the natural humification of lignin may be regarded as oxidative degradation with N-incorporation (Scheme 5).

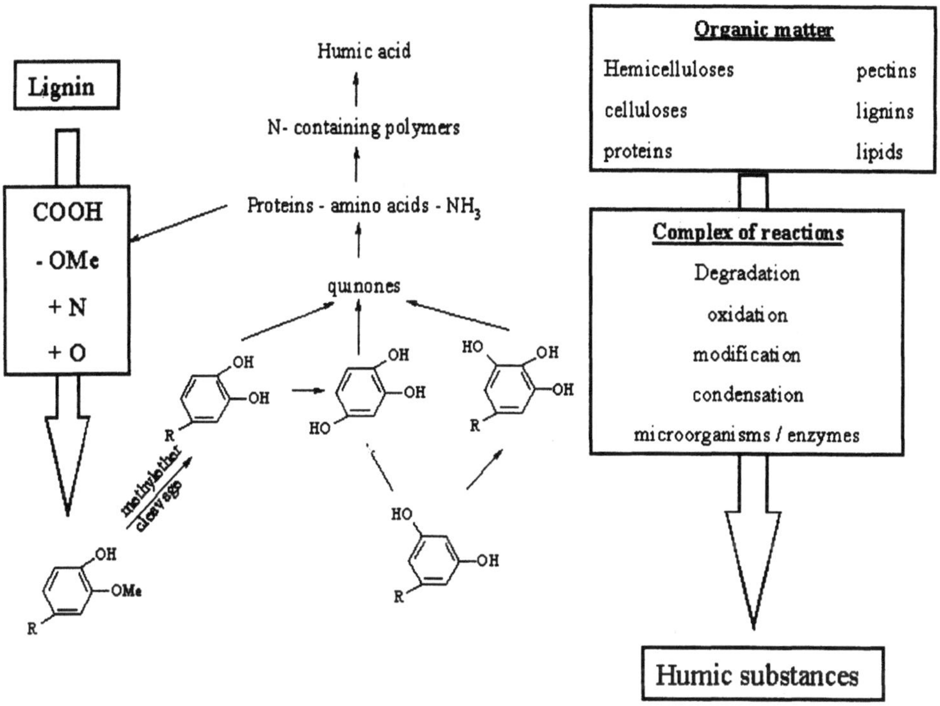

Scheme 5. Reaction scheme for the natural humification of lignin (according to Flaig, 1988).

Natural humic substances are very complex in structures and are highly differentiated. Early research suggested a number of structural models (Schnitzer and Khan, 1972; Hayes *et al.*, 1989). The types of nitrogen bonds have been uncovered to only ~50 percent. Nitrogen has been detected in the soil as ammonium ions, amino groups, amino acids, amino saccharides as well as bases of nucleic acids (Hayes *et al.*, 1989; Schnitzer, 1985).

For further structural elucidation, ^{15}N-NMR and the Py-GC/MS have been used in recent investigations. The accuracy of the NMR method in identifying single forms of N bonds, however, is limited by the low natural ^{15}N-content of the material, the small gyromagnetic ratio of the ^{15}N nucleus and the diversity of various possible N-compounds. Relatively simple ^{15}N NMR spectra show a main signal for amide/peptide and a few small signals for amines, indoles and pyrroles (Preston, 1996) and for aliphatic and aromatic amines in ^{15}N-labeled humic acid, with only ca. 5% of the total area corresponding to pyrroles (Zhuo and Wen, 1992).

The spectra of unlabelled material (Knicker *et al.*, 1993) comprise very small signals attributable probably to NH in guanidine and aniline derivatives, free amino groups as well as substituted amines. The very broad signal for amide/peptide may have superimposed minor signals of other types of N bonds in e.g. heterocycles, quinonimines, melanoides, Maillard products, etc.

However, numerous N-containing substances have been detected by means of Py-GC/MS of humic substances (Schulten and Schnitzer, 1998). The latest structural model suggested a more complete structure of the humic substance (Schulten and Schnitzer, 1998). The model is an embodiment of a macromelocule with aromatic units being connected via aliphatic side chains of various lengths. In an elementary composition of $C_{305}H_{299}N_{16}O_{134}S_1$, it contains 5 aliphatic and 21 aromatic COOH groups, 17 phenolic and 17 alcoholic OH groups, 7 quinone and ketone C=O groups as well as 3 OCH_3 groups and 16 N functions. S is aromatically bonded in the form of benzothiazole. On the basis of the Py-FIMS and Curie-point Py-GC/MS results of humic acids and the whole soils, structural units corresponding to pyrazole, indole, pyrrole, benzothiazole, pyridine, nitriles, aliphatic amines, aromatic amines and acetamide could be deduced.

The N lignins prepared using our NAO method compared well with the natural humic substances in terms of N_{tot} and the C/N ratio. Scheme 6 summarises the reaction scheme involved in the artifical humification of lignins using our AO and NAO processes.

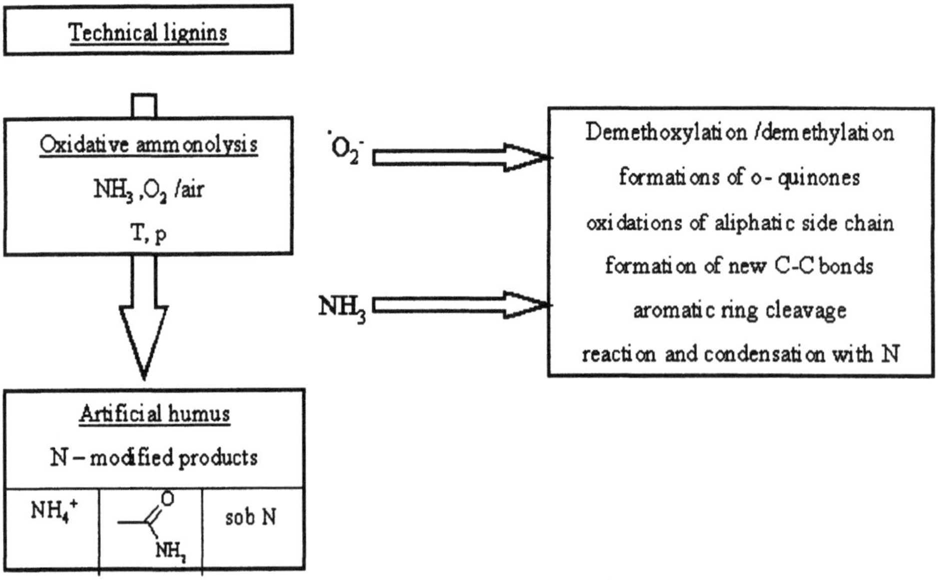

Scheme 6. Reaction scheme for the artificial humification of lignin.

The similarity of basic reactions between oxidative ammonolysis and the natural humification permits a discussion of analogies despite distinct differences. Oxidative ammonolysis in vitro employs harsher conditions that drastically shorten the reaction time in contrast to the humification process that takes place in nature. The biogenic organic matter is converted in nature. In vitro, however, the parent material is technical lignin which has already been altered during pulping and contains only low quantities of admixed substances.

Analogies are found in the macromolecular building principle of the aliphatically linked aromatic units, in the O-functional groups and the N-modification in the form of amide, nitrile and amine functionality including heterocyclic structures. Table 12 shows both the similarities of the C and N contents (thus, the C/N ratios) as well as the N distribution (% of N_{tot}), and the differences in the aromatic structure expressed by the contents of OCH_3 and O-functional groups, the latter detectable by cation exchange capacity (CEC). The artificial humus (N lignins) contains fewer acid groups, thus their cation exchange capacity being lower than that of the natural humic substances.

Table 12. Structural properties of artificial and natural humus in comparison

Humus	C%	N_{tot}%	NH_4^+-N[a]	Amide-N[a]	sob N[a]	OCH_3%	CEC[b]	MW (g/mol)
Artificial	49 - 59	3 - 5	10 - 41	14 - 21	44-73	10 - 12	140 - 180	$\geq 10^4$
Natural[c]	41 - 62	1 - 5	10 - 25	21 - 45	ca. 50	0.9 - 1.8	180 - 500	10^4-2×10^5

[a] % of N_{tot}.
[b] meq./100 g.
[c] Schnitzer and Khan, 1972,1978; Flaig *et al.*, 1975; Stevenson, 1982; Aiken *et al.*,1985; Hayes *et al.*, 1989.

5. APPLICATIONS

Nitrogen-rich N lignins (N_{tot} 10 - 24%) have been tested as organomineral N fertilizers by many research groups; the pioneering work by Flaig and coworkers is to be emphasized once more in this connection. The work performed in the former Soviet Union is outlined in surveys presented by Kalninsh *et al.* (1978) and Sapotnickii (1981). The investigations were conducted on various soils with agricultural crop plants (cf. also Ishibashi *et al.*, 1978; Fiedler and Schmidt, 1979, 1981) and also on species of woody plants (Fiedler and Schmidt 1983). The results of the past decade have been presented by a number of researchers (Meier *et al.*, 1994; Ramirez-Cano *et al.*, 1993; Ramirez-Cano *et al.*, 2001).

The humic substitutes developed by us at the Tharandt Institute have several essential advantages over the hitherto known N lignins. Their humus-like structure is a feature by which their application for the improvement of soils is based. The ameliorative properties of our artificial humus substitutes, their mineralization behavior and influence on plant growth have been examined in numerous pot and lysimeter experiments over the past few years. These examinations took place in a phytochamber and a greenhouse at Dresden University of Technology as well as in greenhouses and in a lysimeter at the Reasearch Institute for Post-mining Landscapes, Finsterwalde (Katzur, 2001).

The particularly favorable effects of our artificial humus substitutes (N lignins) are shown in their abilities to function as excellent organic fertilizers. They are active over several growing seasons (Figure 9), analogous to the natural humus containing various types of N bonds). These N-bonds, specifically, are:

- ammonium-N which is immediately available for the plants
- medium-term, plant-available amide-N
- long-term, plant-available, organically bonded N.

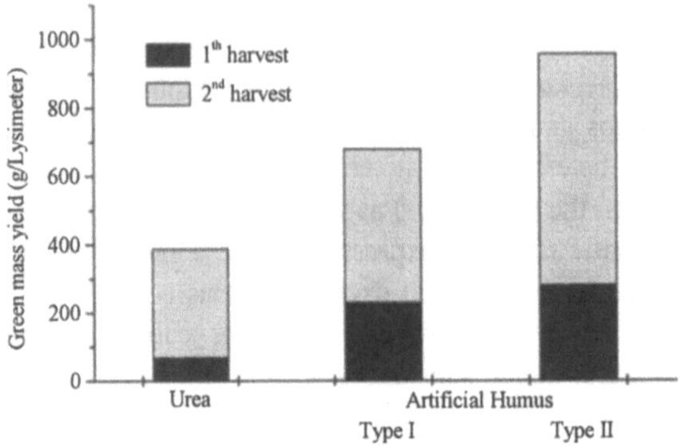

Figure 9. Long term effect of the artificial humus.

Our artificial humus substitutes also improve the nutrient relations and promote soil-microbial processes in the near-surface soil substrate layers, resulting in favorable conditions for plant growth and a high productive evaporation. The latter, coupled with the time-phased N release from the various types of bonds, leads to a markedly lower N leaching as compared to mineral fertilizers (Figure 10).

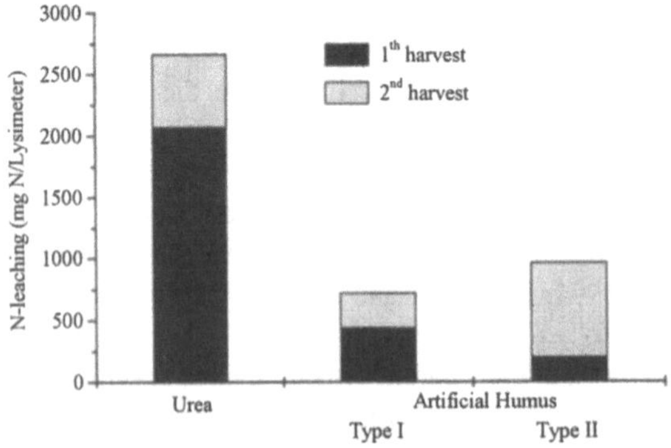

Figure 10. Decrease of N-leaching by the artificial humus.

A multitute of functional groups and a relatively high cation exchange capacity of our artificial humus substitutes also give them a great immobilization potential for noxious, inorganic and organic substances (complexation, salification). Thus, the output of substances along with the seepage water can be drastically reduced in the rehabilitation of problematic sites, if the artificial humus is used instead of the conventional mineral fertilizers (Figure 11).

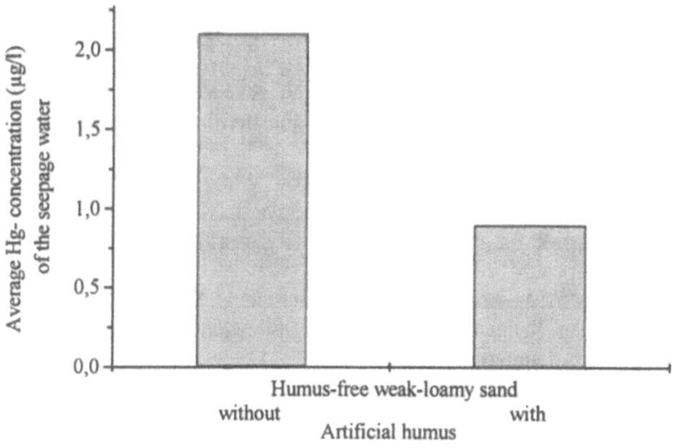

Figure 11. Immobilization of toxic heavy metal Hg by the artificial humus.

6. ACKNOWLEDGEMENT

We are indebted to the Bundesministerium für Bildung und Forschung of Germany for supporting the research on Organocell lignin..

7. REFERENCES

Aiken, G. R., McKnight, D. M., Wershaw, R. L., and MacCarthy, P., eds., 1985, *Humic Substances in Soil, Sediment and Water,* John Wiley & Sons, New York, Section three.

Alber, W., Fiedler, H.-J., Fischer, F., Pruzina, K.-D., Schiene, R., Schmidt, U., and Wienhaus, O., 1977, Verfahren zur Verwertung von Lignin oder ligninhaltigem Material, DDR-Patent 133 788.

Allan, G. G., 1971, Modification reactions, in: *Lignins - Occurence, Formation, Structure and Reactions,* K. V. Sarkanen and C. H. Ludwig, eds., Wiley-Interscience, New York-London-Sidney-Toronto, pp. 511-574.

Bratzler, K., and Aalrust, P., 1962, Vorrichtung zur Gewinnung von organischen stickstoffreichen Düngemitteln aus Zellstoffkocherablaugen oder ähnlichen Stoffen, Deutsches Patent 1 230 045.

Bratzler, K., and Thormann, K., 1962, Verfahren zur Gewinnung von organischen stickstoffreichen Düngemitteln aus Zellstoffkocherablaugen oder ähnlichen Stoffen, Deutsches Patent 1 247 353.

Capanema, E. A., Chen, C.-L., Gratzl, J., Kirkman, A., and Gracz, H., 1998, Studies on the mechanism of oxidative ammonolysis of technical lignin, *Proceedings of* 5th *European Workshop on Lignocellulosics and Pulp* (Aveiro), pp. 397-400.

Capanema, E., Balakshin, M.,Yu., Chen, C.-L., Gratzl, J., Kirkman, A., and Gracz, H., 1999, Effect of temperature on the rate of oxidative ammonolysis and structures of N-modified lignins, ^{15}N-Functionalization of kraft lignin and ist characterization by FTIR, Py-GC-MS and NMR, *Proceedings of 10*[th] *International Symposium of Wood and Pulping Chemistry, Vol. III* (Yokohama), pp. 404-409.

Caro, N., Frank, A. R., and Heimann, H., 1929, Verfahren zur Herstellung von kernnährstoffhaltigen organischen Düngemitteln, Deutsches Patent 527 313.

Caro, N., and Frank, A. R., 1930, Verfahren zur Herstellung von stickstoffhaltigen, organischen Düngemitteln, Deutsches Patent 545 924.

Caro, N., and Frank, A. R., 1930, Verfahren zur Herstellung von kernnährstoffhaltigen, organischen Düngemitteln aus rezenten und fossilen Pflanzenkörpern, Deutsches Patent 545 923.

Caro, N., and Frank, A. R., 1931, Verfahren zur Herstellung leicht assimilierbarer organischer Stickstoffdünger aus rezenten und fossilen Pflanzenkörpern, Deutsches Patent 559 254.

Chudakov, M. I., Antipova, A. L., Egorov, A. E., Sokolova, I. V., Milovanov, A. V., Kondratev, B. N., Saidov, Ch. M., Pavlov, A. A., Ivanova, E. K., and Davydov, P. S., 1968a, Manufacturing methods for fertilizers based on lignin, USSR Patent 223 823 (Russ).

Chudakov, M. I., Antipova, A. L., Egorov, A. E., Kondratev, B. N., Pavlov, A. A., Sokolov, A. M., Ivanova, E. K., and Davydov, P. S., 1968b, Production of nitrogen containing lignin fertilizers, *Gidroliz. lesokhim. prom.* (3), pp. 6-7 (Russ).

Chudakov, M. I., Antipova, A. L., Samsonova, A. P., Mekler, N. A., Egorov, V. A., Basin, D. M., Gurevich, Ya. E.,Voropaev, I. S., Martynenko, K. D., Shiryaev, A. M., Stakhorskaya, L. K., and Kuznetsova, V. I., 1969, Manufacturing process of a substrate for the production of feed yeast, USSR Patent 299 541 (Russ).

Davis, R. O. E., and Scholl, W., 1933, Process for ammoniating organic material, U. S. Patent 2 027 766.

Deineko, I. P., 1989, Chemical conversion of lignin during wood delignification with oxygen, *Author's Abstract of a Doctoral Dissertation*, LTA Leningrad (Russ).

Deineko, I. P., and Kolotov, S. I., 1989, Kinetics of wood delignification with oxygen. 4. Oxidative ammonolysis of wood, *Khim. Drev.* (2), pp. 25-32 (Russ).

Dence, C. W., 1971, Halogenation and nitration, in: *Lignins - Occurence, Formation, Structure and Reactions*, K. V. Sarkanen and C. H. Ludwig, eds., John Wiley & Sons, New York-London-Sidney-Toronto, pp. 373-432.

Duiker, J. A., 1959, Verfahren zur Erzeugung von stickstoffhaltigen Düngemitteln aus salzsaurem Lignin, Österreich. Patent 213 416.

Ehrenberg, C., 1929, Herstellung einer für die Konservierung von Wirtschaftsdüngern geeigneten Substanz, Deutsches Patent 508 258.

Ehrenberg, C., 1930, Verfahren zur Herstellung einer für die Bindung von Ammoniak und Konservierung von Wirtschaftsdünger geeigneten Masse, Deutsches Patent 561 316.

Ehrenberg, C., and Heimann, H.,1929, Herstellung von kernnährstoffhaltigen, organischen Düngemitteln aus rezenten und fossilen Pflanzenkörpern, Deutsches Patent 507 320.

Erasmus, P., 1928, Verfahren zur Herstellung stickstoffhaltiger Verbindungen aus Kohlenhydraten, Deutsches Patent 514 510.

Fiedler, H.-J., and Schmidt, S., 1979, Zur Düngewirkung von AO-Ligninen, *Zellst. Pap. (Leipzig)* **28**(5), pp. 217-219.

Fiedler, H.-J., and Schmidt, S., 1981, Untersuchungen zur Eignung von N-Ligninen als Düngemittel, I, Wirkung von AO-Ligninen auf Weidelgras, *Wiss. Z. Tech. Univ, Dresden* **30**(1), pp. 197-201.

Fiedler, H.-J., and Schmidt, S., 1983, Zur Wirkung von N-Ligninen auf Gehölze, *Wiss. Z. Tech. Univ.,Dresden* **32**(4), pp. 195-200.

Fischer, F., Schiene, R., Wienhaus, O., Miletzky, F., and Pruzina, K.-D., 1983, Möglichkeiten zur Herstellung stickstoffhaltiger Produkte aus verhefter Ammoniumbisulfitablauge, *Wiss. Z. Tech. Univ. Dresden* **32**(3), pp. 179-84.

Fischer, F., Schiene, R., Wienhaus, O., Pruzina, K.-D., Miletzky, F., and Kutnevich, A. M., 1985, Possibilities for producing N-containing products from ammoniumbisulfite spent liquor after previous yeast production, *Khim. Drev.* (2), pp. 65-72 (Russ).

Fischer, K., and Schiene, R., 1990, Erkenntnisse zur Verwertung von Lignin, *Aus der Arbeit von Plenum und Klasse der AdW der DDR*, 7, pp. 23-48.

Fischer, K., Schiene, R., Miletzky, F., and Katzur, J., 1990, Chemical modification of technical lignins by means of oxydative ammonolysis, *Proceedings of 1st European Workshop on Lignocellulosics and Pulp* (Hamburg-Bergedorf), pp. 159-166.

Fischer, K., 1991, N-Lignins - possibilities and chances, *Proceedings of International Forum for Sulfur-Free Lignins* (Schwarzsee, Switzerland), pp. 1-5.

Fischer, K., Schiene, R., and Katzur, J., 1992, Nitrogen modified lignins - a new source of humus, *XVth Internat. Conference in Association with the Royal Society of Chemistry, Volume II* (Lisboa), pp. 353-357.

Fischer, K., Schiene, R., Zier, N., and Krusche, K., 1993, Structure and properties of Organocell lignins and N-modified Organocell lignins, *Proceedings Cell, Meeting of the American Chemical Society*, (Denver/USA, March/April 1993), p. 134.

Fischer, K., Schiene, R., Krusche, K., Zier, N., and Katzur, J., 1995, Artificial humus prepared from lignin, *Proc. Int. Chem. Congress of Pacific Basin Societies* (Honolulu, Hawai), p. 417

Fischer, K., Katzur, J., Schiene, R., and Liebner, F., 2001, N-modifiziertes Lignin - hochwertiger Humusersatzstoff und Langzeit-Düngemittel, *Das Papier* **55**, in press.

Flaig, W., Hingst, G., and Wesselhoeft, P., 1959, Verfahren zur Herstellung von stickstoffreichen Ligninprodukten, Deutsches Patent 1 745 632.

Flaig, W., and Hingst, G., 1960, Verfahren zur Herstellung von stickstoffreichen Ligninprodukten, Österreich. Patent 233 027.

Flaig, W., and Söchtig, H., 1973, Wirkung organischer Bodensubstanzen und Ertragssicherung, *Landbauforschung Völkenrode* **23**(1), pp. 19-28.

Flaig, W., and Söchtig, H., 1974, Ein Beitrag zur umweltfreundlichen Technik durch Verwertung der Sulfitablaugen der Zellstoffindustrie als organischer Stickstoffdünger, *Neth. J. Agric. Sci.* **22**, pp. 255-261.

Flaig, W., 1976, Die organische Bodensubstanz als nachliefernde Stickstoffquelle für die Ernährung der Pflanze und einige Modelle zur technischen Verwirklichung, *Landbauforschung Völkenrode* **26**(2), pp. 117-121.

Flaig, W., Beutelspacher, H., and Rietz, E.,1975, in: *Soil Components* Vol.1, J. E. Gieseking (ed.), Springer Verlag, New York, pp. 1-211.

Flaig, W., 1988, Generation of model chemical precursors, in: *Humic Substances and Their Role in the Environment*, Frimmel, F. H. and Christman, R. F., eds., Wiley - Interscience, Chichester-New York-Brisbane-Toronto-Singapore, pp. 75-9.

Forostyan, Yu. N., and Kovalchuk, B. V., 1972, The conversion of hydrolysis lignin from husks of sunflower seeds with ammonia, *Khim. prirod. soed.* **8**(1), pp. 136-138 (Russ).

Franz, A., and Palm, A., 1930, Verfahren zur Herstellung organischer Düngemittel, Deutsches Patent 561 487.

Gonzalez, C., Alvarez, R., and Coca, J., 1992, Nitrogenous humic fertilizers by ammoniation of humic material obtained from kraft black liquor, *Water, Air and Soil Pollution* **61**, pp. 191-199.

Grosskinski, O., and Klempt,W., 1949, Verfahren zur Herstellung eines an löslichen Ammoniakverbindungen angereicherten Humusdüngers, Deutsches Patent 857 200.

Grosskinski, O., and Klempt, W., 1949,Verfahren zur Gewinnung von Humusdünger, Deutsches Patent 883 609.

Grosskinski, O., and Klempt, W., 1950, Verfahren zur Herstellung von stickstoffreichen Humusdüngemitteln, Deutsches Patent 870 565.

Haider, K., 1996, *Biochemie des Bodens*, Ferdinand Enke Verlag, Stuttgart, pp. 20-64.

Hayes, M. H. B., MacCarthy, P., Malcolm, R. L., and Swift, R. S., eds., 1989, *Humic Substances II - In Search of Structure*, John Wiley & Sons Ltd., Chichester, West Sussex, UK.

Hingst, G., Aalrust, P., Bratzler, K., and Schäfer, H., 1962, Verwendung von Produkten, die durch Umsetzung von Zellstoffkocherablauge mit Ammoniak und Sauerstoff hergestellt worden sind, als Düngemittel, Deutsches Patent 1 302 961.

Hingst, G., Bratzler, K., Schäfer, H., and Aalrust, P., 1963, Process for producing nitrogen-rich organic materials especially for use as fertilizers, Can. Patent 702 812.

Ioffe, L. O., and Sergeeva, V. N., 1972, Process for the delignification of vegetable raw material, USSR Patent 344 054 (Russ).

Ioffe, L. O., Sergeeva, V. N., and Klepech, E. A., 1975, Influence of the essential varying factors of the pulping by oxidative ammonolysis over the quality of pulp and spent liquor, *Khim. Drev.* (5), pp. 60-65 (Russ).

Ishibashi, H., Tanoue, S., Minamata, K., Kudo, M., Khozu, S., Harada, I., and Iwakuni, Y., 1978, Verfahren zur Herstellung von organischen Düngemitteln, Deutsches Patent 28 11 235.

Kaganskii, I. M., Mazhara, M. P., Kharlamov, V. V., and Sapotnitskii, S. A., 1968, Process for the production of nitrogenous fertilizers from wastes of the pulp industry, USSR Patent 213 902 (Russ).

Kalninsh, A. Ya., Trushkin, V. A., Chudakov, M. I., Mozheiko, L. N., Gelfand, E. D., Raskin, M. I., and Telysheva, G. M., 1978, *Utilization of lignin and lignin derivatives in the agriculture*, Riga Zinatne (Russ).

Katzur, J., 2001, Manuscript in preparation.

Kazarnovskii, A. M., 1974, Investigations on process of oxidative ammonolysis of hydrolysis lignin, *Author's Abstract of a Doctoral Dissertation*, Institute of the Pulp and Paper industry Leningrad (Russ).

Kazarnovskii, A. M., and Chudakov, M. I., 1972, Identification of 2-methylimidazole in products of the hydrolysis lignin oxidative ammonolysis, *Khim. Drev.* 11, pp. 107-109 (Russ).

Kazarnovskii, A. M., and Chudakov, M. I., 1973, Oxidative ammonolysis of hydrolysis lignin, *Gidroliz. lesokhim. prom.* 7, pp. 8-10 (Russ).

Kazarnovskii, A. M., and Chudakov, M. I., 1976, Problem concerning the possible mechanism of the oxidative ammonolysis of lignin, *Khim. Drev.* (4), pp. 80-85 (Russ).

Kazarnovskii, A. M., Antipova, A. W., and Ivanova, N. A., 1976, The kinds of nitrogen bonds in the AM lignin and their determination, *Khim. Drev.* (6), pp. 49-52 (Russ).

Kim, Y. K., Plain, W. M., and Hatfield, J. D., 1981, Fertilizer from the oxidative ammoniation of sawdust, *Ind. Eng. Chem. Prod. Res. Dev.*, **20**(2), pp. 205-212.

Kleinert, Th., 1952, Verfahren zur Herstellung von organisch gebundenen Stickstoff enthaltenden Dünge- und Bodenverbesserungsmitteln aus Sulfitablaugen, Österreich. Patent 177 429.

Knicker, H., Fruend, R., and Luedemann, H.-D.,1993, The chemical nature of nitrogen in native soil organic matter, *Naturwissenschaften* **80**, pp. 219-221.

Kopnin, B. N., and Ioffe, L. O., 1975, Kinetics of the delignification of vegetable raw material by oxidative ammonolysis, *Khim. Drev.* (1), pp. 82-87 (Russ).

Kort, M. J., 1970, Reactions of free sugars with aqueous ammonia, *Advances in Carbohydrate Chemistry and Biochemistry* **25**, pp. 311-349.

Krusche, K., Zier, N., Fischer, K., Schiene, R., and Katzur, J., 1994, Investigations on N-modification of Organosolv lignins, *3rd European Workshop on Lignocellulosics and Pulp, Stockholm,* Proceedings, pp. 337-340.

Lapierre, C., Monties, B., Meier, D., and Faix, O., 1994, Structural investigation of kraft lignins transformed via oxo-ammoniation to potential nitrogenous fertilizers, *Holzforschung* **48**, pp. 63-68.

Martinez, A. de J., Delgado, E., Camacho, A., Gonzalez, V., Allan, G., and Zuniga, V., 1999, Catalytic ammoxidation of kraft lignin in an fluidized bed reactor, *10th International Symposium of Wood and Pulping Chemistry, Main Symposium Yokohama, June 7-10,* pp. 549-551.

Mazhara, M. P., Sapotnitskii, S. A., Vakhrushev, Yu. A., and Korosteleva, Z. N., 1969a, Investigation of the technological parameters of the oxidative ammonolysis of lignosulfonates, *Khim. prom. Ukr.* **58**(2), pp. 13-16 (Russ).

Mazhara, M. P., Sapotnitskii, S. A., Vakhrushev, Yu. A., and Korosteleva, Z. N., 1969b, Kinetics of the oxidative ammonolysis of lignosulphonates, *Khim. prom. Ukr.* **58**(2), pp. 58-61 (Russ).

Meier, D., Zuniga-Partida, V., Ramirez-Cano, F., Hahn, N.-C., and Faix, O., 1994, Conversion of technical lignins into slow-release nitrogenous fertilizers by ammoxidation in liquid phase, *Bioresource Technology* **49**, pp. 121-128.

Mekler, N. A., Raskin, M. N., and Chudakov, M. I., 1974, Investigation of the oxidative degradation of lignin in alkaline medium, in: *Chemistry and Utilization of Lignin,* Zinatne Riga, pp. 342-348 (Russ).

Melms, F., and Schwenzon, K., 1967, *Verwertungsgebiete für Sulfitablauge*, VEB Deutscher Verlag für Grundstoffindustrie, Leipzig.

Miletzky, F., 1985, Über die Oxidation von Sulfitablaugen der Zellstoffindustrie mit molekularem Sauerstoff im alkalischen, insbesondere im ammoniakalischen Milieu, *Dissertation Technische Universität Dresden.*

Miletzky, F., and Schiene, R., 1985, Über die oxydative Ammonolyse von Sulfitablauge (V) - Untersuchungen zur Kinetik, *Zellst. Pap. (Leipzig)* **34**(5), pp. 186-188.

Miletzky, F., Schiene, R., Alber, W., and Fischer, F. 1987, Über die oxydative Ammonolyse von Sulfitablauge. Normaldruckverfahren zur Herstellung von AO-Lignin, *Zellst. Pap. (Leipzig)* **36**(1), pp. 4-5.

Miletzky, F., Schiene, R., Fischer, K., and Fischer, F., 1990, Obtaining N-functionalized products from technical lignins, *Proceedings of 1st European Workshop on Lignocellulosics and Pulp* (Hamburg-Bergedorf), pp.179-186.

Murashkevich, T. U., Skrigan, A. I., and Belkova, A. Y., 1970, Absorption of nitrogen from ammonia water by hydrolysis lignin stored for different periods, From *Ref. Zh., Khim.*, No. 19, abstr. P25 in: *Chemical Abstracts* **74** (1971), 88818j.

Nettesheim, C., 1958, Verfahren zur Herstellung von stickstoffreichen Humusdüngemitteln mit hohem Ammoniakstickstoffgehalt, Deutsches Patent 1 061 803.

Potthast, A., 1994, Charakterisierung der Stickstoffbindungsverhältnisse in N-modifizierten Ligninen unter Anwendung spektroskopischer Methoden, speziell der ^{15}N-NMR-Spektroskopie, *Diplomarbeit Technische Universität Dresden.*

Potthast, A., Schiene, R., and Fischer, K., 1996, Structural investigations of N-modified lignins by ^{15}N-NMR spectroscopy and possible pathways for formation of nitrogen containing products related to lignin, *Holzforschung* **50**(6), pp. 554-562.

Potthast, A., Schiene, R., and Fischer, K., 1997, Investigation on structure and formation of N-modified lignins, Book of Abstracts, *213th ACS-Meeting, San Francisco, April 13-17, 1997,* CELL 113 B.

Preston, C. M., 1996, Applications of NMR to soil organic matter analysis: history and prospects, *Soil Science* **161**, pp.144-166.

Pruzina, K.-D., 1979, Oxidative Ammonolyse von Ammoniumbisulfitablauge in einer Kleintechnischen Mehrzweck-Versuchsanlage. Untersuchungen zum Reaktionsablauf, *Dissertation Technische Universität Dresden.*

Ramirez-Cano, F., Meier, D., Faix, O., and Zuniga, V., 1993, Ammoxidation of technical lignins in a liquid-phase reaction, Proceedings of 7th *International Symposium of Wood and Pulping Chemistry, Vol. II.* (Beijing), pp. 975-983.

Ramirez-Cano, F., Ramos-Quirarte, A., Faix, O., Meier, D., Gonzalez-Alvarez, V., and Zuniga-Partida, V., 2001, Slow-release effect of N-functionalized kraft lignin tested with Sorghum over two growth periods, *Bioresource Technology* **76**, pp. 71-73.

Rautavaara, T., Die Verwendung von Sulfitablauge für die Bodenverbesserung und Düngemittelherstellung, in: *Verwertungsgebiete für Sulfitablauge*, F. Melms, K. Schwenzon, Hrsg., VEB Deutscher Verlag für Grundstoffindustrie, Leipzig 1967, pp. 228-238.

Sapotnitskii, S. A., 1981, *Utilization of Spent Sulfite Liquors*, Lesnaya Promyshlennost, Moskva, Chapter 5.7. Modification, pp. 144-152 (Russ).

Schiene, R., Pruzina, K.-D., Wienhaus, O., and Fischer, F., 1979, Über die oxidative Ammonolyse von Sulfitablauge (I). *Zellst. Pap. (Leipzig)* **28**(4), pp. 163-166.

Schiene, R., Miletzky, F., Fischer, F., and Alber, W., 1985, Verfahren und Anlage zur Herstellung von Zwischenprodukten aus Sulfitablauge, DDR Patent 235 250 A1.

Schiene, R., Miletzky, F., Fischer, K., Krusche, K., Zier, N., and Götze, T., 1992, Chemical modification of Organosolv lignins, *Proceedings of 2nd European Workshop on Lignocellulosics and Pulp* (Grenoble), pp. 237-238.

Schiene, R., Scheller, D., and Fischer, K., 1996, Unpublished results.

Schnitzer, M., and Khan, S. U., 1972, *Humic Substances in the Environment,* Marcel Dekker, Inc., New York

Schnitzer, M., and Khan, S. U., 1978, *Soil Organic Matter*, Elsevier Scientific Publishing Company, Amsterdam-Oxford-New York

Schnitzer, M., 1985, in: *Humic Substances in Soil, Sediment and Water,* Aiken, G. R., McKnight, D. M., Wershaw, R. L., and MacCarthy, P. (eds.), John Wiley & Sons, New York, pp. 303-325.

Scholl, W., and Davis, R. O. E., 1933, Ammoniation of peat for fertilizers, *Ind. Eng. Chem.* **25**, pp. 1074-1078.

Schulten, H.-R., and Schnitzer, M., 1998, The chemistry of soil organic nitrogen: a review, *Biol Fertil Soils* **26**, pp. 1-15.

Söchtig, H., and Flaig, W., 1973, Herstellung und Wirkung eines organischen Stickstoffdüngers aus Sulfitablaugen der Zellstoffindustrie, *Landwirtschaftliche Forschung,* 30/II. Sonderheft, *Stand und Leistung agrikulturchemischer und agrarbiologischer Forschung,* XXVII, pp. 17-24.

Stevenson, F. J., 1982, *Humus Chemistry - Genesis, Composition, Reactions*, Wiley, New York.

Sulan, S., Krkoska, P., Borisek, R., and Viluda, J., 1975, Oxidative ammonolysis of sulfite lignin, *Pap. Celul.* **30**(3), pp. 68-70 (Czech).

Sulan, S., Viluda, J., and Krkoska, P., 1978, Effect of reducing agents on the oxidative ammonolysis of sulfite spent liquor, *Pap. Celul.* **33**(5), pp. V39-V41 (Czech).

Suvorov, B. V., Rafikov, S. R., and Kagarlitskii, A. D., 1965, The oxidative ammonolysis of organic compounds, *Uspekhi khimii* **34**(9), pp. 1526-1549 (Russ).

Varela, G., Vidrio, E., Delgado, E., Ramos, J., Zuniga, V., Lopez-Dellamary, F., Ramirez, F., Gonzalez, V., Faix, O., and Meier, D., 1999, ^{15}N-Functionalization of kraft lignin and ist characterization by FTIR, Py-GC-MS and NMR, *Proceedings of 10th International Symposium of Wood and Pulping Chemistry, Vol.II* (Yokohama), pp. 178-181.

Wiesner, P., 1971, Organischer Stickstoffdünger N-Lignin, ein wertvolles Depot-Düngemittel, *Wochenblatt für Papierfabrikation* **99**, pp. 740-743.

Zakis, G. F., Neiberte, B. Ya., and Sergeeva, V. N., 1973a, Effect of persulfates on lignin II. Nitrogen fixation by lignin in the oxidation by persulfate in an ammonia medium. *Khim. Drev.* 13, pp. 85-92 (Russ).

Zakis, G. F., and Neiberte, B. Ya., 1973b, Effect of persulfates on lignin III. Identification of the bond type of readily cleavable nitrogen in N-containing oxidized lignin by high-frequency conductometry, *Khim. Drev.* 13, pp. 93-100 (Russ).

Zakis, G. F., Neiberte, B. Ya., Sergeeva, V. N., and Melke, A. A., 1974, Effect of persulfates on lignin V. Characterization of N-containing oxidized lignin, *Khim. Drev.* 15, pp. 115-125 (Russ).

Zakis, G. F., Neiberte, B. Ya., and Sergeeva, V. N., 1976, Oxidation of lignin by an air/ammonia mixture at elevated temperature, *Khim. Drev.* (4), pp. 86-90 (Russ).

Zakis, G. F., and Neiberte, B. Ya., 1978a, Formation of N-containing derivatives by simultaneous action of ammonia and oxidant on lignin, *Khim. Drev.* (6), pp. 3-18 (Russ).

Zakis, G. F., and Neiberte, B. Ya., 1978b, Study of nitrogen-containing polymers obtained while treating lignin and simple phenols with a persulfate-ammonia solution, II. Alkylation of the N atom, *Khim. Drev.* (6), pp. 82-86 (Russ).

Zhuo, S., Wen, Q., Du, L., and Wu, S., 1992, The nitrogen form of non-hydrolyzable residue of humic acid, *Chin. Sci. Bull.* **37**, pp. 508-511.

Zier, N., 1996, Strukturelle Merkmale eines Organosolv-Lignins bei Variation der Aufschlußparameter, *Dissertation Technische Universität Dresden.*

Zier, N., Schiene, R., and Fischer, K., 1997, Structural characterization of an Organosolv lignin by analytical pyrolysis and wet chemical degradation methods, *J. Anal. Appl. Pyrolysis* **40-41**, pp. 525-541.

Zuniga P.V., Martinez, A. de J., Delgado, E., Camacho, A., Gonzalez, V., and Allan, G., 1992, Ammoxidation of lignocellulosic materials in a fluidized bed reactor, *Proceedings of 2nd Workshop of Lignocellulosic and Pulp* (Grenoble), Extended Papers, pp. 235-236.

PULPING CATALYSTS FROM LIGNIN

- THE DIELS-ALDER STEP

Donald R. Dimmel, Joseph J. Bozell, David G. von Oepen, and Michael C. Savidakis*

1. INTRODUCTION

Certain anthraquinone (AQ) compounds can be used in existing kraft processes to increase pulping rates and pulp yields. Such catalysts could also be the focus for an industry-wide evolution of pulping from high to low to sulfur-free methodology. Catalytic pulping holds the promise of addressing problems with plant bleach effluents by producing pulps that are more easily bleached than standard kraft pulps by modifying the cooking process to reduce the amount of residual lignin present in the pulp prior to bleaching.[1]

In 1989 we first reported a new approach for an inexpensive synthesis of dimethylanthraquinone.[2] The approach involves conversion of an inexpensive pulping by-product, lignin, to a useful AQ-type pulping catalyst using a two-stage process. In the first stage, a low molecular weight (LMW) lignin fraction is removed from a whole lignin-containing source by solvent extraction. The LMW lignin is then converted to AQ or AQ-like products in two chemical steps. First, oxidation of the LMW lignin gives a mixture of two key intermediates, monomethoxybenzoquinone (**1**, MMBQ) and dimethoxybenzoquinone (**2**, DMBQ). In the second step, isoprene is added to **1** and **2** to give a mixture of 2,6- and 2,7-dimethylanthraquinone (**3**, DMAQ) and other quinone products (Equation 1). While other dienes can be used in the second step, DMAQ is a more active pulping catalyst than nonmethylated AQ or more fully methylated AQ.[3]

* Donald R. Dimmel, David, G. von Oepen, and Michael C. Savidakis, Institute of Paper Science and Technology, 500 10th Street N.W., Atlanta, GA, 30338; Joseph J. Bozell, National Renewable Energy Laboratory, 1617 Cole Blvd., Golden, CO, 80401.

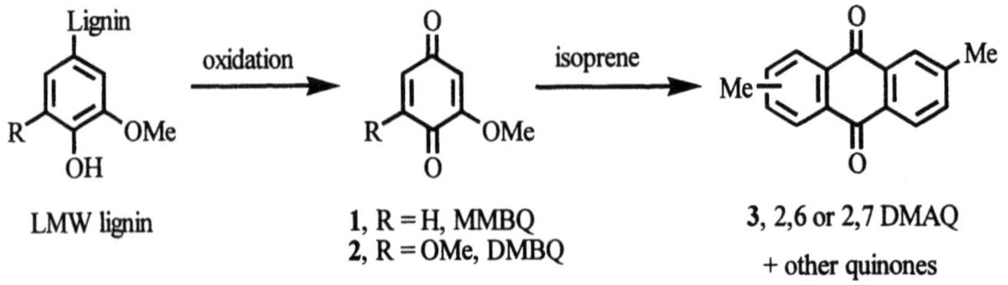

[1]

Much progress has been made on optimizing the initial oxidation step, however, the yield of **1** and **2** is still rather low.[4] Good yields are needed in the Diels-Alder step if the catalyst synthesis is to be commercially feasible. This paper describes our efforts to optimize the Diels-Alder conversion of **1** and **2** to an AQ-based pulping catalyst mixture.

2. BACKGROUND

The reaction between **1** and **2** and a diene can lead to a wide array of products. For this reason, we envisioned that the reaction would give a mixture of AQ, AQ-like materials, and other quinones. This combination of quinones then acts as the pulping catalyst. As the quinone and diene components are varied, description of the various intermediates and final products can become confusing. For example, the initial products of diene addition are nonaromatic and have been categorized as either mono- or bisadducts, depending on how many molecules of diene such as butadiene add to the starting quinone (Figure 1).

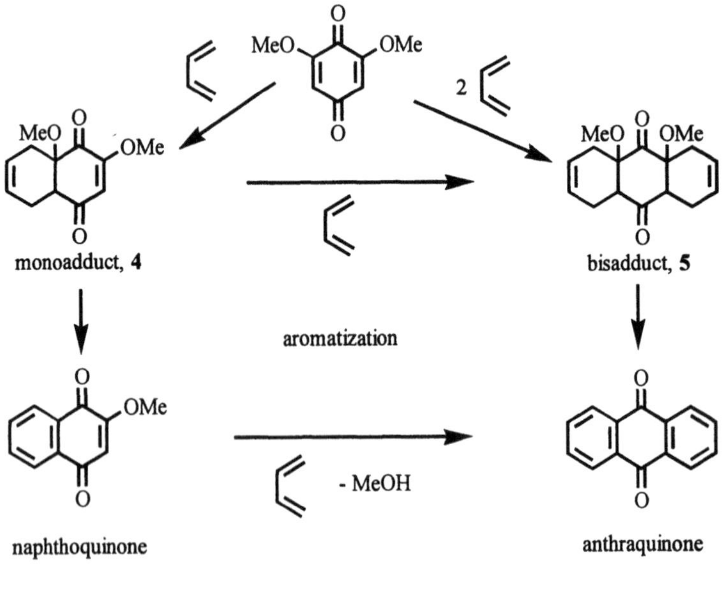

Figure 1. Definition of mono- and bisadducts from Diels-Alder addition.

The most efficient approach for the synthesis of a pulping catalyst would be the simultaneous addition of two molecules of diene such as butadiene to the benzoquinone, initially to give bisadduct **5**, followed by aromatization of **5** to give AQ (Figure 1). Synthesis of anthraquinones by this type of reaction is known.[5]

An alternate path is possible. Diene addition can also take place in a stepwise manner to give monoadduct **4**, followed by addition of a second molecule of diene to give **5**. We felt this path would most likely be successful because of the methoxy substitution on the benzoquinones obtained from lignin oxidation. In general, para-benzoquinones that contain electron-withdrawing substituents readily react with simple dienes to afford the corresponding adducts under mild reaction conditions. Those containing electron donating substituents require more vigorous reaction conditions and usually give monoadducts.[6] The same aromatization processes that convert **5** into AQ would also convert **4** into naphthoquinone derivatives. These derivatives can also undergo Diels-Alder addition to afford AQ (Figure 1).

The catalyst mixture becomes more complex when butadiene is replaced by isoprene. In this case, Diels-Alder reaction/aromatization gives two isomeric DMAQ molecules (Equation 2) via intermediate bisadducts **6**.

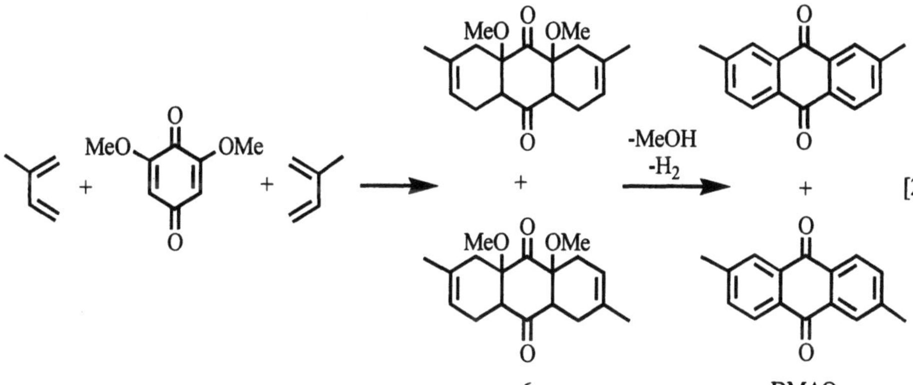

6 DMAQ [2]

Finally, it is also possible for the initial mono and bisadducts to undergo partial aromatization to give compounds such as **7** and **8** (Equation 3).

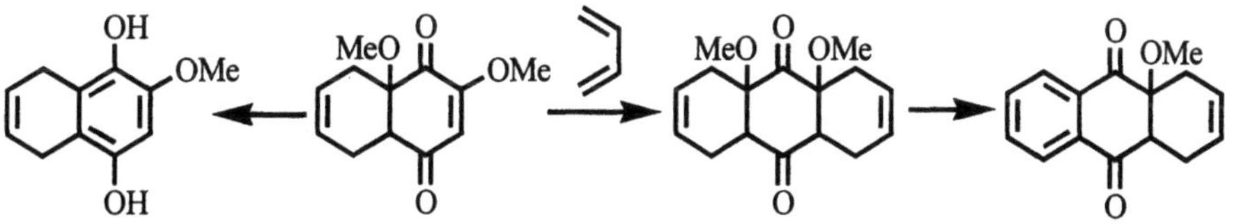

7 **8** [3]

Because of the wide variety of products possible from this process, considerable effort has been devoted to clarify the course of the reaction, identify the cyclization products, and optimize the conditions necessary for the maximum production of AQ-like materials. For simplicity, nonaromatic compounds resulting from the addition of two molecules of any diene to the starting benzoquinone will be referred to generically as bisadducts. Nonaromatic compounds resulting from the addition of a single molecule of diene to the benzoquinone will be referred to as monoadducts. The fully aromatic compounds are generally referred to as AQ or AQ analogs. In most cases described in the following sections, only a single regioisomer will be shown for reactions involving isoprene.

3. DIELS-ALDER REACTIONS IN HOMOGENEOUS SOLUTIONS

Our initial experiments examined the reactivity of benzoquinone (**9**), a reactive quinone expected to readily form bisadducts, as well as MMBQ (**1**) and DMBQ (**2**) with isoprene in toluene solution (Figure 2). Benzoquinone undergoes reaction with isoprene as expected to give a mixture of mono- and bisadducts in excellent yield (98%) with the major product of the reaction being a mixture of bisadduct regioisomers (mono/bisadducts = 30/68).

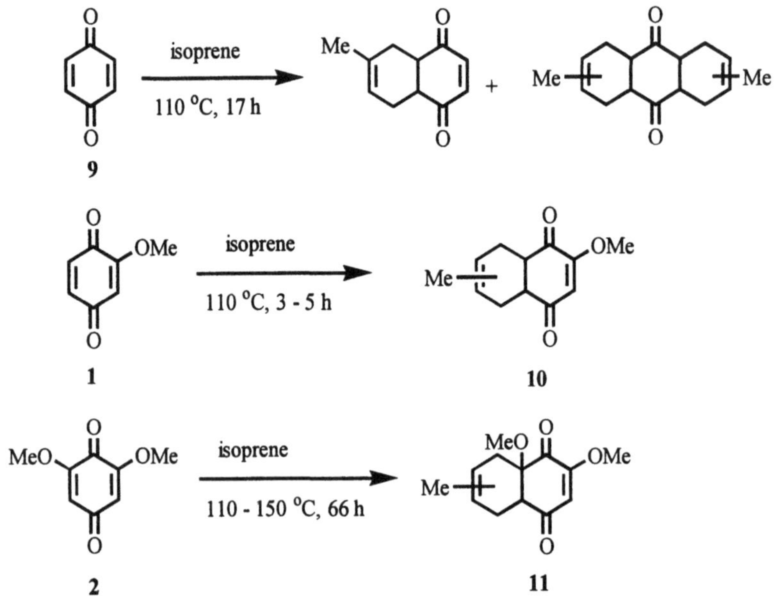

Figure 2. Products observed from the reaction of various quinones with isoprene in homogeneous solutions.

The presence of methoxyl groups on the quinone changed the reactivity. **1** underwent a rapid and high yield reaction with one equivalent of diene to give monoadduct **10** (75% yield). Addition of a second equivalent of diene was not observed at moderate temperatures. **2** also underwent a slower reaction with a single equivalent of diene to give monoadduct **11** in high yield (88%). The DMBQ result was important since it demonstrated that the presence of a methoxyl group on a double bond does not preclude a Diels-Alder addition. Indeed, the result from the reaction of DMBQ suggested that a second addition should be possible since the electronic nature of the remaining olefin should not differ significantly from the first. This initial monoadduct **11** is quite stable, formed in high yield, and could serve as a useful purification point in the lignin-to-catalyst sequence.

We have performed a large number of experiments examining the reaction between DMBQ (**2**) and isoprene. The optimum process treats **2** with excess (about 20x) isoprene in MeCN solvent in a sealed bomb at 140-150 °C. The monoadduct was obtained in 90% yield or greater as a roughly 1:1 mixture of regioisomers **11** in 2.5 hours (Equation 4). ^{1}H NMR and GC/MS confirmed the product structure. The presence of regioisomers is difficult to discern in the ^{1}H NMR; however, GC/MS reveals two closely spaced peaks with almost identical MS spectra and molecular weight.

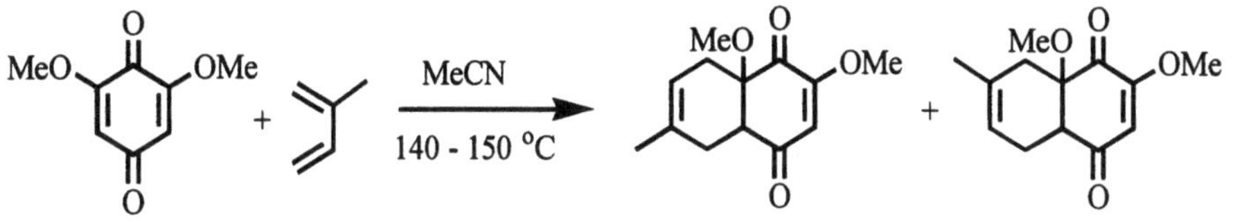

2 **11** [4]

We also investigated the effect of differing amounts of isoprene on the reaction. Decreasing the amount of isoprene caused a decrease in the rate of reaction. In the presence of a 4.2 to 5-fold excess of isoprene, the conversion of **2** after 2.5 h dropped to 70%. At a 1.9-fold excess of isoprene, the conversion of **2** was 47% after 2.5 h. However, the conversion can be increased to > 95% by allowing the reaction to proceed for 17 h at a 5:1 molar ratio of isoprene/DMBQ. We have performed this reaction on a large scale to prepare 6-7 g of the material. It is preferable to run the reaction for a longer period of time rather than at a higher isoprene concentration to minimize the formation of the Diels-Alder dimer of isoprene (Equation 5), which can be a significant side reaction that results in nonproductive consumption of the diene.

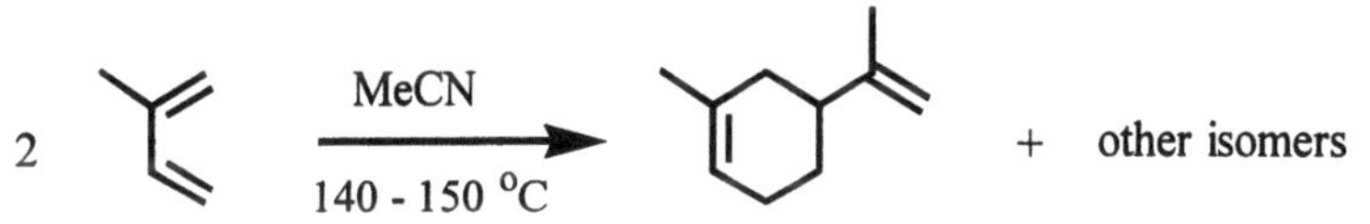

[5]

The monoadducts from MMBQ and DMBQ failed to undergo further reaction with the diene to give a bisadduct of any kind. Since addition of the first molecule of isoprene to DMBQ occurred rapidly and in high yield, we consider this to be primarily a steric problem. We have observed that unsubstituted benzoquinone (**9**) exhibits a distinctly different rate of addition of the diene. The first addition to give a monoadduct was complete in 2-4 hours while addition of the second required heating overnight. Consideration of the three-dimensional structure of the adducts provides a possible explanation. Figure 3 shows the predicted transition state conformation for two modes of addition of a diene to DMBQ.

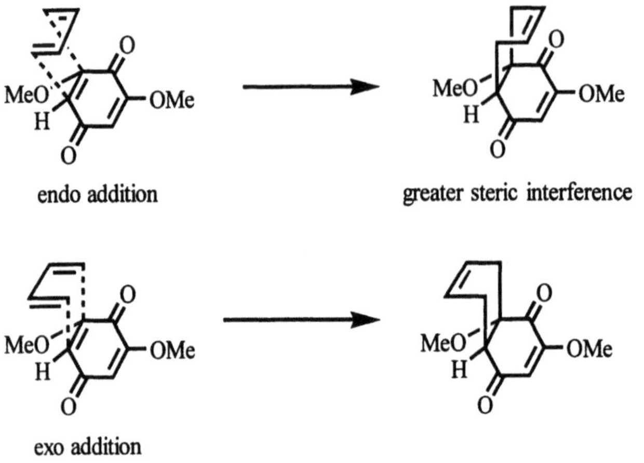

Figure 3. Stereochemistry of the transition state of the Diels-Alder addition.

The incoming diene has the choice of adding in either an endo or exo mode. From a steric standpoint, the exo-mode is preferred since there is less crowding. However, a number of reported examples exist in which endo-addition is observed, despite the greater steric hindrance. This observation has been attributed to a number of factors, including secondary orbital interactions in the transition state, and has been formalized as the Alder rule.[7] Certainly, once the first addition is complete, the resulting product can undergo a ring flip to the more stable exo-conformation. However, some fraction of the mixture will continue to exist in the endo-conformation. The second equivalent of diene will also attempt to adopt an endo-conformation in the transition state. There will be higher steric interference to the second addition regardless of the conformation of the monoadduct. This hypothesis is supported by simple molecular mechanics calculations on the monoadduct obtained from either MMBQ or DMBQ and isoprene (Figure 4). The models (hydrogens omitted) clearly show the bent nature of the monoadducts suggesting steric interference toward introduction of a second molecule of the diene.

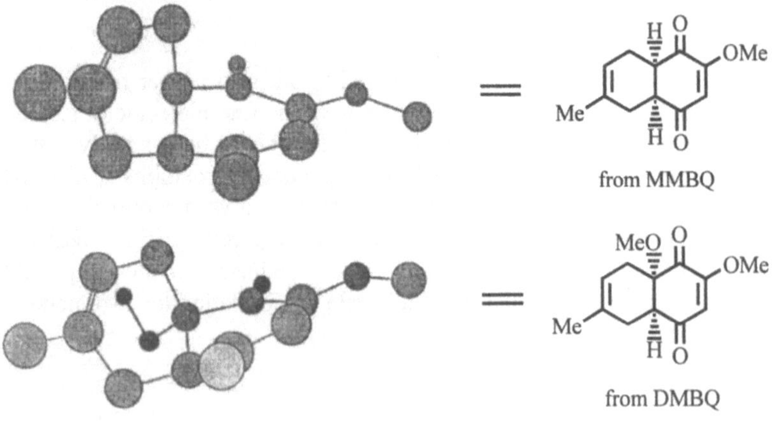

Figure 4. Molecular modeling of the monoadducts from MMBQ and DMBQ, respectively.

4. DIELS-ALDER REACTIONS IN THE PRESENCE OF LEWIS ACID PROMOTERS

If the first addition of diene gives a sterically hindered intermediate product, removal of the hindrance should make the second addition easier. A simple method of removing the hindrance is aromatization of the newly added ring to give methoxynaphthoquinone (**12**) (Figure 5). This transformation would require elimination of MeOH. The intermediate triene would be expected to undergo rapid oxidation to the quinone.

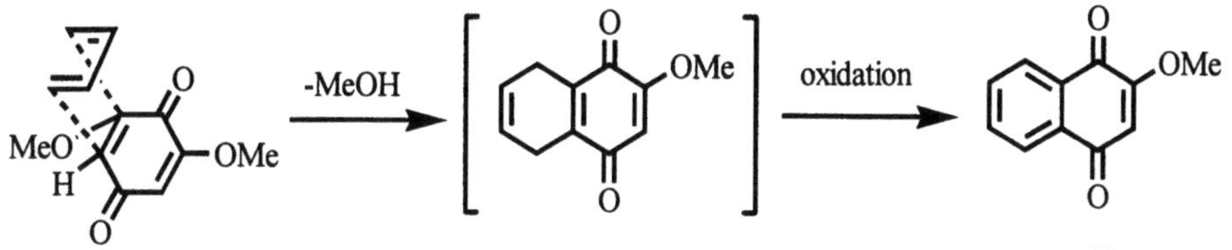

12

Figure 5. Removal of steric hindrance by aromatization.

We envisioned that addition of a Lewis acid could promote the initial loss of MeOH from the monoadduct. In addition, the presence of a Lewis acid could also have a promoting effect on the initial Diels-Alder reaction. The literature reports many examples of the catalysis of Diels-Alder reactions in the presence of Lewis acids.[8] From a process standpoint, an added promoter should be readily available and inexpensive. Silica gel or alumina fulfills this requirement and the use of these materials as Diels-Alder promoters is widely reported.[9] We felt that a one-pot Diels-Alder addition and aromatization could be carried out *in situ* by adding either promoter to the diene / quinone mixture. Since the diene is still present, a second Diels-Alder/aromatization sequence should result, eventually leading to an AQ product (Figure 6).

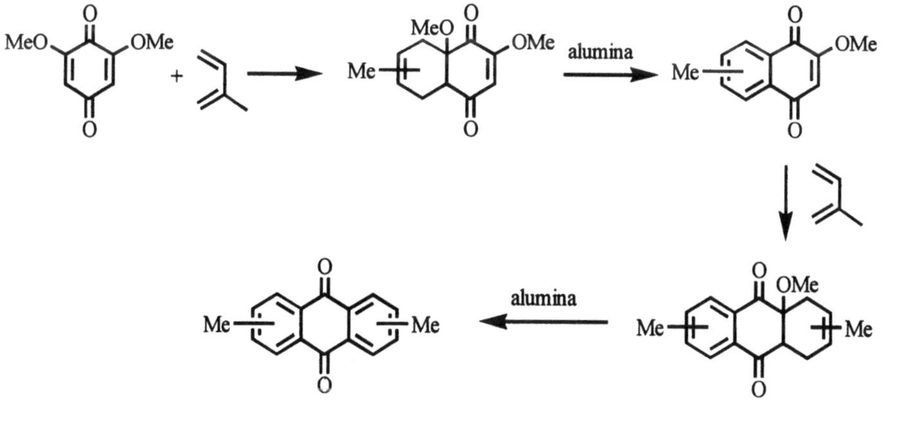

Figure 6. Envisaged one-pot reaction of DMBQ with isoprene in the presence of alumina.

Initial reactions with both alumina and silica gel were promising. Promotion of the Diels-Alder reaction was observed upon mixing of MMBQ and isoprene in the presence of silica gel for 4 hours at room temperature. A mixture of two aromatized Diels-Alder adducts, **13** and **14**, was obtained as major products (Equation 6).

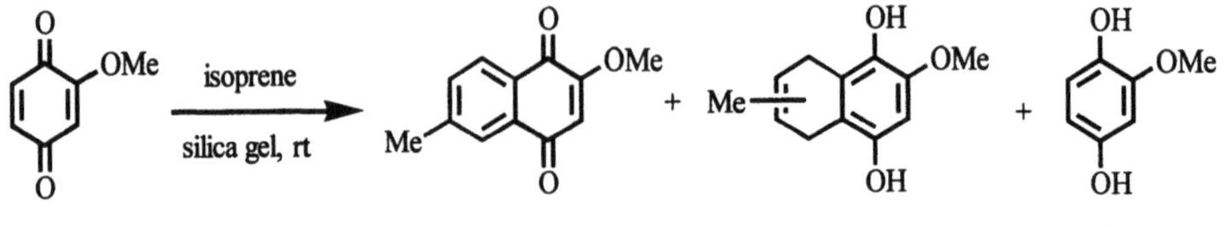

[6]

The diene added to the nonmethoxylated double bond. Although an exact determination of yield was not made for this reaction, the results indicated that sequential Diels-Alder cyclization and aromatization of the intermediate was possible, although the appearance of hydroquinone **14** and a small amount of reduced starting material was unexpected. The reaction was repeated with the less reactive quinone, DMBQ, and a 20% yield of the corresponding naphthoquinone was achieved upon heating at 100 °C, reflecting the lower reactivity of the double bonds in DMBQ toward Diels-Alder addition. Importantly, GC/MS indicated the presence of a small amount of pure DMAQ in the mixture. We also found that the monoadduct isolated from the Diels-Alder reaction of DMBQ and butadiene underwent a facile aromatization when heated in the presence of alumina (Equation 7).

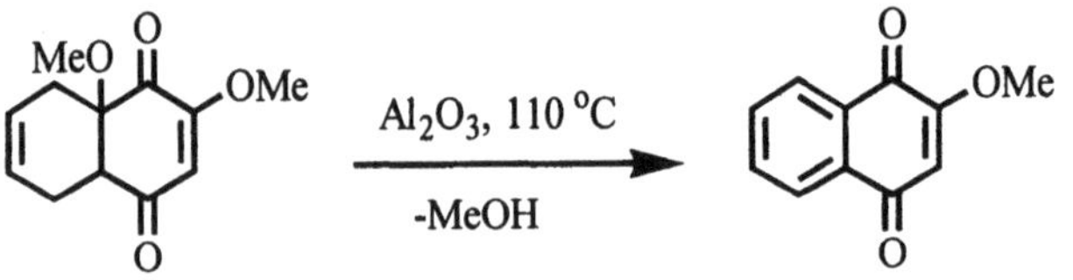

[7]

These results, achieved under very mild conditions, prompted an extensive investigation into the optimization of the reaction.

4.1. Reactions on Alumina

A large number of reactions were performed using DMBQ and isoprene as the Diels-Alder reactants, and various types of alumina as the Lewis acid promoter, in the absence of solvent. The reactions were performed by mixing solid DMBQ and the promoter, adding excess diene, and heating the reaction in a sealed stainless steel vessel at 125 °C for 14-18 h. The products were isolated by washing the solid support with solvent and analyzing by gas chromatography.

The reactions gave a complex mixture of quinone adducts (see Experimental section for a description of the GC/MS analysis); however, the majority of the reaction mixture contained four primary products: starting DMBQ, dimethoxyhydroquinone (**15**), and DMAQ regioisomers (**3**).

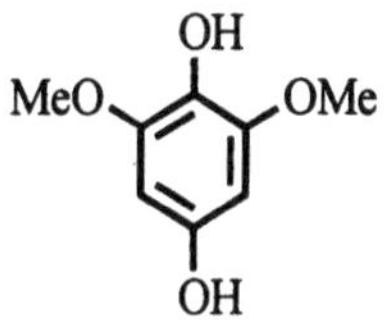

15

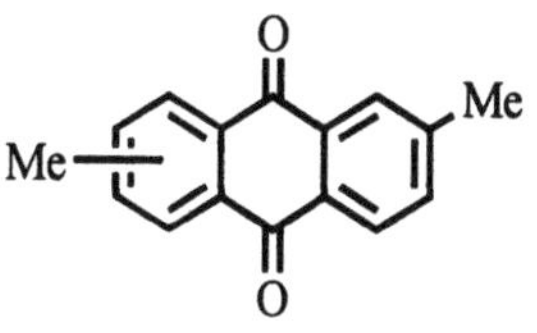

3

The effects of alumina type (basic, neutral, or acidic), alumina activity (Brockmann levels I through V, prepared by addition of varying amounts of water to the solid alumina), alumina amounts (between 250-1000 mg/mmole DMBQ), isoprene amounts (2-20 mmol/mmol DMBQ), and the presence of small amounts of solvent (toluene, acetonitrile, chloroform) were investigated. The large amount of data was also subjected to principal component analysis in order to deconvolute the information and discern correlations.[10] Some general trends were noted:

- Pure DMAQ (**3**) can be prepared using this methodology in chemical yields of 30-40%.
- DMAQ can be prepared directly from DMBQ and isoprene without the intermediacy of any AQ-like materials (i.e., naphthoquinones), which is important from the standpoint of streamling the process.
- Acidic alumina appears to be most effective for promoting the diene addition. The most effective activity level for the alumina is less clear. Brockmann level IV appears to be the best.
- The optimum amount of alumina required (mg/mmol of quinone) is not clear. The product yields varied widely as a function of this parameter.
- The reaction proceeds most efficiently in the presence of a tenfold excess of isoprene. This observation could be a drawback because of the increased possibility of producing isoprene dimer.

4.2. Reactions on Silica Gel

In the presence of silica gel, the major products were generally DMAQ, hydroquinone **14** (Equation 6), dimethylanthrahydroquinone (**16**), and hydroquinone **17**. The presence of hydroquinones has additionally been established by treating an organic solution of the reaction mixture with aqueous KOH. A colored fraction is extracted into the aqueous layer, consistent with the presence of hydroquinones.

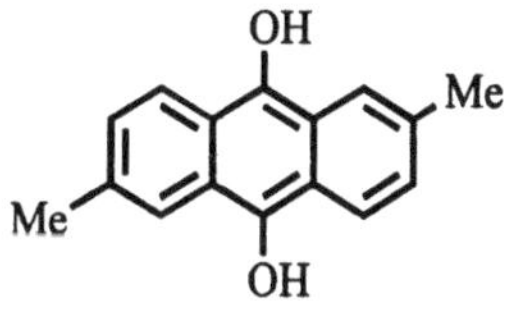

16

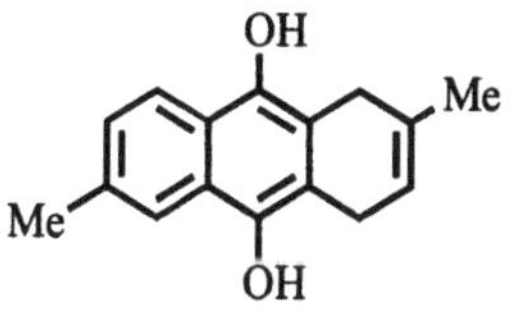

17

The relative proportion of AQ-like materials is about the same as that obtained using alumina as the promoter, but the amount of isoprene dimer is diminished, offering an advantage for the use of silica gel.

Several additional trends were noted from reactions performed in the presence of silica gel. When a large (500 mg/mmole substrate) amount of promoter was used, the difference in reactivity between alumina and silica gel appeared to be minimal. However, reactions with limited (50 mg/mmole substrate) amounts of promoter showed much clearer distinctions. At this level, reactions with silica gel gave only monoadduct, as observed by GC/MS and ^{1}H NMR. Since DMBQ also underwent this reaction in the absence of silica gel, we concluded that the presence of silica gel is ineffective for the further aromatization of the product at these low levels. In contrast, added alumina (basic or acidic, 50 mg/mmole substrate) clearly promoted further reaction of the first-formed adduct to give intermediate **18** and/or the aromatization product via loss of MeOH. Alumina appeared to be a much more active catalyst under these conditions (Equation 8).

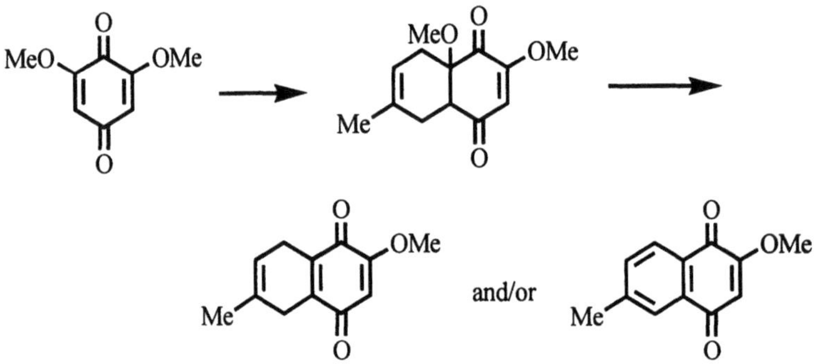

18

[8]

We have also investigated the possibility of catalysis of the reaction by trace metals deposited on the walls of the stainless steel reactor. A Teflon lined reactor was prepared and parallel reactions were performed. We observed minimal difference in the reactions performed in the lined vs. the unlined reactor. At this point, we concluded that the reactor was not promoting the transformation. The silica gel promoter used in the reaction can be recycled into subsequent reactions. After three uses, silica gel still effectively promoted the formation of aromatized Diels-Alder monoadducts, e. g., **13**, but the relative amount of bisadducts dropped considerably.

4.3. Stepwise Reaction Processes

Because the normal conditions employed required the use of a large excess of isoprene, we normally observed significant amounts of isoprene dimer in the reaction product mixture. This product represented a nonproductive consumption of the starting material that would be economically detrimental to the overall process. Accordingly, we examined methods to diminish the amount of this product. Our approach was based on the very efficient reaction observed between DMBQ and isoprene in MeCN in the absence of a promoter (Equation 4). The resulting monoadduct was isolated and then subjected to a second Diels-Alder reaction with isoprene in the presence of silica gel and in MeCN solvent. Under these conditions, the amount of isoprene present was reduced to a 5- to 8-fold excess, with a concomitant reduction (but not elimination) of the isoprene dimer. Importantly, this approach offered cleaner reaction mixtures and a much higher proportion of DMAQ than previous approaches. The primary products in the mixture as determined by GC/MS or ^{1}H NMR were DMAQ (**3**), **13**, hydroquinone **14** and **18** (more likely present as its tautomer, **19**).

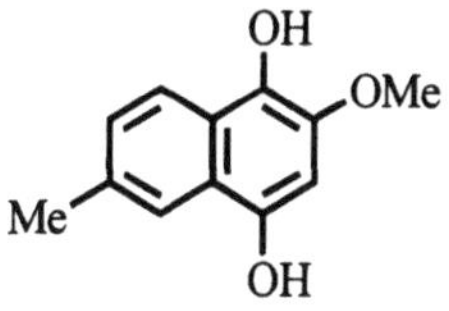

19

The same reaction sequence was also carried out using MMBQ as a starting material. However, formation of the corresponding AQ derivative was not observed. Instead, the major product of a stepwise process was hydroquinone **19**.

4.4. Discussion

The most general result of the Diels-Alder investigation is that DMBQ can be converted to a mixture that contains a significant fraction of AQ-like materials. However, the reaction also produces a number of side products. Some could serve as catalysts, while others may be unreactive. Ideally, conditions should be such that AQ-like materials are the major reaction products. We have investigated several mechanistic aspects of the reaction in order to better understand the process and design experiments to increase the relative amounts of AQ-like materials. Two important observations can be made that have a bearing on the reaction path. First, Diels-Alder reactions proceed to AQ-like materials much more completely with DMBQ than with MMBQ in the presence of promoters. The rationale for this observation is based on the transformations of each quinone during the Diels-Alder step. The formation of the initial monoadduct from either MMBQ or DMBQ is efficient and proceeds in high yield. Neither monoadduct proceeds to AQ-like materials in the absence of Lewis acid promoters.

However, even in the presence of promoters, only DMBQ can be easily converted to AQ-like materials. The primary difference between the two substrates is the ability of the monoadduct from DMBQ to lose a molecule of MeOH during the reaction to give compound **18** with considerably less steric hindrance to further reaction (Equation 8). The monoadduct (**10**) from MMBQ does not have this ability because it imposes more steric hindrance to the approach of a second molecule of the diene due to the bending of the first ring above the plane of the quinone. Simple molecular mechanics calculations support this hypothesis (Figure 7). A more efficient, second addition of the diene is possible from compound **18**, giving higher levels of AQ-type products.

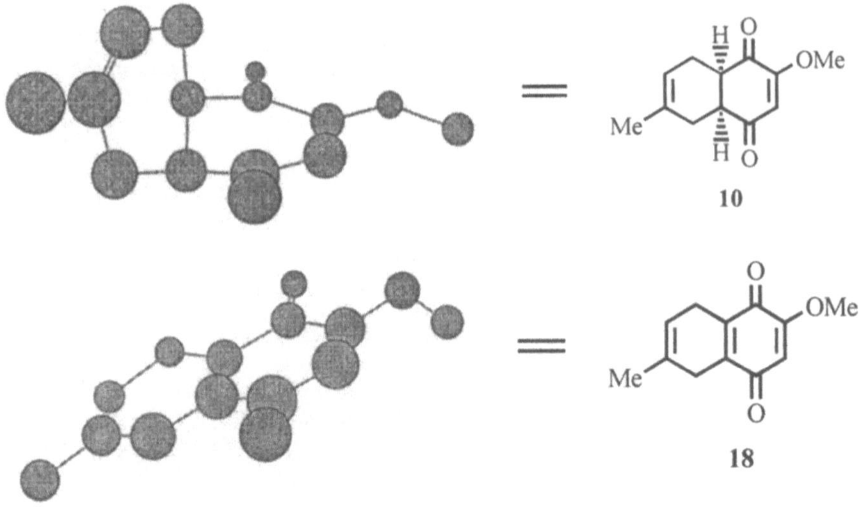

Figure 7. Results of molecular mechanics calculation on the monoadduct (**10**) from MMBQ and isoprene, and on compound **18** derived from DMBQ and isoprene.

A second important mechanistic observation is the presence of hydroquinone derivatives in all of the Diels-Alder reaction mixtures. This is not a problem if one obtains only AQ and anthrahydroquinone (the reduced version of AQ), since both are reported to function as pulping catalysts. However, formation of hydroquinones from any of the starting materials or reaction intermediates is a problem, since conditions do not exist to convert them back into proper Diels-Alder intermediates. If one examines Equation 8, formation of **18** is reasonable by MeOH elimination, promoted by alumina or silica gel. However, formation of the methoxynaphthoquinone (**13**) is more difficult to explain. This product is formed by a dehydrogenation process and requires a hydrogen acceptor in the reaction medium. A mechanistic scheme that appears to account for all the intermediates observed in the reaction is shown in Figure 8.

DMBQ undergoes reaction with isoprene even in the absence of promoter to give the stable and isolable monoadduct **11**. When a promoter is present, **11** is induced to lose MeOH, giving quinone **18**. Our original hypothesis assumed that **18** was the intermediate that was subsequently converted to AQ-like materials by virtue of its flat topology and similar olefin electronics. However, we were never able to observe high concentrations of **18**, even at short reaction times. The presence of hydroquinone derivatives in the reaction mixture suggested an alternate path for **18** was available. We now believe that **18** undergoes two different but rapid conversions. The first is a simple tautomerization, giving hydroquinone **19**. The second, which leads to an important AQ precursor, is a disproportionation, giving a second hydroquinone **14** and methoxymethylnaphthoquinone (MMNQ) **13**. **13** undergoes further Diels-Alder reaction to bisadduct **20**. This material is structurally analogous to **11** and undergoes loss of MeOH to give **21**. This compound, structurally analogous to **18**, can undergo the same pair of tautpmerization/ disproportionation sequences, completing identification of the product slate observed (Figure 8).

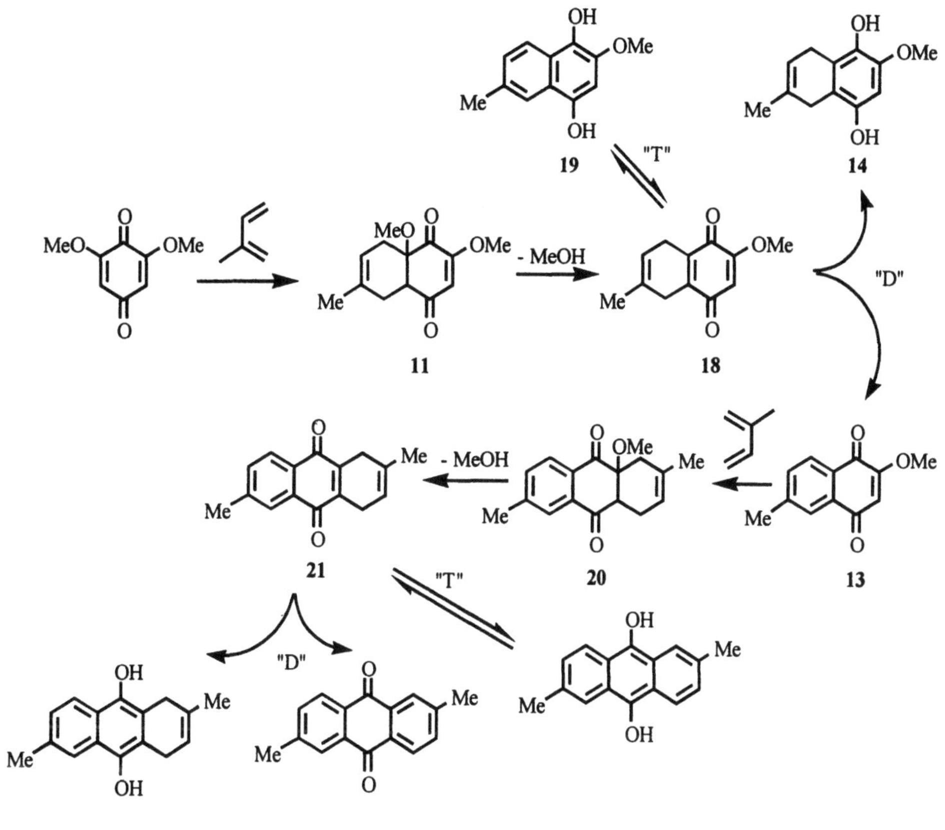

Figure 8. Summary of reaction paths available to DMBQ and derivatives in the Diels-Alder sequence, "T" = tautomerization and "D" = disproportionation.

This hypothesis is supported by an experiment designed to form **18** exclusively (Equation 9).

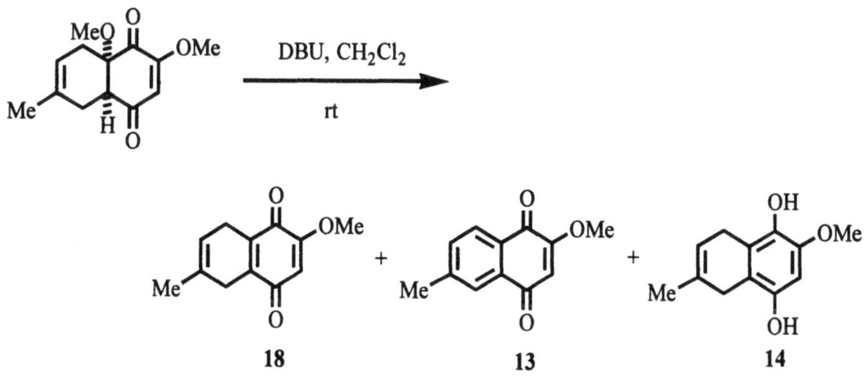

[9]

Treatment of the Diels-Alder monoadduct with DBU (1,8-diazabicyclo[5.4.0] undec-7-ene), a strong, hindered, non-nucleophilic base, caused a rapid (< 5 minutes) consumption of the starting material. ^{1}H NMR spectroscopy showed the presence of considerable amounts of **13**. GC results also indicated that the triene **18** was a minor component in this reaction. In addition, the NMR sample for this reaction deposited an off-white precipitate after standing for a few days. This precipitate was identified as hydroquinone **14**, a compound seen in several other reactions. The disproportionation product is suggested to occur as shown in Figure 9. The initial elimination product undergoes initial base promoted elimination of MeOH to give the expected product (**18**). However, reaction of this product with base induces a hydride transfer to a second elimination product (**18**), giving MMNQ (**13**) and the cyclohexadienone shown. This product can undergo tautomerization followed by proton loss to give the hydroquinone (**14**).

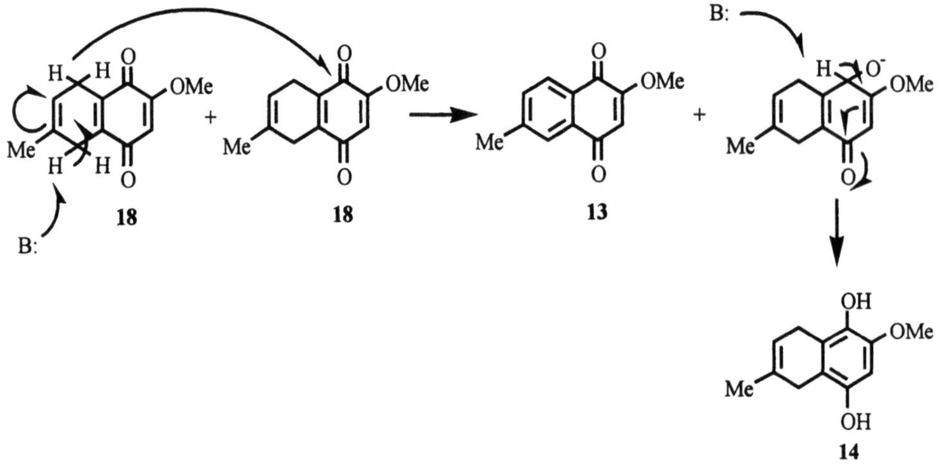

Figure 9. Formation of compounds **13** and **14** via disproportionation of **18**.

5. DIELS-ALDER REACTIONS IN AQUEOUS SOLUTIONS

Changes in solvent can often have dramatic effects on the course of Diels-Alder reactions. Breslow and others have shown that the rate of Diels-Alder reactions between hydrophobic reactants can be greatly enhanced by carrying out these reactions in aqueous solvents.[11] It has been postulated that this rate enhancement occurs due to entropy effects, i.e., micellar catalysis. The Diels-Alder reaction of isoprene and DMBQ was examined in water and the results were compared to those obtained from several other solvents (Table 1).

Water proved to be the most effective solvent for directly providing a good yield of bisadducts and DMAQ. Analysis of the crude reaction mixture obtained from the aqueous DMBQ Diels-Alder reaction by GC/MS allowed for the tentative identification of several of the components (see Experimental section). Several attempts to isolate specific fractions of nonaromatic bisadducts failed; several components of the mixture are not stable to chromatography and were aromatized to DMAQ. The ^{1}H NMR spectra of the crude reaction products agree with proposed product compositions. As the polarity of the reaction mixture decreased, the quantity of bisadducts and DMAQ in the product mixture also decreased. In many cases, the major product was MMNQ. The modest yields of these products from reactions carried out in methanol and water suggested that increasing the molar ratio of isoprene to DMBQ should result in an increase in the amount of DMAQ formed; however, this was not observed.

Table 1. Diels-Alder reactions of DMBQ and isoprene in various solvents[a]

Solvent	Isoprene/DMBQ molar ratio	DMBQ	MMNQ	Mono-adducts	Bis-adducts	DMAQ
Neat	3:1	31	2	0	0	0.5
Hexane	3:1	16	8	0	0	0
Ethyl ether	3:1	20	9	0	0	0
Ethyl acetate	3:1	15	11	0	0	0
Acetone	3:1	24	9	0	0	0
Isopropanol	3:1	4	17	10	0.9	0.4
Butanol	3:1	0	9	11	2	0
Acetic acid	3:1	5	8	6	12	0
Methanol	3:1	17	28	0	0	14
Methanol	6:1	3	28	8	3	18
Methanol	10:1	1	29	16	7	8
Water	3:1	0	4	0	11	22
Water	6:1	0	2	9	20	17
Water	10:1	0	0	0	18	21

[a] Reactions were carried out in steel bombs (4.0 mL capacity) at 160 °C for 16-17 h.

5.1. Optimization of Aqueous Diels-Alder Reactions

The preliminary aqueous Diels-Alder reactions gave a mixture of the nonaromatized bisadduct and DMAQ in 30-40% yield. If the modest yield was a result of micellar catalysis of the cycloaddition, then more efficient micelle formation should result in an increase in the amount of bisadducts and DMAQ. To test this assertion, we performed the aqueous Diels-Alder reaction while applying ultrasonic agitation (Table 2). Good yields of monoadducts and MMNQ were observed at moderate temperature; the yields were much higher than with simple stirring at these temperatures. However, the maximum conversion to the bisadduct was only 5% at 90 °C and in no cases was any AQ formed. High temperatures appear to be required for the formation of bisadducts in water. Unfortunately, the ultrasonic equipment needed to carry out these high temperature and high pressure experiments was not available. The implementation costs for such a process on an industrial scale would probably be substantial, so no further investigation into this type of process was undertaken. Rideout and Breslow have shown that the addition of lithium chloride (LiCl) to aqueous Diels-Alder reaction of cyclopentadiene and methyl vinyl ketone causes a dramatic increase in the rate of cycloaddition.[12] They attribute this effect to the creation of a more polar reaction medium which causes the hydrophobic micelles to become even smaller. The reduced volume has a net effect of reducing the entropy change in going from two molecules to one, thus lowering the activation energy and increasing the rate of reaction. For this reason, we chose to examine the effects of LiCl on the cycloaddition of isoprene to DMBQ (Table 3). The presence of LiCl caused the yield of DMAQ to decrease by a factor of 2. A plausible explanation for the decrease is that the salt causes a reduction in the solubility of DMBQ in the reaction medium, thereby reducing the number of hydrophobic micelles that contained both reactants.

Table 2. Diels-Alder reactions of DMBQ and isoprene using ultrasonic mixing[a]

Temperature (°C)	Time (h)	DMBQ	MMNQ	Mono-adducts	Bis-adducts	DMAQ
22	24	54	0	0.3	0	0
70	23	25	0.4	65	0	0
90	29	0.8	21	66	5	0
90[b]	25	13	6	53	0	0

[a] All product amounts in %.
[b] Performed without ultrasonic mixing.

Table 3. Aqueous Diels-Alder reactions of DMBQ and isoprene in the presence of lithium chloride[a]

Trial	Solvent	DMAQ (%)
1	H_2O	24
2	H_2O	28
3	H_2O	33
4	4.86 M aqueous LiCl	14
5	4.86 M aqueous LiCl	15

[a] Reactions carried out with 5 equivalents of isoprene at 185 °C for 24 hours.

As a possible way to increase the yield of DMAQ in the aqueous Diels-Alder reaction, we examined the effects of the addition of oxidants to the DMBQ/isoprene mixture. The goal was to convert more of the monoadducts to aromatic naphthoquinones, which are key components in generating bisadducts. In general, the addition of polyoxometalates and other oxidants led to lower total combined yields of DMAQ and nonaromatic adducts. There were a few exceptions, such as V 205; however, the small yield improvements probably would not offset the cost of the reagent. Several aqueous Diels-Alder reactions were also conducted in the presence of metal salts, but the yields of DMAQ did not improve. At this point we focused our efforts on the optimization of the high temperature aqueous Diels-Alder reaction (Table 4). In general, increasing the molar ratio of isoprene to benzoquinone and the reaction time causes an increase in the combined yield of bisadducts and DMAQ; however, changing the reaction temperature has the most dramatic effect on the yield. Low temperatures (146 °C) required long reaction times (75 hr) to obtain significant product yields; high temperatures (180 °C) gave only DMAQ. Our optimum reaction conditions, which happened to correspond to a larger scale reaction, gave the desired pulping catalyst mixture in 57% yield. While the aqueous experiments listed in Table 4 gave modestly good DMAQ yields with large levels of isoprene, subsequent larger scale experiments often provided only ~ 30% DMAQ yield. The yield rose to ~ 40% after chromatography, apparently the result of reaction component aromatization. The observed large differences in DMAQ yields through the course of our aqueous Diels-Alder reaction study indicated that there may be subtleties that we do not understand.

Table 4. Optimized aqueous Diels-Alder reactions[a]

Isoprene/DMBQ molar ratio	Temperature (°C)	Time (h)	Bisadducts	DMAQ	Total
7:1	146	75	6	28	34
8:1	146	75	12	22	34
9:1	146	75	20	22	42
3:1	160	16	11	22	33
6:1	160	17	20	17	37
10:1	160	17	18	21	39
10:1	167[b]	8	19	38	57
6:1	170	23	26	22	48
8:1	170	2	4	20	24
8:1	170	17	15	30	45
8:1	170	20	26	23	49
8:1	170	41	21	21	42
5:1	180	24	0	39	39
5:1	185	24	0	28	28

[a] All product yields in %.
[b] Reactions employed more DMBQ (6.0 mmol) in a larger (60 mL) vessel heated for 8 h then cooled to 60 °C over 16 h.

6. DIELS-ALDER REACTION WITH HIGH BISADDUCT YIELD

In subsequent studies we have established that the reaction of unsubstituted benzoquinone with isoprene in ethanol solvent provides the bisadduct shown in Figure 2 in 85% isolated yield.[5] We have shown that this compound is an excellent pulping catalyst and a potential candidate for commercialization.[13]

7. EXPERIMENTAL

7.1. Methods and Materials

Gas chromatography was carried out on a Hewlett-Packard 5890 chromatograph with a flame ionization detector, using one of two methods: 1) 10 m 50% crosslinked phenol methyl silicone gum phase HP-17 macrobore capillary column; helium carrier gas flow was 8 mL/min, injection temperature 250 °C, detector temperature was 280 °C, with a temperature program of 150 °C for 1 min, followed by ramping at 5 °C/min to 275 °C, and holding at 275 °C for 5 min; 2) 0.5 mm x 30 m DB-l column, injection temperature 250 °C, detector temperature was 280 °C, ramping at 5 °C/min between 50-250 °C, and holding at 250 °C until all products had come off the column. Quantitative analysis of the components is difficult because few of the Diels-Alder products are commercially available to determine response factors. Therefore, all bisadducts were assumed to have a response factor equal to that of the DMAQ internal standard. All monoadducts were assumed to have a response factor equal to the MMNQ internal standard. Standard mixtures of the two internal standards gave very good correlations. GC/MS was carried out on a 5890 Hewlett-Packard gas chromatograph using a 35 m DB-5 column and were measured at 70 eV.

7.2. Product Analysis and Identification for Diels-Alder Reactions in Homogeneous Solution and in the Presence of Lewis Acid Promoters

A great majority of the products in the Diels-Alder reaction mixture have been identified by a combination of ^{1}H NMR and GC/MS. GC proved valuable for the rapid monitoring and analysis of the reaction mixtures obtained from both promoted and unpromoted reactions. Table 5 summarizes the structures, retention times, and molecular weights of the products observed in these reactions. Two different instruments were used to measure the GC and the GC/MS. Because of differences in column length and carrier flow rate, the R_t for the two instruments differed slightly. For the signal at R_t 6.7 minutes, the GC/MS revealed a clear molecular ion at m/ e 136. The overall spectrum was very similar to signals at R_t 7.49 and 9.35 minutes. However, the signal at 6.7 minutes was a minor product in the spectrum and its exact structure is not yet known. In general, the assignment of specific structures to the isoprene dimers was based only on presumed steric restrictions that could exist for the Diels-Alder reaction. In some cases, two structures were assigned. The GC/MS with its longer column was sometimes able to resolve the peak into two distinct components. The compound at R_t 28.7 minutes was probably present as a mixture of regioisomers; however, GC/MS was not able to resolve the peak.

7.3. Product Analysis and Identification for Diels-Alder Reactions in Aqueous Medium

GC/MS analysis of crude reaction mixture formed in the aqueous Diels-Alder reaction of DMBQ showed the presence of many of the same products seen in nonaqueous reactions. In addition, several nonaromatic bisadducts were observed, but could not be isolated since they were not stable to chromatography. Some GC/MS assignments were confirmed by ^{1}H NMR. The product mixture contained three naphthoquinones (**12**, **13**, and **19**) and a gamut of AQ-related components, ranging from the expected simple bisadduct **6** to DMAQ. Molecular ions were observed for components corresponding to the structures shown in Figure 10; however, none of these structures were conclusively characterized.

7.4. General Procedure for Diels-Alder Reactions in Homogeneous Solution

A small stainless steel reaction vessel was charged with the starting quinone, isoprene (1.5 - 2.0 x molar excess), and solvent. The mixture was heated at the reaction temperature overnight. Solvent removal and chromatography gave the Diels-Alder products.

Table 5. GCMS results and structural assignments

GC R_t (min)	GCMS R_t (min)	Structure	GCMS MW
6.70	10.15	Isoprene dimer (?)	136
7.49	11.08	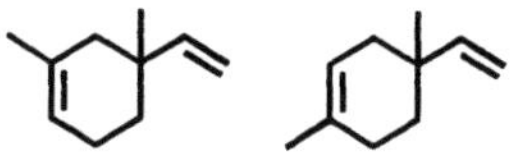	136
9.35	11.08	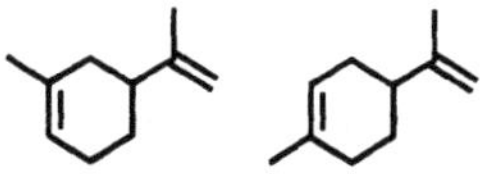	136
10.6		Isoprene dimer (?)	136
22.04	27.78	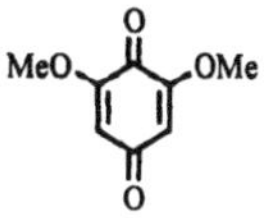	168
28.7	36.4	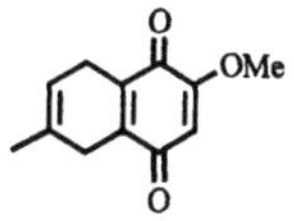	204
29.04	36.6	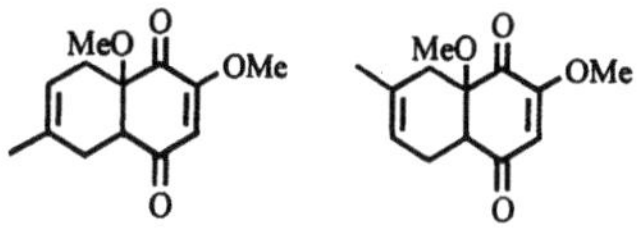	236
29.4	37.5	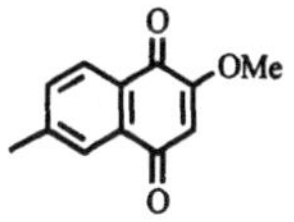	202
31.9	39.7	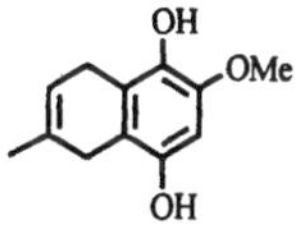	206
33.8	41.9	Unidentified	244
34.5	43.6	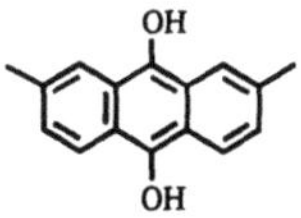	238
35.98	45.95	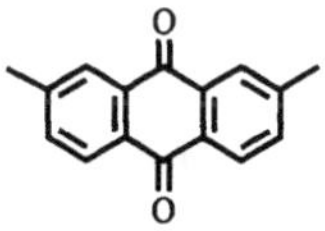	236
40.34	51.1	Unidentified	242
41.03	52.5	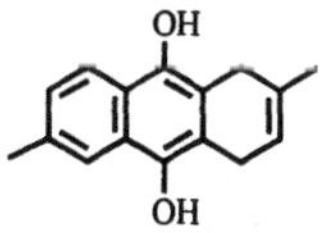	240

7.5. General Procedure for Diels-Alder Reactions in the Presence of Lewis Acid Promoters

The starting quinone was mixed with solid silica gel or alumina. The solid mixture was placed in a small stainless steel reaction vessel. Isoprene was added and the mixture was heated at the reaction temperature overnight. Solvent removal and chromatography gave the Diels-Alder products.

7.6. General Procedure for Diels-Alder Reactions in Aqueous Medium

The starting quinone was mixed with 3.5 mL of deionized water in a pressure vessel to give a 0.043 M solution.[14] Several equivalents of isoprene were added and the vessel was sealed and heated with shaking in a sand bath at 160 °C. After the reaction was completed, the vessel was cooled and the reaction mixture extracted with a 3:1 mixture of chloroform and ethanol. The isolated products were analyzed by GC/MS using an internal standard.

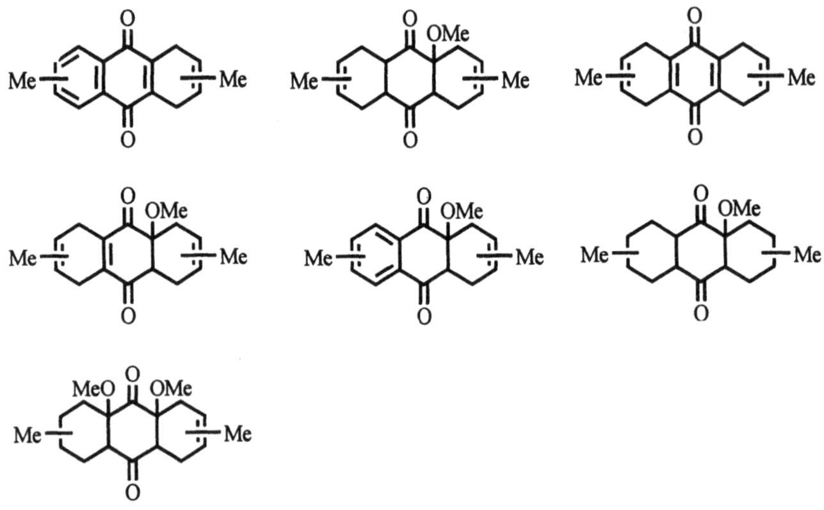

Figure 10. Possible structures for other compounds in product mixtures from aqueous Diels-Alder reactions.

8. REFERENCES

1. J. C. Wozniak, D. R. Dimmel, and E. W. Malcolm, The Generation of quinones from lignin and lignin related compounds, *J. Wood Chem. Technol.* **9**, 491-511 (1989); J. C. Wozniak, D. R. Dimmel, and E. W. Malcolm, Diels-Alder reactions of lignin-derived quinones, *J. Wood Chem. Technol.* **9**, 513-534 (1989).
2. D. R. Dimmel and J. J. Bozell, Pulping catalysts from lignin, *Tappi J.* **74**(5), 239-241 (1991).
3. G. P. Donnini, T. J. Blain, H. H. Holton, and G. W. Kutney, 300 Alkaline pulping additives: structure-activity relationships, *J. Pulp Paper Sci.* TR 134-140 (**1983**); J. C. Wozniak, D. R. Dimmel, and E. W. Malcolm, Lignin-derived quinones as pulping additives, *J. Wood Chem. Technol.* **9**, 535-548 (1989).
4. J. J. Bozell, B. R. Hames, and D. R. Dimmel, Cobalt-Schiff base complex catalyzed oxidation of para-substituted phenolics. preparation of benoquinones, *J. Org. Chem.* **60**, 2398-2404 (1995); D. R. Dimmel, M. R. Karim, M. C. Savidakis, and J. J. Bozell, Pulping catalysts from lignin (5). Nitrogen dioxide oxidation of lignin models to benzoquinones, *J. Wood Chem. Technol.* **16**, 169-189 (1996); D. R.

Dimmel, X. Pan, K. Kuroda, and J. J. Bozell, Pulping catalysts from lignin (6). Nitrogen dioxide oxidation of 5-substituted guaiacyl compounds, *J. Wood Chem. Technol.* **16**, 191-204 (1996); D. R. Dimmel, X. Pan, and J. J. Bozell, Pulping catalysts from lignin (7). Nitrogen dioxide oxidation of a lignin model dimer, *J. Wood Chem. Technol.* **16**, 205-219 (1996); D. R. Dimmel, K. Kuroda, X. Pan, and J. J. Bozell, Pulping catalysts from lignin (8). Nitrogen dioxide oxidation of lignins, *J. Wood Chem. Technol.* **17**, 235-258 (1997); J. J. Bozell, J. O. Hoberg, and D. R. Dimmel, Catalytic oxidation of para-substituted phenols with nitrogen dioxide and oxygen, *Tet. Lett.* **39**, 2261-2264 (1998); J. J. Bozell, J. O. Hoberg, and D. R. Dimmel, Heteropolyacid catalyzed oxidation of lignin and lignin models to benzoquinones, *J. Wood Chem. Technol.* **20**, 19-41 (2000).

5. L. W. Butz and A. W. Rytiva, The Diels-Alder reaction. quinones and other cyclenones, *Org. React.* **5**, 136 -192 (1949).
6. J. March, *Advanced Organic Chemistry, 3rd edition* (Wiley Interscience, New York, 1985), pp. 745-758 and references therein.
7. F. A. Carey and R. J. Sundberg, *Advanced Organic Chemistry - Part A, 2nd edition* (Plenum, New York, 1984) p. 558.
8. Z. Stojanac, R. A. Dickenson, N. Stojanac, R. J. Woznow, and Z. Valenta, Catalyzed orientation reversals in Diels-Alder reactions, *Can. J. Chem.* **53**, 616-618 (1975); J. S. Tou and W. Reusch, Selective catalysis of Diels-Alder reactions of 2-methoxy-5-methyl-l,4-benzoquinone, *J. Org. Chem.* **45**, 5012-5014 (1980); J. B. Hendrickson and V. Singh, Catalysis and regioselectivity of quinone Diels-Alder reactions, *J. Chem. Soc. Chem. Comm.* 837-838 (1983); E. C. Angell, F. Fringuelli, M. Guo, L. Minuti, A Taticchi, and E. Wenkert, Diels-Alder reactions of cycloalkenones 14. Endo diastereoselectivity of 2-cyclohexenones in reactions with cyclopentadiene, *J. Org. Chem.* **53**, 4325-4328 (1988).
9. P. Laszlo and J. Lucchetti, Catalysis of the Diels-Alder reaction in the presence of clays, *Tet. Lett.* 25, 1567-1570 (1984); R. M. Dessau, Catalysis of Diels-Alder reactions by zeolites, *J. Chem. Soc. Chem. Comm.* 1167-1168 (1986); J. Ipaktschi, Diels-Alder reactions in the presence of zeolite, *Z. Naturforsch.* **41**b, 496-498 (1986); V. V. Veselovsky, A. S. Gybin, A. V. Lozanova, A. M. Moiseenkov, W. A. Smit, and R. Caple, Dramatic acceleration of the Diels-Alder reaction by adsorption on chromatography adsorbents, *Tet. Lett.* **29**, 175-178 (1988).
10. K. R. Bebee and B. R. Kowalski, An introduction to multivariate calibration and analysis, *Anal. Chem.* **59**, 1007a-1017a (1987).
11. R Breslow, Hydrophobic effects on simple organic reactions in water, *Acc. Chem. Res.* **24**, 159-164 (1991), and references therein; W. Blokzijl, M. J. Blandamer, and J. B. F. N. Engberts, Diels-Alder reactions in aqueous solutions. Enforced hydrophobic interactions between diene and dienophile, *J. Am. Chem. Soc.* **113**, 4241-4246 (1991).
12. D. C. Rideout and R. Breslow, Hydrophobic acceleration of Diels-Alder reactions, *J. Am. Chem. Soc.* **102**, 7816-7817 (1980).
13. D. R. Dimmel, E. Althen, M. Savidakis, C. Courchene, and J. J. Bozell, New quinone-based pulping catalysts, *Tappi J.* **82**(12), 83-89 (1999); D. R. Dimmel, Method for the delignification of lignocellulosic material by adding a dialkyl substituted octahydroanthra-quinone, U.S. patent: 6,156,155 (2000).
14. B. N. Brogdon and D. R. Dimmel, Fundamental study of relative delignification efficiencies (1). Conventional pulping systems, *J. Wood Chem. Technol.* **16**, 261-283 (1996).

ACETYLATION OF LIGNIN AND PHOTOSTABILIZATION OF LIGNIN-RICH MECHANICAL WOOD PULP AND PAPER

Magnus Paulsson and Rune Simonson*

1. INTRODUCTION

Enhanced production and efficient utilization of lignocellulosic products are issues of critical importance to both industry and society. Benefits from the manufacture of mechanical wood pulp include high yield and low cost. During the past decade, mechanical pulping processes have been developed that can produce high-performance pulps at significantly lower capital and operating costs compared with kraft pulping. Although the production of mechanical pulp is environmentally benign, the use of mechanical pulps is restricted severely by their tendency to undergo color reversion (yellowing) upon exposure to daylight and/or long-term storage at ambient temperatures. This discoloration is cited as the principal reason for limited commercial applications of mechanical wood pulps in high-quality printing papers (Cockram, 1989; Ford and Sharman, 1996).

It is generally accepted that the photoyellowing initially is a surface phenomenon due mainly to the photooxidation of lignin via a series of radical reactions (Forman, 1940; Nolan *et al.*, 1945; Leary, 1968; Kringstad, 1969). Although significant progress has been made in recent decades in understanding and elucidating the mechanism of photoyellowing (cf. Gratzl, 1985; Heitner, 1993; Leary, 1994; Davidson, 1996; Forsskåhl, 2000), the course of the reactions is still not fully understood and requires further study. The complexity and insufficient knowledge of the precise reaction pathways leading to discoloration have made it difficult to develop stabilizing treatments that meet all the needs of the paper industry.

A large number of methods have been suggested to stabilize lignin-rich pulps against light-induced yellowing. The proposed treatments can be divided into two categories. The first category is characterized by the addition of inhibitors, such as ultraviolet absorbers (which reduce UV-light absorption by pulp), quenchers (which inactivate excited molecules), and antioxidants (which trap intermediate radicals). The second category

* Chalmers University of Technology, Department of Forest Products and Chemical Engineering, SE-412 96 Göteborg, Sweden.

includes treatments that function by suppressing the formation of chromophores by chemical modification of functional groups in the lignin moiety (cf. Paulsson and Ragauskas, 1998a). The chemical modification of potential chromophoric or leucochromophoric groups in lignin includes the reduction of α-carbonyl groups, hydrogenation of ring-conjugated ethylenic groups, and acylation or alkylation of hydroxyl groups. One of the most effective treatments tested so far has been the chemical modification of hydroxyl groups in softwood or hardwood lignins through etherification or esterification reactions. A variety of alkylation and acylation such as acetylation, methylation, benzoylation, and hydroxyethylation have been tested. Many researchers have shown that these treatments not only increase the photostability of the pulps but often brighten and in some cases, also photobleach the pulps.

Although successful photostabilization has been achieved by derivatizing lignin phenolic hydroxyl groups in wood pulp materials, no treatment has yet become technically or economically feasible due to the large amounts of chemicals required for successful protection and to the deterioration in strength properties accompanying the treatment. The strength reduction is thought to be due to the fact that hydroxyl groups in carbohydrates also participate in alkylation or acylation reactions (Castellan *et al.*, 1992; cf. Manchester *et al.*, 1960; Lorås, 1968). Furthermore, many of the chemicals used (*e.g.*, dimethylsulfate for methylation) are highly toxic and the use of these chemicals on a commercial scale is therefore questionable.

Acetylation was early on reported to increase the photostability of jute fibers (Peill, 1946; Callow, 1947; Callow, 1952) and later found to photostabilize and, in some cases, also brighten groundwood pulps (GWP) (Manchester *et al.*, 1960; Leary, 1968; Lorås, 1968; Ek *et al.*, 1992; Hirashima and Sumimoto, 1996; Itoh *et al.*, 1997), thermomechanical pulps (TMP) (Paulsson *et al.*, 1995; Hirashima and Sumimoto, 1996; Itho *et al.*, 1997; Agarwal, 1998; Paulsson and Ragauskas, 1998b; Agarwal, 1999), and chemithermomechanical pulps (CTMP), (Paulsson *et al.*, 1995; Hirashima and Sumimoto, 1996; Paulsson and Ragauskas, 1998b). Acetylation also increases the photostability of milled wood lignins (Agarwal *et al.*, 1999; Heitner *et al.*, 1999), wood (Dunningham *et al.*, 1992; Plackett *et al.*, 1992; Dawson *et al.*, 1995; Hon, 1995; Torr *et al.*, 1996; Chang and Chang, 2000), fiberboard (Feist *et al.*, 1991), and wood meal (Ota *et al.*, 1996). Furthermore, acetylation can also retard the heat-induced discoloration of different types of lignin-containing pulps (Paulsson and Ragauskas, 1998c).

Among the various chemical modifications proposed for increasing the dimensional stability of papermaking fibers (see reviews by Goheen, 1958; Ward *et al.*, 1965; Ward, 1973), the most promising method, so far, is acetylation. Acetylation has been shown to improve the dimensional stability of hardboards (Klinga *et al.*, 1965; Klinga and Tarkow, 1966) and low density fiberboards (Rowell *et al.*, 1995). Acetylation has also been reported to increase the dimensional stability of paper made from unbleached or bleached chemical pulps (Harrison, 1944; Stamm and Beasley, 1961; Herdle and Griggs, 1965; Klinga and Back, 1966; Ward, 1973), paper made from mechanical wood pulps (Paulsson *et al.*, 1994), cellulose rag stock (Bletzinger, 1943), and jute fabric (Callow, 1952). The prerequisite is, however, that the paper product, instead of the pulp, is acetylated or that the acetylated pulp is beaten prior to the manufacture of the paper. The reason for this is that the acetylated fibers lose their ability to become hydrated since hydrophilic hydroxyl groups are replaced with more hydrophobic acetyl groups during the acetylation treatment. The solid matrix of paper is held together by interactions at fiber contact points, interactions that are believed to be hydrogen bonds (Nissan, 1967; Nissan, 1976; Bates *et al.*, 1999). It is, therefore, not possible to produce a strong sheet of paper from acetylated wood fibers without introducing new hydrophilic surfaces (*e.g.*, by beating). However, this will lower the dimensional stability obtained with acetylation.

Acetylation is thought to be a single-site reaction that is acid- or base-catalyzed. During acetylation with acetic anhydride, acetic acid is formed as a byproduct. Examples of catalysts that have been used are pyridine, zinc chloride, dimethylformamide, urea-ammonium sulfate, magnesium perchlorate, trifluoroacetic acid, boron trifluoride, sodium acetate, potassium hydrogen phosphate, potassium acetate, and γ-rays (cf. Rowell, 1982). However, acetylation can also be performed without any catalyst.

Lignin has been reported to be more reactive than holocellulose in jute fibers (Callow, 1951) as well as in Southern pine wood (Rowell, 1982). Furthermore, the order of reactivity of isolated cell-wall components from pinewood species towards uncatalyzed acetylation with acetic anhydride has been reported to be lignin > hemicelluloses >> holocellulose, whereas the reactivity of cellulose was almost negligible (Rowell *et al.*, 1994).

In this paper, the uncatalyzed acetylation of lignin and its effect on the photostability of different types of lignin-rich mechanical wood pulps will be discussed. The mechanism of photoyellowing will also be commented.

2. EXPERIMENTAL

2.1. Lignin Model Compounds

The lignin model compounds (**1**)-(**19**) used for studying the reactivity of softwood lignin toward acetylation are shown in Scheme 1. They were: the *erythro* (**1**) and *threo* (**2**) forms of 1-(3,4-dimethoxyphenoxy)-2-(2-methoxyphenoxy)-1,3-propanediol, the *erythro* (**3**) and *threo* (**4**) forms of 1-(4-hydroxy-3-methoxyphenyl)-2-(2-methoxyphenoxy)-1,3-propanediol, 3-hydroxy-1-(4-hydroxy-3-methoxyphenyl)-2-(2-methoxyphenoxy)-1-propanone (**5**), (*E*)-3-(3,4-dimethoxyphenyl)propenal (**6**), (*E*)-3-(3,4-dimethoxyphenyl)-2-propen-1-ol (**7**), 1,2-dihydroxybenzene (**8**), 1,2-dihydroxy-4-methylbenzene (**9**), 1,4-dihydroxybenzene (**10**), 1,4-dihydroxy-2-methoxybenzene (**11**), 3,5-di-*tert*-butyl-1,2-benzoquinone (**12**), 4-methyl-1,2-benzoquinone (**13**), 3-methoxy-5-methyl-1,2-benzoquinone (**14**), *trans*-2,3-dihydro-7-methoxy-2-(5-methoxy-3,4-benzoquinonyl)-3-methyl-5-propylbenzofuran (**15**), 2,6-dimethoxy-1,4-benzoquinone (**16**), 6-methoxy-2-methyl-1,4-benzoquinone (**17**), *trans*-4-hydroxy-3-methoxystilbene (**18**), and 3,3'-dimethoxystilbenequinone (**19**). More information on the preparation of the lignin model compounds can be found in Paulsson *et al.* (1996a; 1996b).

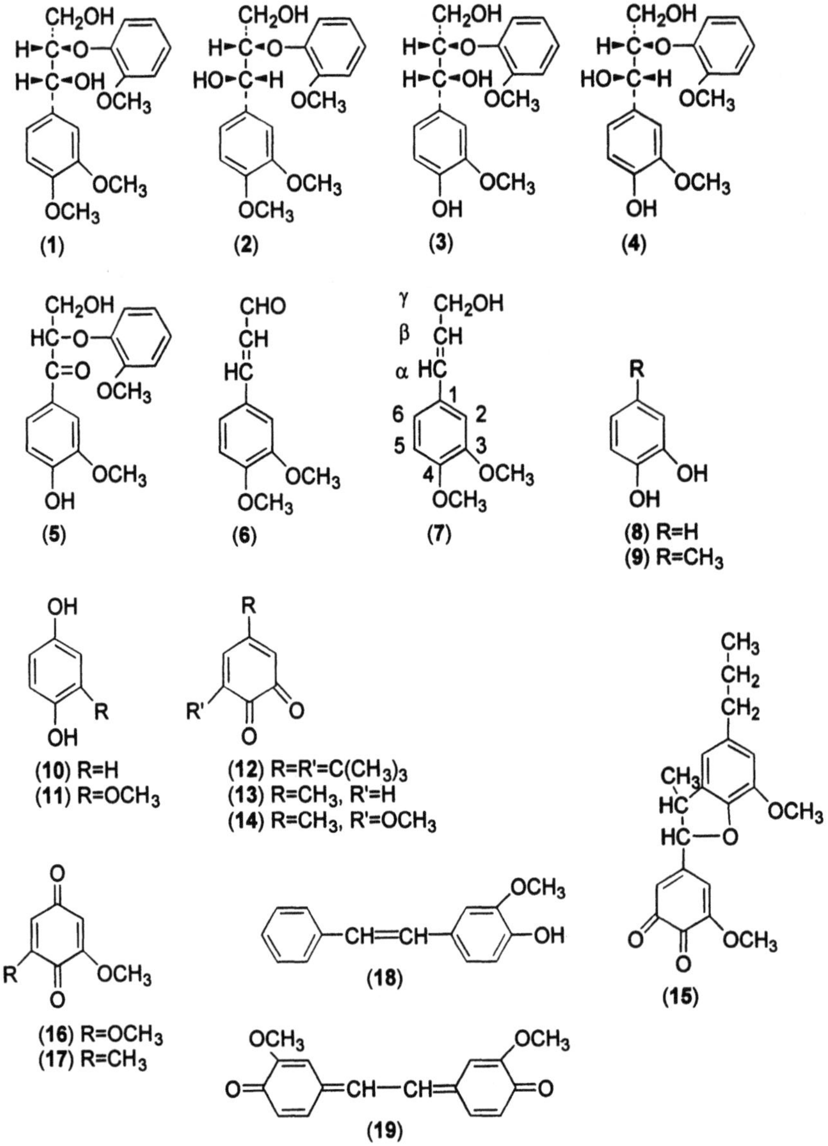

Scheme 1. Lignin model compounds examined in the acetylation experiments (Paulsson *et al.*, 1996a; Paulsson *et al.*, 1996b). Denotations of the carbons in the phenylpropane unit are shown for compound (7).

2.2. Pulp and Paper Samples

Different commercially produced softwood and harwood mechanical pulps have been examined in order to evaluate the mechanical, optical, and aging characteristics of the acetylated pulps. The softwood pulps examined were the following: unbleached spruce (*Picea abies*) thermomechanical pulp (TMP), dithionite-bleached spruce (*Picea abies*) TMP, hydrogen peroxide-bleached spruce (*Picea abies*) TMP, and hydrogen peroxide-bleached spruce (*Picea abies*) chemithermomechanical pulp (CTMP). The two hardwood pulps examined were a hydrogen peroxide-bleached aspen (*Populus tremula*) CTMP and a hydrogen peroxide-bleached aspen (*Populus tremuloides*) alkaline peroxide mechanical pulp (APMP). The pulps were, in all cases, obtained as dried samples and used as received to make handsheets according to SCAN-M 5:76 (or TAPPI Test Method T 205 om-88). The paper sheets were then conditioned at 23 °C and 50% relative humidity according to SCAN-P 2:75 (or TAPPI Test Method T 402 om-88) before modification with acetic anhydride.

2.3. Acetylation Procedure

2.3.1. Lignin Model Compounds

In a typical acetylation experiment, the lignin model compound (ca. 15 mg) was dissolved in acetic anhydride (ca. 3 ml) and the solution was heated at 100 °C for different periods of time. After cooling, the excess of acetic anhydride was decomposed by the addition of methanol (10 ml) and the solvent was removed by film evaporation. The residual acetic acid was removed by azeotropic distillation with toluene. The product was dried in *vacuo* over P_2O_5.

2.3.2. Papers Made from Lignin-Rich Mechanical Wood Pulps

Handsheets of mechanical wood pulps were acetylated at 80 - 110 °C (normally 100 °C) using acetic anhydride without any catalyst or solvent. The procedure was as follows: Conditioned handsheets were placed in a preheated glass container with an excess of acetic anhydride. After the reaction time was completed, the handsheets were immersed in water to stop the acetylation reaction. The acetylated paper sheets were then thoroughly washed with water to remove remaining chemicals and thereafter pressed for five minutes at 400 kPa. The handsheets were dried on press plates and then conditioned at 23 °C and 50% relative humidity according to SCAN-P 2:75 (or TAPPI Test Method T 402 om-88).

2.4. Chemical Analysis

2.4.1. Lignin Model Compounds

The reactivity of the model compounds was followed by thin layer chromatography (TLC) performed on silica gel plates (Merck, Kiselgel 60 F_{254}), using toluene-dioxane-acetic acid (90:25:4) as the eluent. Spots were made visible by spraying with formalin-H_2SO_4 (1:9) and subsequent heating (140 °C). The reaction products were identified with ^{1}H NMR spectroscopy. The ^{1}H NMR spectra were recorded at 400 MHz with a Varian XL-400 (VXR-5000) instrument (temperature, 293 K or 300 K; solvent, deuteriochloroform; internal reference, $(CH_3)_4Si$). ^{1}H NMR data for lignin model compounds (**1**)-(**19**) and their acetate derivatives are given in Paulsson *et al.* (1996a; 1996b). Acetylation rates of non-phenolic and phenolic arylglycerol β-aryl ether lignin model compounds (**1**)-(**4**) were also followed by quantitative high performance liquid chromatography (HPLC) analysis. The HPLC analysis was performed on a Nucleosil C_{18} column (4.6 × 250 mm,

10μ particle size) using UV-detection (λ, 280 nm). The mobile phase consisted of water(*w*)-methanol(*m*)-acetonitrile(*a*), with a linear gradient between *w-m-a* (81:13:6, vol.%) to *m-a* (55:45, vol.%) during 35 minutes (flow rate, 1 ml/min.). UV/VIS absorption spectra were recorded with a Shimadzu UV-256FW spectrophotometer.

2.4.2. Papers Made from Lignin-Rich Mechanical Wood Pulps

The acetyl content was calculated from the amount of acetate liberated after saponification with sodium hydroxide, as described in Paulsson *et al.* (1996c). The phenolic hydroxyl content was determined according to the procedure described by Lai *et al.* (1990), which is a slight modification of the method originally developed by Adler *et al.* (1958). The lignin content was determined as Klason lignin and as acid-soluble lignin measured spectrophotometrically at 205 nm (ε = 110 l/g cm).

2.5. Accelerated Light-Induced Yellowing

The paper samples were subjected to an accelerated light-induced yellowing in a SUNTEST CPS (Heraeus HANAU, Hanau, Germany) light-aging tester equipped with a xenon burner and filters (ultraviolet and window-glass), which eliminate radiation of wavelengths below 310 nm. The spectral distribution of the transmitted light was similar to that of average indoor daylight (cf. Paulsson and Ragauskas, 1998b). The temperature was kept close to room temperature by means of a cooling fan. The irradiance was controlled by an optical sensor that compensated for possible main voltage fluctuations and burner aging. Unacetylated paper sheets were used in all experiments as controls.

2.6. Optical Measurements

Brightness and color changes according to the CIELAB color scale (L*, a*, and b*) were measured using an Elrepho 2000 spectrophotometer. The optical properties were averaged from two to six sheets (three measurements per sheet of paper). The reflectance of a single sheet of paper (60 g/m^2) over a completely black, nonreflecting surface (over a hollow black body, reflectance < 0.5%) and the reflectance over a stack of paper (high enough to inhibit any transparence of light) were recorded. The specific light scattering coefficient (s, at 457 nm) and the specific light absorption coefficient (k, at 457 nm) were then calculated using the Kubelka-Munk theory. The post color (PC) number (at 457 nm) was calculated for the acetylation treatment (PC_1) and for the light-induced aging (PC_2) (Giertz, 1945; cf. Janson and Forsskåhl, 1989). The sum of PC_1 and PC_2 represents the total effect of the acetylation and photoaging treatment: $PC = PC_1 + PC_2$.

3. RESULTS AND DISCUSSION

3.1. Acetylation of Lignin Model Compounds

Lignin cannot be separated from lignocellulosic materials unless its structure is modified. It is therefore difficult to determine the reaction rates of lignin structures present in mechanical wood pulps or other lignin-containing materials. As a consequence, our approach was to study the reactivity of model compounds representative of structures that are suggested to be important for the discoloration and brightening reactions of lignin. The lignin model compounds subjected to acetylation using acetic anhydride at 100 °C are shown in Scheme 1 and the identified reaction products are shown in Scheme 2 (see also Table 3).

The acetylation of lignin model compounds of the β-guaiacyl ether type (**1**)-(**5**) was studied to obtain a basis for judging how fast and in which order different hydroxyl groups in lignin are acetylated. The course of the sequential acetylation of the hydroxyl groups was followed by TLC and HPLC analyses and the products were identified by ^{1}H NMR spectroscopy (Paulsson *et al.*, 1996a).

The reactivity of the two diastereomeric forms, (**1**) and (**2**), of a non-phenolic arylglycerol β-guaiacyl ether model was in general similar; the hydroxyl group in the γ-position (see Scheme 1 for denotation) was most easily acetylated whereas the α-hydroxyl group was comparatively resistant to acetylation. This is somewhat surprising since such benzylic hydroxyl groups (sometimes referred to as "the reactive group of lignin") are usually the reactive sites in many types of reactions in wood chemistry, *e.g.*, in most pulping reactions. It is possible that the steric environment is important for acetylation (α-hydroxy is sterically more hindered than γ-hydroxy). The phenolic hydroxyl group in the arylglycerol β-aryl ether models (**3**) and (**4**) was rapidly acetylated. The other hydroxyl groups (α-OH and γ-OH) in these models were acetylated in a mode similar to that of the corresponding hydroxyl groups in the non-phenolic compounds (**1**) and (**2**); the γ-OH with a rate comparable with that of the phenolic-OH and the α-OH, subsequently in a rather slow reaction. The 2-aryloxypropiophenone (**5**) reacted in a manner similar to that of model compounds (**3**) and (**4**) in the sense that both the phenolic hydroxyl group and the γ-hydroxyl group were completely derivatized in a short period of time. Figure 1 shows the conversion rate of compounds (**1**) and (**5**) during acetylation.

The behavior of the end-groups of the lignin macromolecule during photoyellowing is not clear. There are various data in the literature regarding the role as well as the consumption and/or generation of end-groups during photoaging.

Coniferaldehyde end-groups are considered to be one of the major colored structures in lignins (Pew and Connors, 1971). Several investigations have shown that most, but not all of the coniferaldehyde groups are eliminated during hydrogen peroxide-bleaching (Hirashima and Sumimoto, 1987; Pan *et al.*, 1992; Pan *et al.*, 1994; Agarwal and McSweeny, 1995; Schmidt and Heitner, 1995) and during sulfite pretreatment in a CTMP process (Heitner and Min, 1987). Photooxidation of coniferaldehyde groups has been suggested as an explanation for the photobleaching of high-yield pulps and for the decrease in absorbance of irradiated lignin containing materials observed in the UV-region with a maximum at 360 nm (cf. Wang *et al.*, 1993; Agarwal and McSweeny, 1995; Schmidt and Heitner, 1995). It is also possible that coniferaldehyde end-groups can be produced through photooxidation of coniferyl alcohol groups.

Acetylation of the softwood lignin coniferaldehyde model (**6**) yields the acylal (**6a**). The formation of the acylal is a slow reaction; the product could be detected after 40 minutes of reaction time and the yield was approximately 20% after 4 hours (^{1}H NMR spectroscopy).

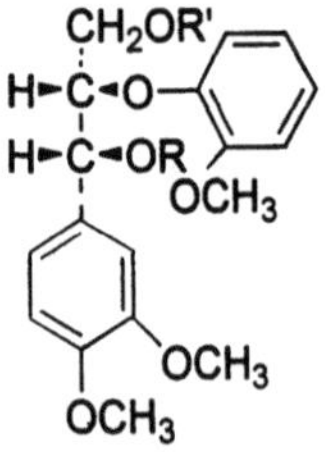

(**1a**) R=H, R'=$COCH_3$
(**1b**) R=R'=$COCH_3$

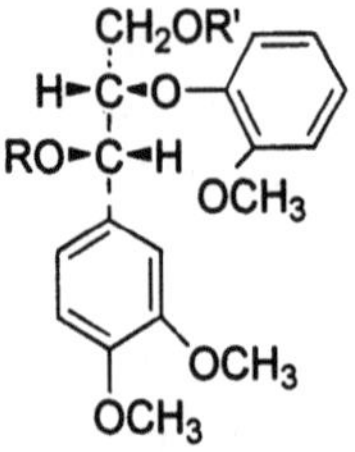

(**2a**) R=H, R'=$COCH_3$
(**2b**) R=R'=$COCH_3$

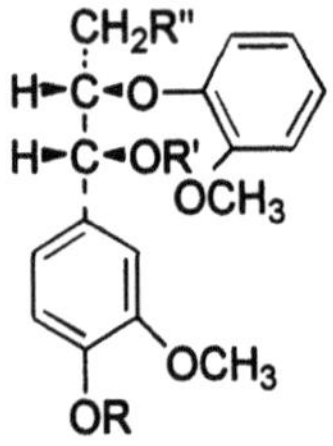

(**3a**) R=$COCH_3$, R'=R''=H
(**3b**) R=R'=H, R''=$COCH_3$
(**3c**) R=R''=$COCH_3$, R'=H
(**3d**) R=H, R'=R''=$COCH_3$
(**3e**) R=R'=R''=$COCH_3$

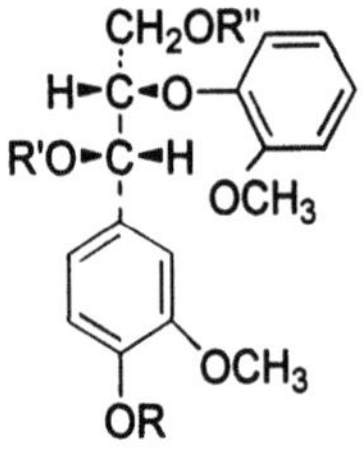

(**4a**) R=$COCH_3$, R'=R''=H
(**4b**) R=R'=H, R''=$COCH_3$
(**4c**) R=R''=$COCH_3$, R'=H
(**4d**) R=H, R'=R''=$COCH_3$
(**4e**) R=R'=R''=$COCH_3$

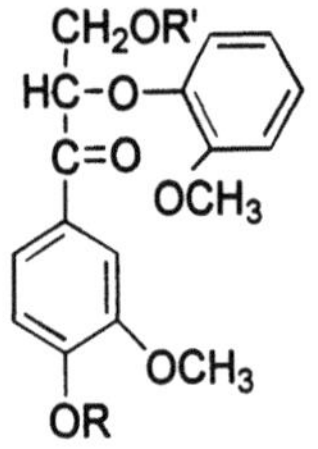

(**5a**) R=$COCH_3$, R'=H
(**5b**) R=R'=$COCH_3$

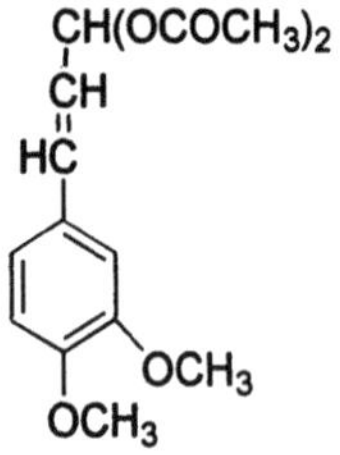

(**6a**)

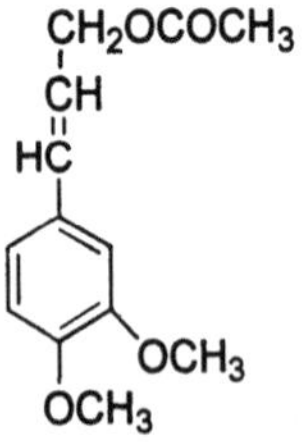

(**7a**)

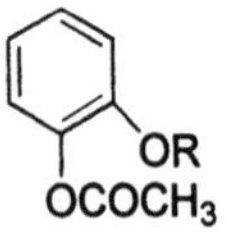

(**8a**) R=H
(**8b**) R=$COCH_3$

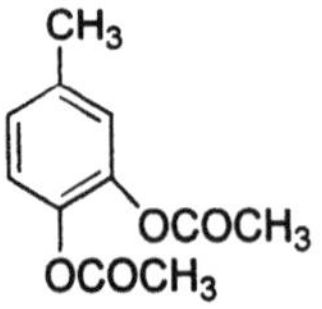

(**9a**)

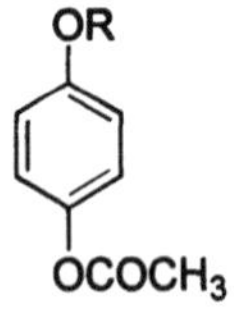

(**10a**) R=H
(**10b**) R=$COCH_3$

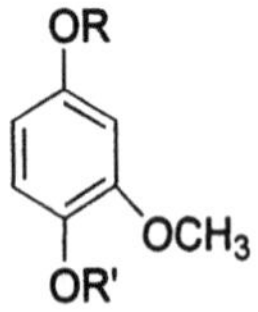

(**11a**) R=H, R'=$COCH_3$
(**11b**) R=$COCH_3$, R'=H
(**11c**) R=R'=$COCH_3$

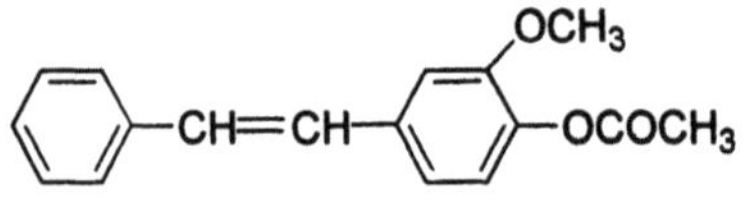

(**18a**)

Scheme 2. Acetylation products identified by ^{1}H NMR spectroscopy (Paulsson *et al.*, 1996a; Paulsson *et al.*, 1996b).

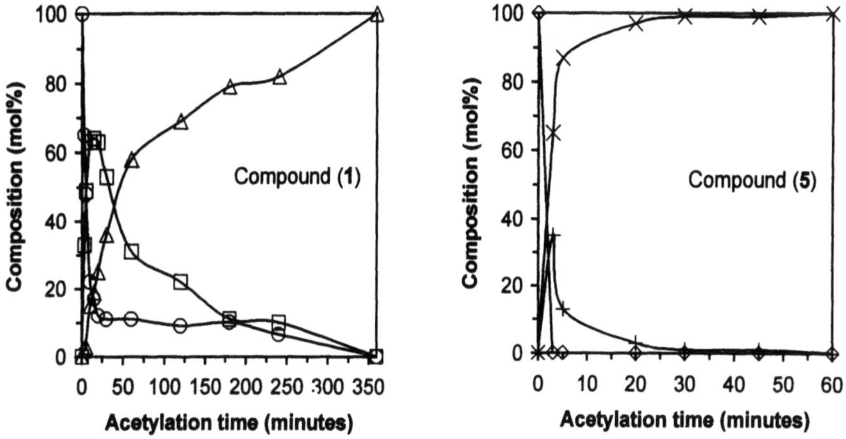

Figure 1. Composition of the acetylation mixtures from the reaction of the model compounds (**1**) and (**5**). Legends: (o), (**1**); (□), (**1a**); (Δ), (**1b**); (◊), (**5**); (+), (**5a**); (x), (**5b**).

Coniferyl alcohol end-groups are known to be present in high-yield pulps. According to several reports (*e.g.*, Gellerstedt and Zhang, 1992; Pan *et al.*, 1992; Pan *et al.*, 1994; Agarwal *et al.*, 1995), bleaching operations do not completely remove this type of structural units. It is also possible that they are formed from β-O-4 structures during high-yield pulping at elevated temperatures (Westermark *et al.*, 1995). They are decomposed to some extent upon irradiation (Pan *et al.*, 1992; Agarwal and McSweeny, 1995). Simultaneously, the number of vanillin end-groups increases and this may be explained as being associated with the cleavage of the ring-conjugated ethylenic bonds in the coniferyl alcohol units (cf. Gellerstedt and Petterson, 1975).

According to the experiments with lignin model compound (7), coniferyl alcohol units are easily acetylated. After 5 minutes of reaction time, the yield of the monoacetate **7a** was already 60% and complete acetylation was achieved after 20 minutes. Acetylation is expected to lower the reactivity of the coniferyl alcohol groups with respect to oxidative cleavage of the double bond, as well as conversion into coniferaldehyde groups.

Ortho- and *para*-quinones have been recognized as important chromophoric units in mechanical pulps. They are considered to be involved in photoyellowing, photobleaching, and thermal discoloration reactions (Leary, 1968; Kringstad, 1973; Forsskåhl *et al.*, 1981; Lebo *et al.*, 1990; Castellan *et al.*, 1993; Argyropoulos *et al.*, 1995; Schmidt *et al.*, 1995; Zhu *et al.*, 1995). It has been suggested that several chromophoric systems consisting of quinone units and their leucochromophoric counterparts (catechol and hydroquinone structures) are involved in the discoloration reactions of high-yield pulps (Ek *et al.*, 1992; Hirashima and Sumimoto, 1994). It is known that the proportion of catechol structures increases during photoyellowing most likely due to demethylation reactions (Holmbom *et al.*, 1992; Pan *et al.*, 1992). Acetylation of quinones (reactions of the Thiele-Winter type, cf. Finley, 1974) has been put forward as an explanation of the chromophore elimination (brightening) and photostabilization observed in partial acetylation of spruce groundwood pulp (Ek *et al.*, 1992).

Acetylation of catechol (**8**) shows that one of the hydroxyl groups is completely acetylated in a short period of time. The yield of 1,2-diacetoxybenzene (**8b**) was about 60% after 5 minutes of acetylation; small amounts of 2-acetoxy-l-hydroxybenzene (**8a**) (yield, 10%) were present even after several hours. The 4-methyl catechol (**9**) was completely acetylated within 10 minutes, yielding 1,2-diacetoxy-4-methylbenzene (**9a**). The hydroquinone (**10**) was completely acetylated within 45 minutes, giving 1,4-diacetoxybenzene (**10b**); one of the hydroxyl groups was derivatized within 5 minutes. It is suggested that methoxyhydroquinones play a key role in the discoloration of high-yield pulps, especially hydrogen peroxide-bleached pulps (Castellan *et al.*, 1993). Acetylation for 45 minutes of a model representative of such units, 1,4-dihydroxy-2-methoxybenzene (**11**), gave 1,4-diacetoxy-2-methoxybenzene (**11c**) in a quantitative yield. ^{1}H NMR examination of the acetylation mixture obtained after 10 minutes showed the presence of the starting material (**11**) (yield, 20%), 4-acetoxy-l-hydroxy-2-methoxybenzene (**11a**) (yield, 20%), 1-acetoxy-4-hydroxy-2-methoxybenzene (**11b**) (yield, 20%), and (**11c**) (yield, 40%). It follows from the experiments that leucochromophoric groups, such as catechol and hydroquinone units in pulps, can be expected to be derivatized and eliminated in an initial stage of the acetylation treatment.

An *ortho*-quinonoid model compound (**12**) did not react during acetylation (the compound remained unchanged even after seven hours of reaction). However, a more appropriate model compound (**13**), was rapidly decomposed upon acetylation (^{1}H NMR and UV/VIS examination). The composition of the reaction product was very complex; the product was colored, but the absorption properties differed considerably from those of the starting material. A third model compound (**14**) was also rapidly decomposed upon acetylation and a related model compound (**15**) reacted similarly. Compounds (**13**)-(**15**) gave colored reaction products upon acetylation. However, acetylation decreases the absorbance in the visible region considerably. The *para*-quinone model compound (**16**) did not react during acetylation (even after a reaction time of seven hours), whereas the model compound (**17**) was degraded to some extent (less than 15% after seven hours).

Hydroxystilbenes can be produced through the degradation of lignin during mechanical pulping or during alkaline hydrogen peroxide-bleaching (Gellerstedt and Agnemo, 1980; Wu *et al.*, 1991; Gellerstedt and Zhang, 1992). The hydroxystilbene model (**18**) was highly reactive to acetic anhydride and its phenolic hydroxyl group was completely acetylated within 10 minutes. Diguaiacylstilbenes (the term guaiacyl refers to a 4-hydroxy-3-methoxyphenyl residue) have been proposed as the leucochromophores largely responsible for the initial discoloration of high-yield pulps (Gellerstedt and Zhang, 1993) and wood (Morgan and Orsler, 1968). 3,3'-Dimethoxystilbenequinone (**19**) is the colored product obtained when diguaiacylstilbene is oxidized. Acetylation of (**19**) resulted in the introduction of acetyl groups, as judged from the ^{1}H NMR spectra. The absorption in the visible region almost completely disappears after acetylation for a few minutes.

3.2. Acetylation of Paper Made from Lignin-Rich Mechanical Wood Pulps

Acetylation of paper, made from different commercially produced mechanical wood pulps, was performed at 80 to 110 °C (normally 100 °C) using acetic anhydride as the acetylating agent without any catalyst or solvent. The temperature was chosen to minimize the thermal discoloration of the pulps and to maintain good control of the different phases of the reaction. Although acetylation is comparatively slow at these relatively low reaction temperatures, most (~ 80%) of the phenolic hydroxyl groups in the paper samples reacted very rapidly (within 15 minutes at 100 °C), regardless of their origins (Figure 2). This corresponds to a total acetyl content of 5-7% (by mass) depending on the wood species, pulp type and reaction temperature used. If only phenolic hydroxyl groups are acetylated in softwood lignin, the increase in acetyl content should be less than 1 percentage unit, whereas a complete acetylation of all lignin hydroxyl groups corresponds to an acetyl content of about 7.5% (Paulsson *et al.*, 1994). It is therefore obvious that hydroxyl groups other than the phenolic ones are acetylated simultaneously even at rather short reaction times. The remaining phenolic hydroxyl groups are more resistant or less accessible to acetylation.

Table 1 shows the change in optical properties upon acetylation of papers made from different mechanical wood pulps. Generally, acetylation of paper made from unbleached or dithionite-bleached pulps slightly lowered both the light scattering coefficient (s) and light absorption coefficient (k). This results in a small increase in brightness (negative PC_1-number). The change in optical properties, evaluated according to the CIELAB color system, was manifested as a decrease in yellowness (b*), an increase in lightness (L*), and a small decrease in redness (a*). For paper made from hydrogen peroxide-bleached pulps, the acetylation treatment resulted in a decrease in brightness (decreased s-value, increased k-value). Acetylation also increased the yellowish cast of the paper (increased b*). In summary, the change in optical properties that occurred upon acetylation was rather small, at least at low derivatization degrees (*i.e.*, short acetylation times) and the relatively low reaction temperature used.

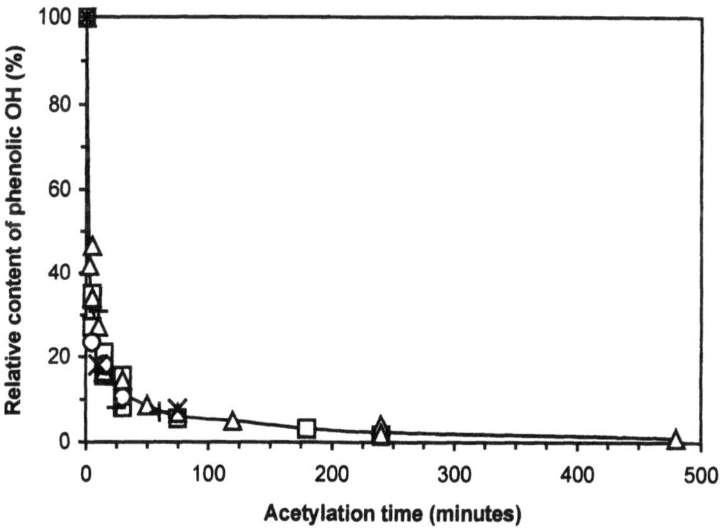

Figure 2. The relative amount of phenolic hydroxyl groups in various mechanical wood pulps (cf. Paulsson *et al.*, 1994). The acetylation temperatures are given within parentheses. Legends: (□), unbleached spruce (*Picea abies*) TMP (100 °C); (Δ), H_2O_2-bleached spruce (*Picea abies*) TMP (100 °C); (o), H_2O_2-bleached spruce (*Picea abies*) CTMP (100 °C); (x), H_2O_2-bleached aspen (*Populus tremula*) CTMP (80°C); (+), H_2O_2-bleached aspen (*Populus tremula*) CTMP (110 °C).

Table 1. Change in optical properties upon acetylation of papers made from different mechanical wood pulps. Acetylation times (minutes) are given in parentheses (cf. Paulsson *et al.*, 1995; Paulsson and Ragauskas, 1998b)

Pulp	Acetyl content (% by mass)	Brightness (%)	L*	a*	b*	PC_1-number[1]
Unbleached spruce (Picea abies) TMP						
Control	1.4	60.2[4]	89.1	0.8	14.2	---
Acetylated[2] (5)	6.1	63.4[4]	90.8	0.3	13.9	-2.6
(15)	9.6	63.9[4]	91.1	0.3	13.8	-3.0
Dithionite-bleached spruce (Picea abies) TMP						
Control	1.4	65.7[5]	92.1	-0.4	13.6	---
Acetylated[2] (5)	6.2	66.5[5]	92.4	-0.5	13.2	-0.52
(15)	10.6	67.5[5]	92.8	-0.6	13.0	-1.1
H_2O_2-bleached spruce (Picea abies) TMP						
Control	0	76.9[4]	92.8	-0.7	7.7	---
Acetylated[2] (5)	4.6	75.6[4]	92.9	-0.8	9.0	0.21
(15)	6.4	75.9[4]	92.8	-0.8	11.2	0.36
H_2O_2-bleached spruce (Picea abies) CTMP						
Control	0.5	78.2[5]	95.4	-1.9	8.8	---
Acetylated[2] (5)	4.6	76.3[5]	95.2	-2.0	10.0	0.64
(15)	9.4	75.6[5]	95.1	-2.1	10.4	0.90
H_2O_2-bleached aspen (Populus tremula) CTMP						
Control	0.9	79.9[4]	95.1	-0.8	7.0	---
Acetylated[3] (10)	7.7	77.2[4]	94.7	-0.8	8.6	0.84
H_2O_2-bleached aspen (Populus tremuloides) APMP						
Control	0.6	84.3[5]	95.8	-1.0	4.3	---
Acetylated[2] (5)	4.1	80.9[5]	95.3	-1.4	6.4	0.87
(10)	6.0	78.5[5]	94.9	-1.3	7.6	1.6

[1] The post color (PC_1) number at 457 nm (due to acetylation) was calculated according to Giertz (1945).
[2] Acetylation temperature, 100 °C.
[3] Acetylation temperature, 110 °C.
[4] Brightness measured with an Elrepho 2000 spectrophotometer. This instrument uses a diffuse light source and measures the reflected light at a 0° viewing angle, *i.e.*, the "Elrepho brightness" cannot directly be translated to TAPPI brightness that is determined with an instrument employing a directional light source (incident angle of 45°) and measures the reflected light at a 0° viewing angle. The difference in measured brightness was, however, small between the two instruments (less than 1 brightness unit).
[5] TAPPI brightness.

Since much of the competitiveness of mechanical pulps is based on a low production cost, a photostabilizing treatment must be inexpensive unless other positive features, such as improved mechanical or optical properties can also be obtained. The influence of acetylation on important mechanical paper properties, such as tensile strength, tearing resistance and wet tensile strength is shown in Table 2. The acetylation reaction was carried out on already formed papers in an attempt to preserve the original strength properties, *i.e.*, to avoid the reduction in paper strength that occurs when pulp is derivatized and then formed into paper. Both the tearing resistance and tensile strength were essentially unaffected by the acetylation treatment, whereas the wet tensile strength was considerably increased. The relative wet tensile strength (% of dry strength) of TMP paper samples was about 25% after a reaction time of 5 minutes. Prior to acetylation, the reference paper sheets showed no measurable wet strength under the testing conditions used. Mechanical wood pulps can therefore be acetylated without losing mechanical properties provided that the paper product rather than the pulp is used.

Table 2. Change in mechanical properties upon acetylation. Acetylation times (minutes) are given in parentheses (cf. Paulsson *et al.*, 1994)

Pulp	Acetyl content (% by mass)	Tensile strength (kN/m)	Tearing resistance (mN)	Relative wet tensile strength[1] (%)
Unbleached spruce (Picea abies) TMP				
Control	1.4	1.71	345	0
Acetylated[2] (5)	6.1	1.79	365	24
(15)	9.6	1.75	310	26
(180)	16.3	1.62	295	38
H_2O_2-bleached spruce (Picea abies) TMP				
Control	0	1.95	280	0
Acetylated[2] (5)	4.6	2.11	295	26
(15)	6.4	2.17	250	33
(120)	12.6	1.97	225	44

[1] Percentage of the dry tensile strength after a soaking time of 30 seconds.
[2] Acetylation temperature, 100 °C.

The equilibrium moisture content decreased with increasing degree of acetylation, which can be explained by the fact that hydrophilic hydroxyl groups are replaced with more hydrophobic acetyl groups so that the amount of bonded water is reduced. This is in accordance with results obtained for acetylated fiber building boards (Bristow and Back, 1969) and for acetylated flakeboards (Rowell *et al.*, 1986). The reduction in equilibrium moisture content can contribute to the retention of the mechanical properties of acetylated paper and can possibly account for the improvement in tensile strength observed after short acetylation times. Tensile strength and tearing resistance depend on the number of hydrogen-bonding sites and on the strength of the individual fibers. Consequently, acetylation of paper seems neither to have diminished the number of hydrogen bonds between fibers nor to have weakened the fiber structure to any great extent. The decrease in equilibrium moisture content observed probably increased the dimensional stability of the paper (fewer free sorption sites, *i.e.*, hydroxyl groups). Together, this may, under wet conditions, protect existing fiber-fiber bonds and thus give wet strength to the acetylated paper.

3.3. Photostabilization of Paper Made from Acetylated Mechanical Wood Pulps

The photoyellowing of papers made from unbleached and bleached lignin-rich mechanical wood pulps is efficiently inhibited by acetylation at a low degree of derivatization (Paulsson *et al.*, 1995; Paulsson *et al.*, 1996c; Paulsson and Ragauskas, 1998b). The change in brightness (Figure 3a) and post color (PC_2) number (Figure 3b) as a result of the accelerated light-induced yellowing of acetylated, unbleached and bleached spruce TMP illustrates the photostability obtained. The photostability of other types of mechanical wood pulps is similar, although the extent of stabilization depends on the degree of acetylation, as well as on the aging conditions applied (Paulsson and Ragauskas, 1998b). Although acetylation efficiently preserves a high brightness value, the structures formed during extensive aging still cause a yellow tint (an increase in b*).

The acetylated unbleached pulp was initially slightly photobleached. Several researchers have reported that methylated and, particularly acetylated lignocellulosic materials can be photobleached during irradiation (Callow, 1947; Callow and Speakman, 1949; Manchester *et al.*, 1960; Andrews and DesRosiers, 1966; Lorås, 1968; Ek *et al.*, 1992). The photobleaching of acetylated materials has been ascribed to the formation of acetyl peroxide or peroxides formed from carbohydrate acetates during aging (Callow and Speakman, 1949; Lorås, 1968). The cause of photobleaching is, however, not fully understood.

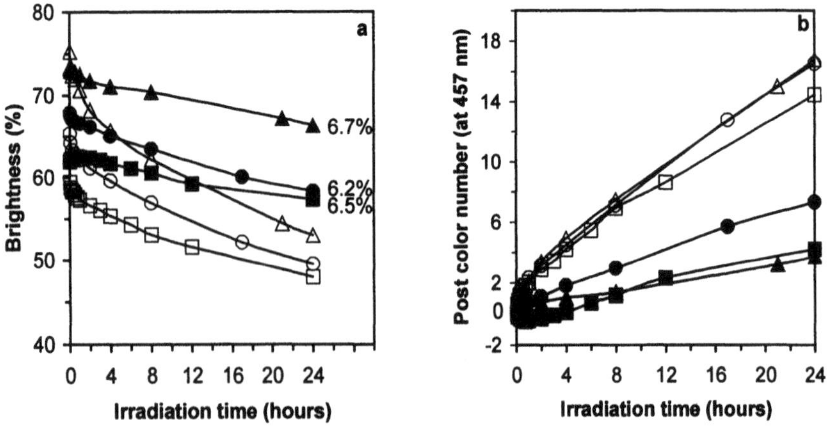

Figure 3. Brightness (a) and post color (PC_2) number (b) as a function of irradiation time for spruce (*Picea abies*) TMPs. Legends: (□), unbleached TMP; (o), dithionite-bleached TMP; (Δ), hydrogen peroxide-bleached TMP. Filled symbols represent acetylated samples (acetyl contents are given in Figure 3a).

The improvement in light stability was closely related to the decrease in the concentration of phenolic hydroxyl groups as a result of the acetylation treatment (Figure 4). Irradiation for a short period of time further decreased the number of free phenolic hydroxyl groups. New phenolic hydroxyl groups were, however, generated upon prolonged irradiation, which suggests that reactions that generate and those that consume phenolic groups coexist and that one or the other of the reactions dominates at different phases of irradiation (Paulsson *et al.*, 1996c).

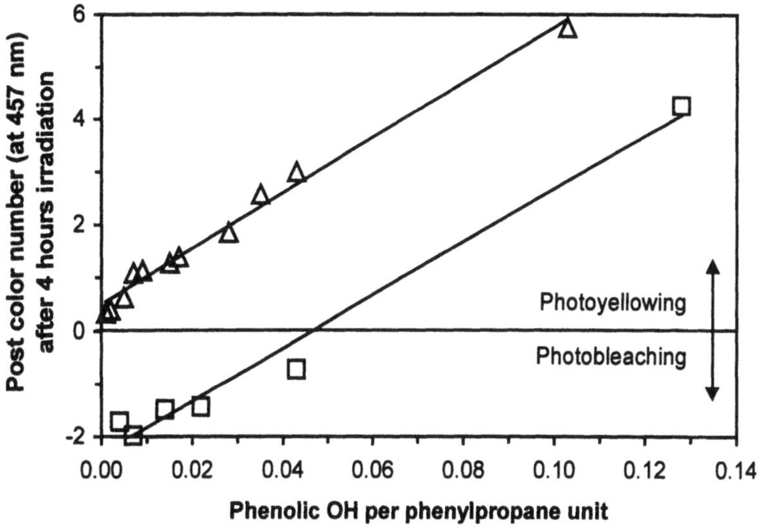

Figure 4. Relationship between the post color (PC_2) number after accelerated light-induced yellowing (for 4 hours) and the concentration of phenolic hydroxyl groups per phenylpropane unit for spruce (*Picea abies*) TMPs. Legends: (□), untreated or acetylated, unbleached TMP; (Δ), untreated or acetylated, hydrogen peroxide-bleached TMP.

3.4. Comparison of Acetylated Papers and Acetylated Model Compounds

Table 3 shows the performance of model compounds representative of different types of lignin structures during acetylation (acetic anhydride, 100 °C, no solvent or catalyst). Formulae for the examined lignin models **(1)**-**(19)** are given in Scheme 1. The readiness of the lignin in the paper to react with acetic anhydride should probably be lower than that of the model compounds for both steric and chemical reasons. It is important to take this into consideration when the results of the model compound studies are compared with those obtained from the lignin-rich papers.

Table 3. Performance of lignin model compounds (1)-(19) during acetylation

Model compound	Representative of	Acetylation time (minutes) 15	60	Notes
1	non-phenolic β-guaiacyl ethers	64 mol% (**1a**) 19 mol% (**1b**)	31 mol% (**1a**) 58 mol% (**1b**)	
2	non-phenolic β-guaiacyl ethers	~75 mol% (**2a**) > 7 mol% (**2b**)	57 mol% (**2a**) 41 mol% (**2b**)	The order of reactivity of the different hydroxyl groups [α, γ, and phenolic (ϕ)] in β-O-4 structures was found to be: ϕ-OH ≥γ-OH>>α-OH
3	phenolic β-guaiacyl ethers	63 mol% (**3c,3d**)[1] 36 mol% (**3e**)	9 mol% (**3c,3d**)[1] 91 mol% (**3e**)	
4	phenolic β-guaiacyl ethers	57 mol% (**4c,4d**)[1] 43 mol% (**4e**)	31 mol% (**4c,4d**)[1,2] 69 mol% (**4e**)[2]	
5	phenolic β-guaiacyl ethers	~ 5 mol% (**5a**) ~ 95 mol% (**5b**)	completely acetylated	
6	coniferaldehyde units	no reaction	a few % acetylated	Forming an acylal product.
7	coniferyl alcohol units	> 90% acetylated	completely acetylated	γ-OH acetylated.
8	catechol units	One OH acetylated within 5 min., 10% left with one free OH after 6 h.		
9	catechol units	Completely acetylated within 10 min.		Most of the catechol and hydroquinone structures are derivatized in the initial phase of acetylation.
10	hydroquinone units	One OH acetylated within 5 min., completely acetylated in 45 min.		
11	hydroquinone units	80% reacted after 10 min., completely acetylated in 45 min.		
12	*ortho*-quinones	No reaction after 7 h. [This is a very stable *ortho*-quinone.]		
13	*ortho*-quinones	Completely decomposed within 15 min.		The *ortho*-quinones were generally more reactive to acetic anhydride than the *para*-quinones. The structural changes that took place have not yet been determined.
14	*ortho*-quinones	Completely decomposed within 15 min.		
15	*ortho*-quinones	Completely decomposed within 15 min.		
16	*para*-quinones	No reaction after 7 h.		
17	*para*-quinones	Less than 15% decomposed after 7.		
18	stilbene units	Completely acetylated within 10 min.		Phenolic-OH acetylated.
19	stilbene-quinones	Acetyl groups are introduced.		The stilbenequinone was rapidly decolored.

[1] The two compounds could not be separated by HPLC (cf. Paulsson *et al.*, 1996a).
[2] Acetylation time, 45 minutes.

The photostabilizing and brightening effects of acetylation are most pronounced in the first few minutes of acetylation of papers made from lignin-rich mechanical wood pulps. Structures important for the yellowing can therefore be found among those that react during this initial phase of acetylation. The rate of acetylation of the phenolic model compounds (**3**), (**4**), (**5**) (β-0-4 type), and (**18**) (stilbene type) shows that the phenolic hydroxyl group is easily derivatized. More than 95% of the phenoiic hydroxyl groups in models (**3**)-(**5**) were acetylated within 5 minutes and the phenolic hydroxyl group in model (**18**) was acetylated within 10 minutes. It is also evident that most of the catechol and hydroquinone structures examined are easily acetylated. These findings are in agreement with observations of acetylated unbleached and hydrogen peroxide-bleached mechanical wood pulps (see Figure 2). The γ-hydroxyl groups in models of the β-0-4 type (**1**)-(**5**) and coniferyl alcohol type (**7**) are comparatively rapidly acetylated, whereas the (benzylic) α-hydroxyl groups in the β-0-4 models seem to be more resistant to acetylation. Acetylation for 60 minutes leaves about 40-60% of the α-OH in the etherified models (**1**) and (**2**) underivatized.

It has been suggested that acetylation affects functional groups other than the hydroxyl groups discussed above. Coniferaldehyde chromophores have been suggested to react with acetic anhydride to yield a derivative of the acylal type (Nakamura and Kitaura, 1957). Acetylation of quinones has been put forward as an explanation for the chromophore elimination (brightening) and photostabilization observed in partial acetylation of spruce groundwood pulp (Ek *et al.*, 1992). Our model experiments show that coniferaldehyde units (**6**) react slowly with acetic anhydride and that this reaction therefore cannot explain the increase in brightness achieved after a moderate extent of acetylation. *ortho*-Quinones (**13**)-(**15**) are, however, rapidly decomposed (and to a large extent decolored) as a result of the acetylation treatment, while *para*-quinones (**16**)-(**17**) were essentially unaffected.

3.5. Some Comments on the Mechanism of Photoyellowing

Schemes 3 and 4 show the proposed reaction pathways for the formation of radicals (phenoxyl, ketyl, *etc.*) and those for the formation of chromophores (initially quinones), respectively.

Acetylation of lignin-rich mechanical wood pulps removes most of the free phenolic hydroxyl groups after a short reaction time (cf. Figure 2). Acetylation is also effective in retarding light-induced yellowing at low derivatization degrees (cf. Figure 3). This suggests that photooxidation of free phenolic hydroxyl groups (reaction pathways **I** and **II**, Scheme 3) is important for the light-induced yellowing of high-yield pulps. Acetylation of the phenolic hydroxyl group in 3-methoxy-acetophenone was found to be effective in retarding light-induced discoloration (Hu *et al.*, 2000). A recent investigation of acetylation rates of lignin model compounds has shown that phenolic hydroxyl groups reacted faster than non-phenolic ones (Haque and Hill, 2000). Furthermore, there is nothing to suggest that α-hydroxyl groups should be more reactive in the papers than in the model compounds studied. The number of underivatized benzylic hydroxyl groups present in the lignin moiety at an acetylation level that inhibits the photoyellowing of high-yield pulps to a large extent can thus be expected to be fairly high. It is therefore difficult to explain the stabilizing effect obtained by acetylation as a protection of the β-O-4 aryl ether bond since it is suggested that most of the degradation of such bonds proceeds via cleavage of arylglycerol β-aryl ether structures (reaction pathway **III** in Scheme 3, cf. Schmidt and Heitner, 1993). The acetylation of γ-hydroxyl and phenolic hydroxyl groups could possibly influence the stability of the β-O-4 aryl ether bond and thereby retard the subsequent breakdown of these lignin units. Further work is needed to evaluate this possibility. However, acetylation of the phenolic hydroxyl group in 2-aryloxy-l-arylpropanone, a model of type (**5**) with CH_3 instead of CH_2OH in the γ-position, could not prevent the cleavage of the β-O-4 aryl ether structure upon irradiation (Hon, 1995). This suggests that reaction pathway **IV** (cf. Girer and Lin, 1972) is still a possible source of phenoxyl radicals (and thereby chromophore formation) upon irradiation of acetylated high-yield pulps. The cleavage of 2-aryloxy-l-arylpropanone structures is probably responsible for only a small part of the total discoloration of high-yield pulps, since the number of such units in lignin is low compared with the number of the phenolic hydroxyl groups. Furthermore, solid-state ^{13}C NMR spectroscopic studies of unirradiated and irradiated, ^{13}C-selectively enriched cell wall-dehydrogenation polymer have shown that the major part of the β-O-4 structures are intact, even after severe accelerated photoaging (Parkås, 2000).

In summary, phenolic hydroxyl groups and γ-hydroxyl groups (in both β-0-4 and coniferyl alcohol structures) together with *ortho*-quinonoid units are easily derivatized during the initial stage of uncatalyzed acetylation. The (benzylic) α-hydroxyl groups in β-0-4 structures are acetylated more slowly and most of these groups are still present after a moderate extent of acetylation. This suggests that the reaction involving arylglycerol β-aryl ether structures (**III**) is a less probable reaction pathway in chromophore formation during photoyellowing than reaction pathways involving photooxidation of free phenolic hydroxyl groups (**I** and **II**) or cleavage of 2-aryloxy-l-arylpropanone structures (**IV**), at least during the initial and most detrimental phase of photoyellowing. However, the importance of structures that can initiate discoloration of lignin-containing pulps and participate in a cyclic process must also be taken into consideration. The significance of the removal of *ortho*-quinonoid units or some other still unknown structures needs to be investigated further.

The photostability of the acetylated papers made from mechanical wood pulps has been evaluated under accelerated aging conditions using a xenon burner as the light source (see experimental). However, light sources with different wavelength distributions influence the extent of stabilization (Paulsson and Ragauskas, 1998b; Paulsson and Ragauskas, 1998d). It is therefore important to verify the photostability obtained by acetylation under realistic long-term aging conditions.

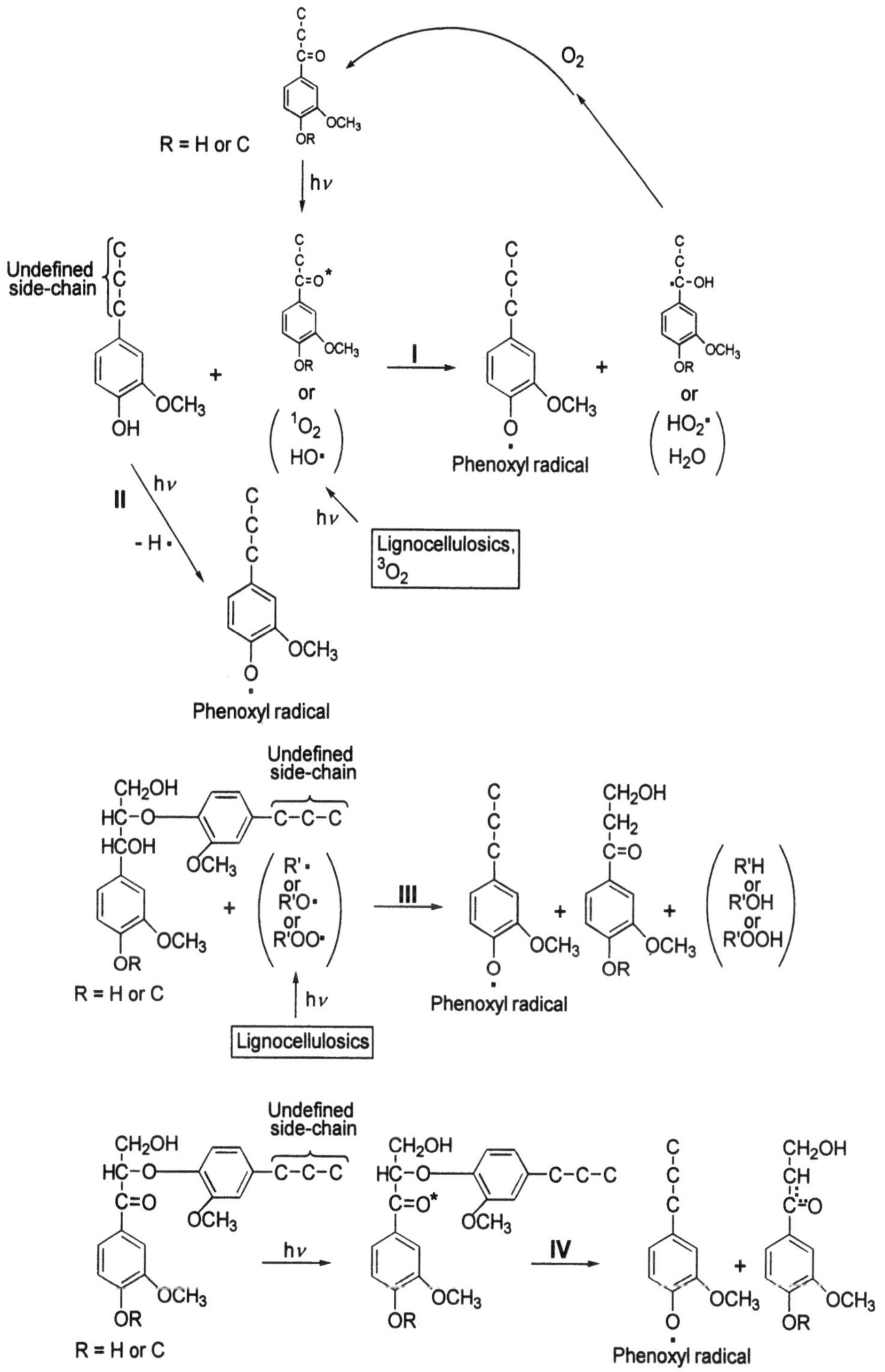

Scheme 3. Proposed reaction pathways (I-IV) in the formation of radicals during photoyellowing (based on references given in Gratzl, 1985; Heitner, 1993; Leary, 1994; Davidson, 1996; Forsskåhl, 2000).

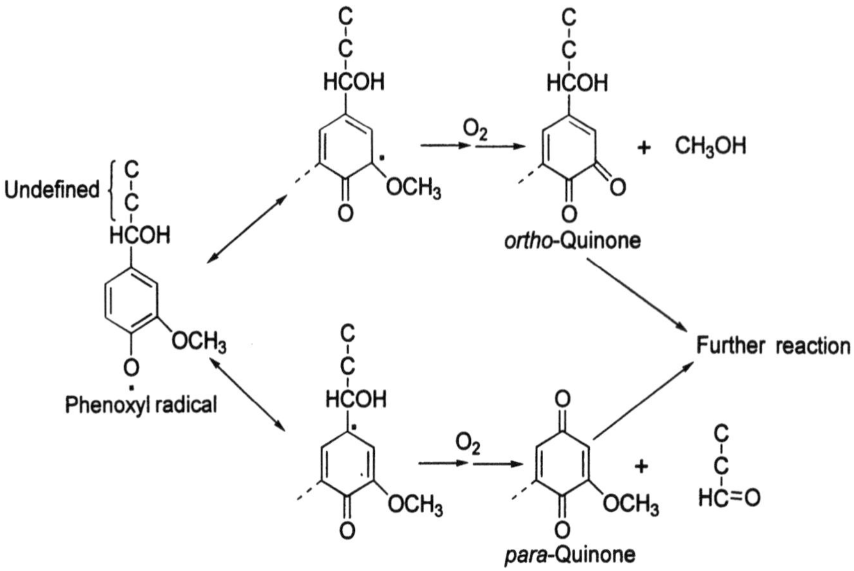

Scheme 4. Proposed reaction pathways in the formation of *ortho-* and *para*-quinones from phenoxyl radicals during photoyellowing (based on references given in Gratzl, 1985; Heitner, 1993; Leary, 1994; Davidson, 1996; Forsskåhl, 2000).

3.6. Commercial Perspective of Acetylation of Mechanical Pulp and Paper

Acetylation has been used as a tool to provide a better understanding of the photoyellowing mechanism of lignin-rich mechanical wood pulps and papers. A better understanding of the mechanism may facilitate the development of a viable photostabilizing treatment, a treatment that should be economically and technically feasible, and environmentally sound.

Acetylation is one of few chemical treatments that can generate a photostable product without affecting other paper properties (optical, mechanical) too negatively and this can be done with a rather low degree of derivatization. Acetylation results in the formation of covalently bonded acetyl groups. Simultaneously, acetic acid is formed as a by-product, but can also be the result of the hydrolysis of acetic anhydride if water is present in the system.

A continuous process for the acetylation of mechanical wood pulp to be used in dimensionally stable fiber composites for building and construction purposes has been developed in Sweden. The process is currently being demonstrated in a pilot plant with a capacity of about 500 kg of acetylated fiber per hour. To fulfil the requirements regarding dimensional stability under wet or humid conditions and also allow exterior use of the product and obtain resistance towards degrading fungi, an acetyl content of about 20% (by mass) on pulp is normally required. In the production of fiberboard, the wood pulp is dried to moisture content of a few percent before being pressed into flat boards. If such a dried wood pulp is acetylated according to the developed process, the consumption of acetic anhydride due to hydrolysis with water is rather limited and the main part of the anhydride consumed is used for the acetylation of the wood fiber. Provided that an efficient recovery of chemicals is performed and the formed (from the reaction) and purchased acetic acid is internally converted to acetic anhydride, the total acetylation cost is estimated to be about US$250 per ton of acetylated wood fiber.

Paper, on the other hand, is normally not dried to very low moisture contents. Acetylation of papers made from wood pulps with a moisture content of *e.g.* 8% (by mass) will result in about 10 times consumption of acetic anhydride due to the reaction with water than what is needed to introduce an acetyl content of 2% (by mass) via acetylation. Even though the acetyl content required for a good photostabilization effect is low, considerable amounts of acetic anhydride need to be handled in a process for the acetylation of papers. Assuming a production level of 100,000 tons of acetylated paper per annum, the acetylation cost is estimated to be about US$85 and US$105 at added acetyl contents of 2% and 5% (by mass), respectively. The technical difficulties associated with the acetylation of papers on a commercial scale, however, still remain to be evaluated.

4. CONCLUSIONS

Acetylation strongly improves the long-term light-stability of papers made from unbleached, dithionite-bleached, and hydrogen peroxide-bleached mechanical wood pulps even at a low degree of derivatization. The improvement in photostability coincided with a reduction in the content of phenolic hydroxyl groups present in the papers. Irradiation for a short period of time decreased the number of free phenolic hydroxyl groups in the acetylated lignin-containing papers, but new phenolic hydroxyl groups were generated upon extensive irradiation. These results suggest the coexistence of reactions that generate and consume phenolic groups and that the reactions dominate at different phases of photoyellowing.

Model compounds representative of structures thought to be important in discoloration and brightening reactions of lignin have been acetylated in order to obtain a better knowledge of the photoyellowing mechanism. The results show that phenolic hydroxyl groups and γ-hydroxyl groups together with *ortho*-quinonoid units are easily derivatized during the initial stage of acetylation. The (benzylic) α-hydroxyl groups (in β-O-4 structures) are acetylated more slowly and most of these groups are still present after a moderate extent of derivatization that, to a large extent, inhibits the photoyellowing of papers made from lignin-rich mechanical wood pulps. This suggests that reactions involving the photooxidation of free phenolic hydroxyl groups could be important for the light-induced discoloration of lignin-containing materials. Cleavage of arylglycerol β-aryl ether structures is a less probable reaction pathway in the formation of phenoxyl radicals, at least in the initial stage of photoyellowing.

5. ACKNOWLEDGEMENTS

Financial support from the Swedish research foundation, Stiftelsen Nils och Dorthi Troëdssons Forskningsfond, is gratefully acknowledged. The authors wish to thank Lic. Eng. Jim Parkås for valuable discussions, and Ms. Brita Ohlsson and Ms. Lena Fogelquist for their adroit technical assistance.

6. REFERENCES

Adler, E., Hernestam, S., and Walldén, I., 1958, Estimation of phenolic hydroxyl groups in lignin, *Sven. Papperstidn.* **61**(18B): 641.

Agarwal, U.P. and McSweeny, J.D., 1995, Photoyellowing of thermomechanical pulps: looking beyond alpha carbonyl and ethylenic groups as the initiating structures, *8th Int. Symp. Wood Pulp. Chem.*, Helsinki, Finland, June 6-9, Vol. 1, 435.

Agarwal, U.P., Atalla, R.H., and Forsskåhl, I., 1995, Sequential treatment of mechanical and chemimechanical pulps with light and heat: a Raman spectroscopic study, *Holzforschung* **49**(4): 300.

Agarwal, U.P., 1998, Assignment of the photoyellowing-related 1675 cm^{-1} Raman/IR band to *p*-quinones and its implications to the mechanism of color reversion in mechanical pulps, *J. Wood Chem. Technol.*, **18**(4):381.

Agarwal, U.P., 1999, On the importance of hydroquinone/*p*-quinone redox system in the photoyellowing of mechanical pulps, *10th Int. Symp. Wood Pulp. Chem.*, Yokohama, Japan, June 7-10, Vol. 1, 694.

Agarwal, U.P., McSweeny, J.D., and Ralph, S.A., 1999, An FT-Raman study of softwood, hardwood, and chemically modified black spruce MWLs, *10th Int. Symp. Wood Pulp. Chem.*, Yokohama, Japan, June 7-10, Vol. 2, 136.

Andrews, D.H. and DesRosiers, P., 1966, The influence of wood components on the bleach response of jack pine groundwood to peroxide, *Pulp Pap. Mag. Can.* **67**(C): T119.

Argyropoulos, D.S., Heitner, C., and Schmidt, J.A., 1995, Observation of quinonoid groups during the light-induced yellowing of softwood mechanical pulp, *Res. Chem. Intermed.* **9**(3-5): 263.

Bates, R., Beijer, P., and Podd, B., 1999, Wet strengthening of paper, in: Papermaking Chemistry, L. Neimo, ed., Fapet Oy, Helsinki, Finland, pp. 288-301.

Bletzinger, J.C., 1943, Effect of acetylation on water-binding properties of cellulose rag stock, *Ind. Eng. Chem.* **35**(4): 474.

Bristow, J.A. and Back, E.L., 1969, The moisture content of fibre building boards, *Sven. Papperstidn.* **72**(11): 367.

Callow, H.J., 1947, Action of light upon jute, *Nature* **159**(4035): 309.

Callow, H.J. and Speakman, J.B., 1949, The action of light on jute, *J. Soc. Dyers Colour.* **65**(12): 758.

Callow, H.J., 1951, Acetylation of cellulose and lignin in jute fibers, *J. Indian Chem. Soc.* **28**(11): 605.

Callow, H.J., 1952, Acetylation of jute, *J. Text. Inst.* **43**(8): T423.

Castellan, A., Nourmamode, A., Noutary, C., and Lachenal, D., 1992, Improvement of the photostability of peroxide bleached chemithermomechanical pulps using cationic phase transfer catalyst and hydroxyl blocking agents, *Cellul. Chem. Technol.* **26**(4): 451.

Castellan, A., Nourmamode, A., Jaeger, C., and Forsskåhl, I., 1993, Photochemistry of methoxy-*p*-benzoquinone and methoxyhydroquinone in solid 2-hydroxypropylcellulose films and on filter paper: UV/Vis absorption and diffuse reflectance spectroscopy studies, *Nord. Pulp Pap. Res. J.* **8**(2): 239.

Chang, S.-T. and Chang, H.-T., 2000, Improved lightfastness of wood by esterification, *6th European Workshop on Lignocellulosics and Pulp*, Bordeaux, France, September 3-6, 257.

Cockram, R.A., 1989, CTMP in fine papers. *Int. Mech. Pulp. Conf.*, Helsinki, Finland, June 6-8, Vol. 1, 20.

Davidsson, R.S., 1996, The photodegradation of some naturally occurring polymers, *J. Photochem. Photobiol., B:* **33**(1): 3.

Dawson, B.S.E., Cowan, F.J., Uprichard, J.M., Cronshaw, D.R., and Coombridge, B.A., 1995, Chemical treatment of *Pinus radiata* against photoyellowing, *Surf. Coat. Aust.* **32**(10): 16.

Dunningham, E.A., Plackett, D.V., and Singh, A.P., 1992, Weathering of chemically modified wood. Natural weathering of acetylated radiata pine: Preliminary results, *Holz Roh- Werkst.* **50**(11):429.

Ek, M., Lennholm, H., Lindblad, G., and Iversen, T., 1992, A study on the mechanism of the photo-yellowing of partially acetylated groundwood pulps, *Nord. Pulp Pap. Res. J.* **7**(3): 108.

Feist, W.C., Rowell, R.M., and Youngquist, J.A., 1991, Weathering and finish performance of acetylated aspen fiberboard, *Wood Fiber Sci.* **23**(2): 260.

Finley, K.T., 1974, The addition and substitution chemistry of quinones, in: The Chemistry of the Quinonoid Compounds, S. Patai, ed., John Wiley & Sons Ltd., Bristol, England, pp. 877-1144.

Ford, M.J. and Sharman, P.M., 1996, HYH spells a good deal for the future, *Pulp Pap. Int.* **38**(10): 29.

Forman, L.V., 1940, Action of ultraviolet light on lignin, *Pap. Trade J.* **111**(24): 266.

Forsskåhl, I., Gustafsson, J., and Nybergh, A., 1981, The photochemical discolouration of methoxy-*p*-benzoquinone in solution, *Acta Chem. Scand., Ser. B:* **35**(6): 389.

Forsskåhl, I., 2000, Brightness reversion, in: Forest Products Chemistry, P. Stenius, ed., Fapet Oy, Helsinki, Finland, pp. 278-332.

Gellerstedt, G. and Pettersson, E.-L., 1975, Light-induced oxidation of lignin. The behaviour of structural units containing a ring-conjugated double bond, *Acta Chem. Scand., Ser. B:* **29**(10): 1005.

Gellerstedt, G. and Agnemo, R., 1980, The reactions of lignin with alkaline hydrogen peroxide. Part V. The formation of stilbenes, *Acta Chem. Scand., Ser. B:* **34**(6): 461.

Gellerstedt, G. and Zhang, L., 1992, Formation and reactions of leucochromophoric structures in high yield pulping, *J. Wood Chem. Technol.* **12**(4): 387.

Gellerstedt, G. and Zhang, L., 1993, Formation of leucochromophores during high-yield pulping and H_2O_2 bleaching, in: Photochemistry of Lignocellulosic Materials, C. Heitner and J.C. Scaiano, eds., ACS Symposium Series 531, American Chemical Society, Washington, DC, USA, pp. 129-146.

Gierer, J. and Lin, S.Y., 1972, Photodegradation of lignin. A contribution to the mechanism of chromophore formation, *Sven. Papperstidn.* **75**(7): 233.

Giertz, H.W., 1945, Om massans eftergulning, *Sven. Papperstidn.* **48**(13): 317.

Goheen, G.E., 1958, Chemical reactions of fibrous cellulose, *Tappi* **41**(12): 37.

Gratzl, J.S., 1985, Lichtinduzierte Vergilbung von Zellstoffen – Ursachen und Verhütung, *Papier* **39**(10A): V14.

Haque, M.N. and Hill, C.A.S., 2000, Chemical modification of model compounds, wood flour and fibre with acetic anhydride, *J. Inst. Wood Sci.* **15**(3): 109.

Harrison, J.J., 1944, Effect of esterification of pulp fibers upon strength properties, hygroscopicity, and hygroexpansivity of paper, *Pap. Trade J.* **119**(5): 28.

Heitner, C. and Min, T., 1987, The effect of sulphite treatment on the brightness and bleachability of chemithermomechanical pulp, *Cellul. Chem. Technol.* **21**(3): 289.

Heitner, C., 1993, Light-induced yellowing of wood-containing papers. An evolution of the mechanism, in: Photochemistry of Lignocellulosic Materials, C. Heitner and J.C. Scaiano, eds., ACS Symposium Series 531, American Chemical Society, Washington, DC, USA, pp. 2-25.

Heitner, C., Manley, R.St.J., Ahvazi, B., and Wang, J., 1999, The effect of acetylation on the photodegradation of lignin, *10th Int. Symp. Wood Pulp. Chem.*, Yokohama, Japan, June 7-10, Vol. 1, 728.

Herdle, L.E. and Griggs, W.H., 1965, Partially acetylated cellulose – its properties and potential applications, *Tappi* **48**(7), 103A.

Hirashima, H. and Sumimoto, M., 1987, Fundamental properties of mechanical pulp lignins II. Behaviors of coniferyl aldehyde type structures in pulp lignin, *Mokuzai Gakkaishi* **33**(1): 1.

Hirashima, H. and Sumimoto, M., 1994, A basic chromophore and leucochromophore in mechanical pulps: possible repetition of photochemical reduction-oxidation-reduction, *Tappi J.* **77**(1): 146.

Hirashima, H. and Sumimoto, M., 1996, Fundamental properties of mechanical pulp lignins IV. Photostabilization of H_2O_2-bleached mechanical pulp sheets, *Mokuzai Gakkaishi* **42**(5): 506.

Holmbom, B., Ekman, R., and Eckerman, C., 1992, Degradation products formed during light and heat treatment of spruce groundwood, *J. Pulp Pap. Sci.* **18**(4): J146.

Hon, D.N.-S., 1995, Stabilization of wood color: is acetylation blocking effective?, *Wood Fiber Sci.* **27**(4):360.

Hu, T.Q., Cairns, G.R., and James, B.R., 2000, Removal of phenolic hydroxyl groups and its effect on photostability, *Holzforschung* **54**(2): 127.

Itoh, K., Adachi, Y., Tsubota, H., Oki, T., and Tachibana, S., 1997, Studies on biological treatment of mechanical pulps (IX). Effect of phenolic hydroxyl groups in mechanical pulps on depression of color reversion of mechanical pulps, *Jpn. Tappi J.* **51**(7): 1090.

Janson, J. and Forsskåhl, I., 1989, Color changes in lignin-rich pulps on irradiation by light, *Nord. Pulp Pap. Res. J.* **4**(3):1 97.

Klinga, L.O., Tarkow, H., and Back, E.L., 1965, Dimensional stabilization of hardboard by combined acetylation and heat treatment, *Sven. Papperstidn.* **68**(17): 583.

Klinga, L.O. and Back, E.L., 1966, Wet strengthening of paper by partial acetylation. A new wet strengthening method, *Sven. Papperstidn.* **69**(3):64.

Klinga, L.O. and Tarkow, H., 1966, Dimensional stabilization of hardboard by acetylation, *Tappi* **49**(1): 23.

Kringstad, K., 1969, Degradation of wood and high-yield pulps by light, *Tappi* **52**(6): 1070.

Kringstad, K.P., 1973, Einige mögliche Reaktionen in der lichtinduzierten Degradierung von ligninreichen Hochausbeutezellstoffen, *Papier* **27**(10A): 462.

Lai, Y.-Z., Guo, X.-P., and Situ, W., 1990, Estimation of phenolic hydroxyl groups in wood by a periodate oxidation method, *J. Wood Chem. Technol.* **10**(3): 365.

Leary, G.J., 1968, The yellowing of wood by light: Part II, *Tappi* **51**(6): 257.

Leary, G.J., 1994, Recent progress in understanding and inhibiting the light-induced yellowing of mechanical pulps, *J. Pulp Pap. Sci.* **20**(6): J154.

Lebo, S.E., Lonsky, W.F.W., McDonough, T.J., Medvecz, P.J., and Dimmel, D.R., 1990, The occurrence and light-induced formation of *ortho*-quinonoid lignin structures in white spruce refiner mechanical pulp, *J. Pulp Pap. Sci.* **16**(5): J139.

Lorås, V., 1968, Bleaching and stabilizing of flash-dried mechanical pulp, *Pulp Pap. Mag. Can.* **69**(2): 57.

Manchester, D.F., McKinney, J.W., and Pataky, A.A., 1960, The brightening of groundwood, *Sven. Papperstidn.* **63**(20): 699.

Morgan, J.W.W. and Orsler, R.J., 1968, The chemistry of colour changes in wood. I. The significance of stilbenes, *Holzforschung* **22**(1): 11.

Nakamura, T. and Kitaura, S., 1957, Lignin color reactions, *Ind. Eng. Chem.* **49**(9): 1388.

Nissan, A.H., 1967, The significance of hydrogen bonding at the surfaces of cellulose network structure, in: Surfaces and Coatings Related to Paper and Wood, R.H. Marchessault and C. Skaar, eds., Syracuse University Press, Syracuse, NY, USA, pp. 221-265.

Nissan, A.H., 1976, H-bond dissociation in hydrogen dominated solids, *Macromol.* **9**(5): 840.

Nolan, P., Van den Akker, J.A., and Wink, W.A., 1945, The "fading" of groundwood by light II. The physical mechanism of fading, *Pap. Trade J.* **121**(11): 101.

Ota, M., Abe, K., and Sekiguchi, T., 1996, Light-induced color change of acetylated wood meal of kiri (*Paulownia tomentosa* Steud.), *Mokuzai Gakkaishi* **42**(2): 216.

Pan, X., Lachenal, D., Lapierre, C., and Monties, B., 1992, Structure and reactivity of spruce mechanical pulp lignins. Part I. Bleaching and photoyellowing, *J. Wood Chem. Technol.* **12**(2): 135.

Pan, X., Lachenal, D., Lapierre, C., Monties, B., and Neirinck, V., 1994, On the behaviour of spruce thermomechanical pulp lignin during hydrogen peroxide bleaching, *Holzforschung* **48**(5): 429.

Parkås, J., 2000, Light-Induced Yellowing Studied with ^{13}C-Enriched CW-DHP and Solid State ^{13}C NMR Spectroscopy, *Lic.Eng. Thesis*, Department of Forest Products and Chemical Engineering, Chalmers University of Technology, Sweden.

Paulsson, M., Simonson, R., and Westermark, U., 1994, Chemical modification of lignin-rich paper. Part 1. Acetylation of paper made from spruce TMP and aspen CTMP. Evaluation of paper properties, *Nord. Pulp Pap. Res. J.* **9**(4): 232.

Paulsson, M., Simonson, R., and Westermark, U., 1995, Chemical modification of lignin-rich paper. Part 2. Photostabilizing by acetylation of paper made from spruce TMP and aspen CTMP, *Nord. Pulp Pap. Res. J.* **10**(1): 62.

Paulsson, M., Li, S., Lundquist, K., Simonson, R., and Westermark, U., 1996a, Chemical modification of lignin-rich paper. Part 3. Acetylation of lignin model compounds representative of β-O-4 structures of the β-guaiacyl ether type, *Nord. Pulp Pap. Res. J.* **11**(2): 109.

Paulsson, M., Li, S., Lundquist, K., Simonson, R., and Westermark, U., 1996b, Chemical modification of lignin-rich paper. Part 4. Elimination of chromophoric and leucochromophoric structures by acetylation, *Nord. Pulp Pap. Res. J.* **11**(4): 220.

Paulsson, M., Simonson, R., and Westermark, U., 1996c, Chemical modification of lignin-rich paper. Part 5. Long-term irradiation of unbleached and hydrogen-peroxide-bleached acetylated spruce TMP, *Nord. Pulp Pap. Res. J.* **11**(4): 227.

Paulsson, M. and Ragauskas, A.J., 1998a, Chemical modification of lignin-rich paper. Part 7. Photostabilizing high-brightness aspen CTMP by combining various classes of additives and acetylation, *Nord. Pulp Pap. Res. J.* **13**(2): 124.

Paulsson, M. and Ragauskas, A.J., 1998b, Chemical modification of lignin-rich paper. Part 8. Effect of light source on the accelerated light-induced yellowing of untreated and acetylated high-yield pulps, *Nord. Pulp Pap. Res. J.* **13**(2): 132.

Paulsson, M. and Ragauskas, A.J., 1998c, Chemical modification of lignin-rich paper. Part 9. Effect of dry heat and moist heat on the accelerated yellowing of untreated and acetylated high-yield pulps, *Nord. Pulp Pap. Res. J.* **13**(3): 191.

Paulsson, M. and Ragauskas, A.J., 1998d, Chemical modification of lignin-rich paper. Part 10. The light-induced yellowing of untreated and acetylated high-yield pulps studied by solid-state UV/VIS diffuse reflectance spectroscopy, *Nord. Pulp Pap. Res. J.* **13**(3): 198.

Peill, P.L.D., 1946, Permanent bleaching of ligno-cellulosic materials, *Nature* **128**(4016): 554.

Pew, J.C. and Connors, W.J., 1971, Color of coniferous lignin, *Tappi* **54**(2): 245.

Plackett, D.V., Dunningham, E.A., and Singh, A.P., 1992, Weathering of chemically modified wood. Accelerated weathering of acetylated radiata pine, *Holz Roh- Werkst.* **50**(4): 135.

Rowell, R.M., 1982, Distribution of acetyl groups in Southern pine reacted with acetic anhydride, *Wood Sci.* **15**(2), 172.

Rowell, R.M., Tillman, A.-M., and Simonson, R., 1986, Vapor phase acetylation of Southern pine, Douglas-fir, and aspen wood, *J. Wood Chem. Technol.* **6**(2): 293.

Rowell, R.M., Simonson, R., Hess, S., Plackett, D.V., Cronshaw, D., and Dunningham, E., 1994, Acetyl distribution in acetylated whole wood and reactivity of isolated wood cell-wall components to acetic anhydride, *Wood Fiber Sci.* **26**(1): 11.

Rowell, R.M., Kawai, S., and Inoue, M., 1995, Dimensionally stabilized, very low density fiberboard, *Wood Fiber Sci.* **27**(4): 428.

Schmidt, J.A. and Heitner, C., 1993, Light-induced yellowing of mechanical and ultra-high yield pulps. Part 2. Radical-induced cleavage of etherified guaiacylglycerol-β-arylether groups is the main degradative pathway, *J. Wood Chem. Technol.* **13**(3): 309.

Schmidt, J.A. and Heitner, C., 1995, Light-induced yellowing of mechanical and ultra-high yield pulps. Part 3. Comparison of softwood TMP, softwood CTMP and aspen CTMP, *J. Wood Chem. Technol.* **15**(2): 223.

Schmidt, J.A., Kimura, F., and Gray, D.G., 1995, IR and UV spectroscopic study of borohydride reduced mechanical pulp during monochromatic and wide band irradiation, *Res. Chem. Intermed.* **21**(3-5): 287.

Stamm, A.J. and Beasley, J.N., 1961, Dimensional stabilization of paper by acetylation, *Tappi* **44**(4): 271.

Torr, K.M., Dawson, B.S.W., Ede, R.M., and Singh, J., 1996, Surface changes on acetylation and exposure to ultraviolet radiation of *Pinus radiata* using X-ray photo-electron spectroscopy, *Holzforschung* **50**(5): 449.

Wang, J., Heitner, C., and Manley, R.St.J., 1993, The photodegradation of milled wood lignin. Part II. The effect of inhibitors, *7th Int. Symp. Wood Pulp. Chem.*, Beijing, P.R. China, May 25-28, Vol. 2, 740.

Ward, K.Jr., Morak, A.J., Murray, M.L., and Thomas, W.B., 1965, The improvement of bonding in paper sheets by chemical modification of the fibers: Some screening tests with cotton linters, *Tappi* **48**(4): 207.

Ward, K.Jr., 1973, in: Chemical Modification of Papermaking Fibers, Marcel Dekker Inc., New York, NY, USA.

Westermark, U., Samuelsson, B., and Lundquist, K., 1995, Homolytic cleavage of the β-ether bond in phenolic β-O-4 structures in wood lignin and in guaiacylglycerol-β-guaiacyl ether, *Res. Chem. Intermed.* **21**(3-5): 343.

Wu, Z.-H., Matsuoka, M., Lee, D.-Y., and Sumimoto, M., 1991, Mechanochemistry of lignin. VI. Mechanochemical reactions of β-1 lignin model compounds, *Mokuzai Gakkashi* **37**(2): 164.

Zhu, J.H., Olmstead, J.A., and Gray, D.G., 1995, Fluorescent detection of *o*-quinones formed in lignin-containing pulps during irradiation, *J. Wood Chem. Technol.* **15**(1): 43.

CATALYTIC MODIFICATION AND PHOTO-STABILIZATION OF LIGNIN FUNCTIONAL GROUPS

Thomas Q. Hu and Brian R. James*

1. INTRODUCTION

The primary components of wood, the main raw material used for papermaking in North America, are the fibrous, hydrophilic carbohydrates (cellulose and hemicelluloses) and the three-dimensional, hydrophobic lignin. For the purpose of making papers, wood, either in the form of wood chips or sawmill residues, must first be reduced to pulps (discrete fibres) by a chemical or mechanical pulping process.[1] In chemical pulping, pulps are produced in a yield of 45-55% through the dissolution of lignin by the pulping chemicals at an elevated temperature (≥ 160 °C). In mechanical pulping, pulps are formed in a yield of 90-98% through the action of mechanical forces that separate the fibres but retain lignin. From 1980 to 1999, the world mechanical pulp production has increased steadily from 27.6 to 35.1 million tonnes, accounting for ~ 23% of the total world wood pulp production.[2] Such a steady increase is expected to continue because of the low cost and high efficiency in material usage involved in the production of mechanical pulps. Indeed, as far as the pulp and paper industry is concerned, one of the most profitable uses of lignin is to use it as a component of pulps and papers.[3]

Besides their low-cost and high-yield advantages, mechanical wood pulps have high opacity that is desirable for the production of printing papers. However, papers made from mechanical wood pulps turn yellow rapidly when exposed to light. Such a yellowing continues to limit the use of mechanical wood pulps mainly to the production of non-permanent papers such as newsprint and catalogue paper.[1]

The light-induced yellowing of mechanical wood pulps and papers is known to be due to the photo-induced oxidation of lignin.[4] Upon exposure to light such as ultraviolet (UV) irradiation, lignin phenols,[5, 6] α-carbonyl-β-O-4-aryl ethers[7, 8] and α-hydroxy-β-O-4-aryl ethers[9, 10] are known to produce phenoxy radicals that can be readily oxidized to the yellow *o*-quinones (Scheme 1).

* Thomas Q. Hu, Pulp and Paper Research Institute of Canada, Vancouver Laboratory, 3800 Wesbrook Mall, Vancouver, British Columbia, V6S 2L9, Canada. Brian R. James, Department of Chemistry, University of British Columbia, 2036 Main Mall, Vancouver, British Columbia, V6T 1Z1, Canada.

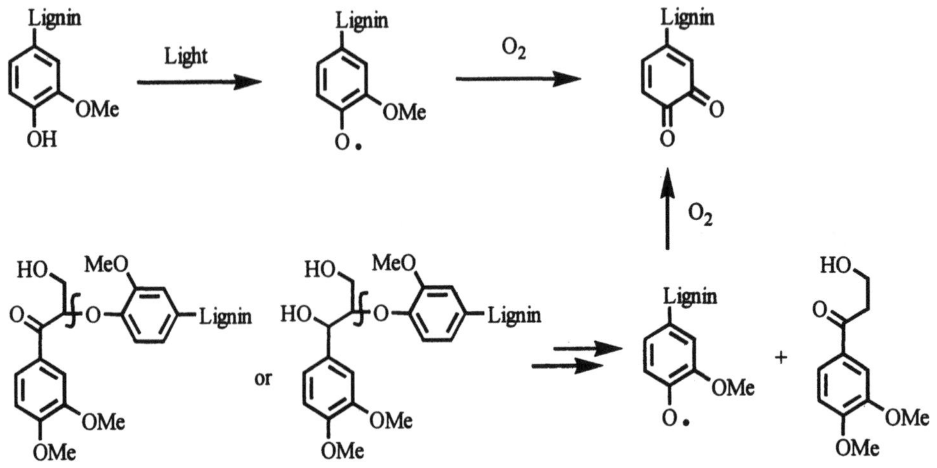

Scheme 1. Formation and subsequent oxidation of lignin phenoxy radicals from various lignin structures.

Since the mid-1960s, many lignin modification methods,[11-19] most notably alkylation and acetylation of lignin phenols[11-16,18] and reduction of lignin α-carbonyl (C=O),[17,18] have been attempted for a possible yellowing inhibition of mechanical wood pulps and papers. Alkylation of lignin phenols requires a large excess of an alkylating agent and is often complicated by the alkali-induced darkening of the pulps during the reaction. Acetylation gives a significant yellowing inhibition, but the reaction needs to be done on pre-formed papers in order to maintain the mechanical strength of the acetylated papers.[16] Reduction of lignin α-carbonyl groups is ineffective because the reduced product, the α-hydroxyl groups, can either be oxidized back to the α-carbonyl[20] or participate in the yellowing via the so-called "ketyl radical" pathway[9, 10] (cf. Scheme 1) when exposed to light. None of the lignin modification methods reported in the literature is capable of selectively and completely inhibiting the yellowing of mechanical wood pulps.

This paper reviews three catalytic lignin modification methods developed in our laboratories over the past few years and the effects of these modifications on the photostability of lignin model compounds and, in some cases, on the properties of isolated milled wood lignin and pulps.

2. CATALYTIC MODIFICATIONS OF LIGNIN FUNCTIONAL GROUPS

Scheme 2 summarizes three catalytic lignin modification methods developed by us. The first and most dramatic one (method A), aiming at a complete yellowing inhibition and bleaching of mechanical wood pulps, is the catalytic, hydrogen (H_2)-hydrogenation of lignin aromatic rings to their cyclohexyl derivatives. The second one (method B) is the catalytic hydrogenolysis of lignin phenols (Ar-OH) to phenyl (Ar-H) groups. The third one (method C), targeting at a partial yellowing inhibition and bleaching, is the catalytic H_2-reduction of lignin α-carbonyl (-C=O) and α-hydroxyl (-CH-OH) groups to α-methylene (-CH_2) groups.

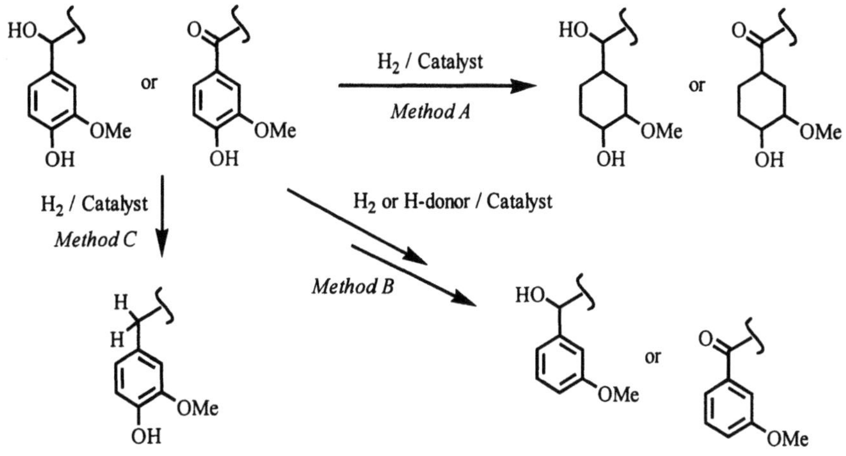

Scheme 2. Methods for the catalytic H_2-modifications of various lignin functional groups.

We are interested in the use of catalytic H_2-modifications of lignin for bleaching and yellowing inhibition of mechanical wood pulps because of the potential of catalytic reactions,[21, 22] particularly homogeneous ones which can be carried out under mild conditions with high selectivity.[23] Their use for the treatment of mechanical wood pulps may allow us to preserve the high-yield advantage of the pulps. Catalytic reactions generally require less energy and produce a minimum amount of waste with high "atom utilization".[22] These are very attractive attributes for the pulp and paper industry where, for example, approximately 73 tonnes/day of bleaching chemicals (hydrogen peroxide, sodium hydroxide and sodium silicate) have been used to produce 550 air-dried metric tonnes (ADMT)/day of bleached chemithermomechanical pulps (CTMP) to an ISO brightness of 80%,[24] based on non-catalytic, "atom inefficient" chemistry. Hydrogen (H_2) is a widely used feedstock in the chemical, foods and refining industries and it is forecast[25] to become a major source of energy such as a clean fuel for fuel-cell vehicles[26] in the near future.

The outcome of our catalytic H_2-modification approaches will depend on whether we can develop catalysts that are: (1) active for the intended H_2-modification of lignin; (2) H_2O-soluble and homogeneous; and (3) economical and applicable to pulps.

3. CATALYTIC H_2-HYDROGENATION OF LIGNIN AROMATICS

Because lignin aromatic rings are the central elements of lignin leucochromophores and chromophores in mechanical wood pulps, their hydrogenation is expected to bleach and inhibit the yellowing of the pulps. Hydrogenation of lignin aromatics has previously been attempted on lignin model compounds by Eachus and Dence.[27] The reaction, however, required the use of high temperatures and pressures (≥ 220 °C, ≥ 1500 psi H_2)[27] and was accompanied by a significant hydrogenolysis/cleavage of the substituent side-chains. Hydrogenation of the aromatic rings in lignin or in pulps without a significant hydrogenolysis and the breakdown of the lignin macromolecules had never been reported prior to our studies. Instead, efforts undertaken elsewhere have been directed towards the conversion of lignin or wood to chemicals and/or gaseous fuels using such a seemingly unavoidable hydrogenolysis.[28-30] It is interesting to note that catalytic hydrogenation of lignin aromatics and its accompanying hydrogenolysis were used over 60 years ago by lignin chemists to assist in the elucidation of lignin structure.[31, 32]

3.1. Catalytic, Selective H_2-Hydrogenation of Lignin Model Compounds

We consider that the hydrogenation of lignin aromatics may be used for the bleaching and yellowing inhibition of mechanical wood pulps, if the reaction can be achieved under mild conditions (i.e. $\leq$ 100 °C and $\leq$ 700 psi H_2) so as to avoid the breakdown/dissolution of lignin and the carbohydrate pulp components. Because of the high stability of aromatic carbon-carbon double bonds, the use of an active transition metal catalyst is deemed necessary if their hydrogenation is to be effected under mild conditions. The first series of catalysts that we studied was based on rhodium (Rh) because of its high, heterogeneous catalytic activity towards the hydrogenation of simple carbocyclic aromatics.[33] Table 1 summarizes data on the hydrogenation of 2-methoxy-4-propylphenol (**1**), a lignin model compound that mimics the structure of softwood lignin phenols, to 2-methoxy-4-propylcyclohexanol (**1a**) and 4-propylcyclohexanol (**1b**) [Eq. (1), Pr = n-propyl] using three of the Rh catalysts studied.[34-37] The chemoselectivity for the hydrogenated product (**1a**) over the hydrogenated and hydrogenolyzed product (**1b**) may serve as an indication of how well the breakdown/dissolution of the lignin macromolecules can be avoided when such a reaction is performed on pulps.

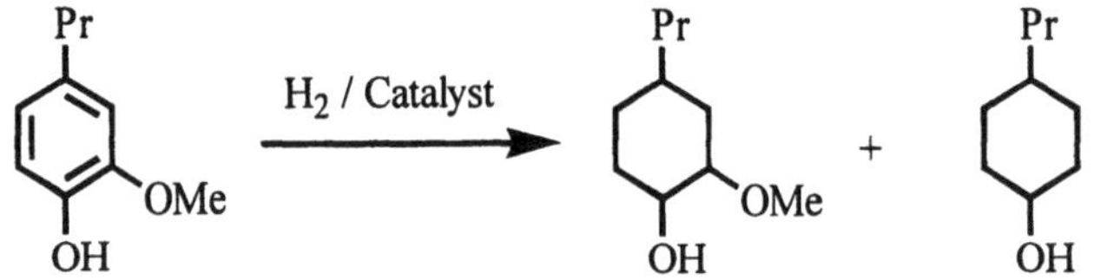

1 **1a** **1b** (1)

Di-μ-chloro-bis(η^4-1,5-hexadiene)dirhodium(I), $[(1,5\text{-}C_6H_{10})RhCl]_2$ (catalyst **I**), reported to be active for the hydrogenation of simple aromatics,[38] was an active catalyst precursor for the chemoselective hydrogenation of **1** (selectivity for **1a** > 91%) at room temperature (~ 20 °C) under 1 atm H_2 in a two-phase, aqueous/hexane medium in the presence of a phase transfer catalyst (PTC) (Table 1).[34,36] No intermediate cyclohexadiene or cyclohexene derivatives were observed. The reactions were not only chemoselective, but also stereoselective with **1a** being largely the *cis*-2-methoxy-*cis*-4-propylcyclohexanol (e.g. ~ 86% all *cis* configuration in **1a** with Bu_4NHSO_4 as the PCT).[34,36] During the reaction, colloidal rhodium was formed which eventually aggregated to give inactive Rh metal.[34,36] The longer the alkyl chain of the phase transfer catalyst tetraalkylammonium salt, the higher the activity of the Rh catalyst and the higher the selectivity for **1a**. This is likely due to the more efficient stabilization of the colloidal Rh(0) catalytic species by the ammonium salt in the organic phase.[39] Stereoselective hydrogenation of *o*-, *m*- or *p*-xylene to mainly its corresponding *cis*, *cis*-dimethylcyclohexane under biphasic conditions using Rh colloid stabilized by hydroxyalkylammonium salt has also more recently been reported by another group.[40] Arene hydrogenation catalyzed by colloidal-Rh has also been accomplished in aqueous/supercritical fluid media.[41]

Other lignin model compounds such as 4-propylphenol and 3,5-dimethoxy-4-propylphenol were also hydrogenated using **I** as the catalyst precursor in the presence of Bu_4NHSO_4. The hydrogenation rate decreased and the stereoselectivity for the all *cis* product increased with an increasing number of methoxy (OMe) substitutents on the phenolic ring.[34, 42]

Selective hydrogenation of **1** to **1a** (selectivity = ~ 92%) could also be achieved in a one-phase, aqueous-^iPrOH medium using **I** as the catalyst precursor in the absence of a PTC, even though the reaction was slower. The use of polyvinylpyrrolidone (PVP)-stabilized Rh colloid, Rh(0)-PVP-$3Et_3NHCl$ (catalyst **II**) prepared *in-situ* from the reaction of PVP, $RhCl_3$ and Et_3N[35] or a rhodium cluster compound, $Rh_6(\mu_3\text{-}O)_4(OH)_{12}(DMF)_n$ (catalyst **III**)[37] also allowed the selective hydrogenation of **1** to be performed in a one-phase, aqueous-ethanol medium (Table 1). In the hydrogenation using **III** as the catalyst precursor, colloidal Rh was also formed during the reaction, and this eventually aggregated to give inactive Rh metal. **II** was stable throughout the hydrogenation without any deposition of Rh metal. All the systems are thought to involve colloidal Rh as the active catalyst.

Interestingly, the presence of water was found to be necessary for the hydrogenation reactions to occur. In the absence of H_2O, less than 1.0% of **1** was hydrogenated using either **I** or **III** as the catalyst (Table 1). Deuterium-labelling experiments showed the incorporation of hydrogen from H_2O into the hydrogenated products.[36] The requirement and involvement of H_2O in the catalytic hydrogenation of lignin model compounds are encouraging, as H_2O is the most desirable solvent for pulp processing.

Table 1. Conversion (%) and selectivity for **1a** in the hydrogenation of **1** using Rh catalysts (**I** – **III**) in various solvents with or without a PTC at room temperature (~ 20 °C) under 1 atm H_2 for 24 h (unless otherwise specified in parentheses)

Catalyst	Solvent/PTC	Conversion (%)	Selectivity for **1a** (%)
$[(1,5\text{-}C_6H_{10})RhCl]_2$ (**I**)[a]	H_2O-hexane/Et_4NCl	43.1	91.4
	H_2O-hexane/Bu_4NHSO_4	52.9	92.4
	H_2O-hexane/$(^nC_6H_{13})_4NCl$	65.0	94.9
	H_2O-hexane/$(^nC_8H_{17})_4NBr$	91.1	96.1
(**I**)[a]	H_2O-^iPrOH	38.1	92.6
Rh(0)-PVP-$3Et_3NHCl$ (**II**)[b]	H_2O-EtOH	100 (160 h)	86.0
$Rh_6(\mu_3\text{-}O)_4(OH)_{12}(DMF)_n$ (**III**)[c]	H_2O-EtOH	57.0 (42 h)	82.0
(**I**)	Hexane/$(^nC_8H_{17})_4NBr$	0.5	25.6
	^iPrOH	0.2	27.1
(**III**)	EtOH	1.0 (20 h)	-

[a] The aqueous phase of the reaction was buffered at pH 7.5 using borate/citrate/phosphate, H_2O/hexane or ^iPrOH = 5 / 15 mL, [Rh] = 3.0 mM, [**1**]/[Rh] = 25, [PTC]/[Rh] = 7 if PTC was used.
[b] Catalyst prepared from $RhCl_3 \cdot 3H_2O$, polyvinylpyrrolidone (PVP) and Et_3N,[35] H_2O/EtOH = 10 / 10 mL, [Rh] = 0.7 mM, [**1**]/[Rh] = 50.
[c] Rh catalyst prepared from $Rh_6(CO)_{16}$,[37] H_2O/EtOH = 4 / 3 mL, [Rh] = 0.8 mM, [**1**]/[Rh] = 60.

Although the three Rh catalyst systems (I – III) all showed good activity for the selective hydrogenation of lignin model compounds such as **1**, they were not suitable for the hydrogenation of lignin in pulps possibly because of the unsuitably large size of the colloids for **II**, and the deposition of black Rh metal in the cases of **I** and **III**.

We have also studied catalytic systems based on ruthenium (Ru). Ru has been shown to possess the second highest activity (after Rh) among the several most common transition metals (Rh, Ru, Pt, Pd, Ni and Co) for the heterogeneous hydrogenation of benzene and isobutylbenzene.[33] Several Ru complexes have also been reported to catalyze the homogeneous hydrogenation of simple aromatic compounds.[43-47]

Table 2 summarizes data on the hydrogenation of **1** using several of the Ru compounds that show good to excellent activities (conversion of **1** ~ 36 - 100%).[36,48] The three Ru(0)-hexamethylbenezene (C_6Me_6) complexes **IV-VI** and two dinuclear Ru(II)-arene complexes **VII** and **VIII** were effective for the selective hydrogenation of **1** but black Ru metal was present at the end of the reactions, indicating that the catalysis was again probably colloidal.

The most active Ru catalytic system (**IX** or **X**) was generated using $RuCl_3$ in the presence of 3.5 or 7.0 equiv. of n-trioctylamine (TOA) in H_2O-^iPrOH at 50 °C under 50 atm H_2. Practically all the substrate was hydrogenated within 24 h even though the molar ratio of [**1**]/[Ru] was 300, i.e. 6 times higher than those used for the Ru catalysts **IV-VIII**. However, Ru colloids were produced which eventually aggregated to give inactive Ru metal. Interestingly, when 1.0 equiv. of triphenylphosphine monosulfonate (TPPMS) was used in conjunction with 3.5 equiv. of TOA, a homogeneous system was observed, but with a smaller conversion (56.4% after 24 h) (Table 2). The use of TOA was essential; in its absence, no conversion of **1** was found.

Table 2. Conversion (%) and selectivity for **1a** in the hydrogenation of **1** using various Ru catalysts (2.0 mM) in 20 mL H_2O-^iPrOH (1:1 V/V) at 100 °C under 50 atm H_2 for 24 h with [**1**]/[Ru] = 50, unless otherwise specified

Catalyst	Conversion (%)	Selectivity for **1a**
$Ru(\eta^6\text{-}C_6Me_6)(\eta^2\text{-}C_2H_4)_2$ (**IV**)	62.6	78.6
$Ru(\eta^6\text{-}C_6Me_6)(\eta^4\text{-}1,3\text{-}C_8H_{12})$ (**V**)	73.2	77.0
$Ru(\eta^6\text{-}C_6Me_6)(\eta^4\text{-}1,3\text{-}C_6H_8)$ (**VI**)	94.7	82.1
$Ru_2Cl_4(1,3,5\text{-}C_6H_3Me_3)_2$ (**VII**)[a]	36.7	82.6
$Ru_2Cl_4(C_6H_6)_2$ (**VIII**)[b]	82.2	63.9
$RuCl_3$ / 3.5 TOA (**IX**)[c]	99.7	85.6
$RuCl_3$ / 7.0 TOA (**X**)[c, d]	> 99.9	87.8
$RuCl_3$ / 3.5 TOA / TPPMS (**XI**)[c, e]	56.4	80.2

[a] [Ru] = 0.72 mM, 80 °C.
[b] [Ru] = 10 mM, 90 °C, solvent H_2O, H_2 pressure 60 atm.
[c] [1]/[Ru] = 300.
[d] [Ru] = 2.0 mM.
[e] [Ru] = 3.8 mM.

3.2. Catalytic H_2-Hydrogenation of Milled Wood Lignin

Catalytic H_2-hydrogenation of milled wood lignin (MWL)[49,50] isolated from a peroxide-bleached spruce thermomechanical pulp (TMP) was performed either at room temperature (~ 20 °C) under 1 atm H_2 using the Rh cluster catalyst (**III**) in aqueous methyl cellosolve, H_2O-$MeOCH_2CH_2OH$,[37] or at 80 °C under 50 atm H_2 using the Ru catalyst (**IX**) in H_2O-^iPrOH-$MeOCH_2CH_2OH$.[36] The extent of aromatic ring hydrogenation was established by analysis of the ^1H NMR integration ratio of the remaining aromatic protons (δ 6-8) to the protons (δ 0.8-2.1) on the hydrogenated rings that are not α to the OH or OMe groups.[37] Similar to the lignin model compounds, no intermediates from the hydrogenation of the aromatic rings were observed; *i.e.*, hydrogenation of the aromatic rings proceeded to give only cyclohexane derivatives. However, in contrast to the hydrogenation of lignin model compounds, reaction with lignin was extremely slow and not all the aromatic rings could be hydrogenated even after a prolonged reaction time (10 days). For example, only about 38% of the aromatic rings in the MWL were hydrogenated after 10 days using the Ru catalyst (Figure 1). In addition, for a MWL that was 52% hydrogenated, the product yield was only 28%. The slow reaction likely results from the low accessibility of some of the aromatic rings in the lignin polymer to the colloidal Ru(0) that is stabilized by TOA. Increased aromatic ring hydrogenation may require cleavage of some of the interlinkages in lignin. GPC analysis of the hydrogenated lignins reveals a linear relationship between the extent of aromatic hydrogenation and the average molecular weight of the hydrogenated lignin (Figure 2). Interestingly, as the extent of aromatic hydrogenation increases, the polydispersity (M_w/M_n) of the lignin decreases, indicating increasing uniformity within the partially hydrogenated lignin.

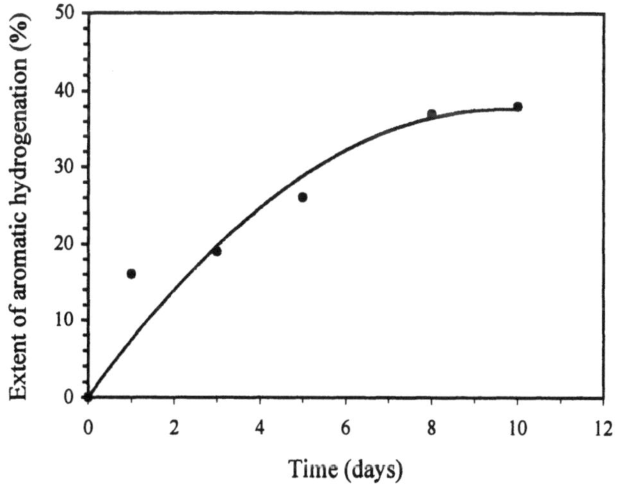

Figure 1. Extent of aromatic hydrogenation of milled wood lignin (MWL) vs. hydrogenation time; MWL = 80 mg, $RuCl_3$ (0.04 mmol), TOA (0.14 mmol), $H_2O/^iPrOH/MeOCH_2CH_2OH$ = ~ 6/14/4 mL.

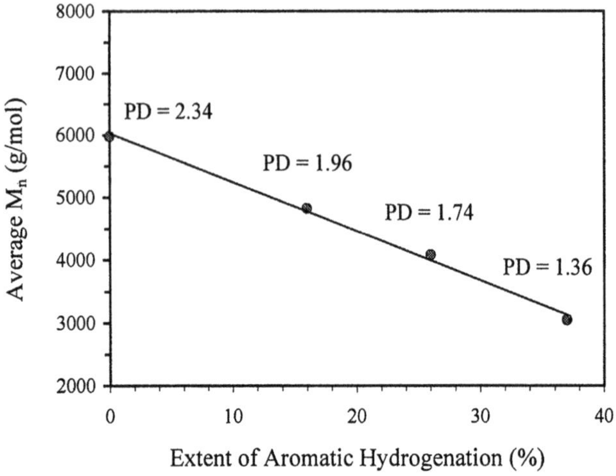

Figure 2. Average molecular weight (M_n) and polydisperisty (PD) of the hydrogenated milled wood lignins at various extents of aromatic hydrogenation using the Ru catalyst.

3.3. Photostability of Hydrogenated Lignin Model Compounds and Lignin

Photostability of the hydrogenated lignin model compounds[51] and lignin[48] was evaluated by impregnating them onto a cellulosic matrix, in the form of Whatman (No. 1) filter papers, and exposing the papers to UV (253 nm) light. Figure 3 shows the %ISO brightness values of the papers impregnated with the lignin model phenol **1**, 1,2-dimethoxy-4-propylbenzene (**2**, see below) and the hydrogenated phenol, *cis,cis*-**1a**.[51] Complete brightness stabilization/photostability was obtained for the paper impregnated with the aromatic hydrogenated compound, while only partial stabilization was achieved with the methylated compound. We have also found partial photostabilization for a partially hydrogenated lignin.[48]

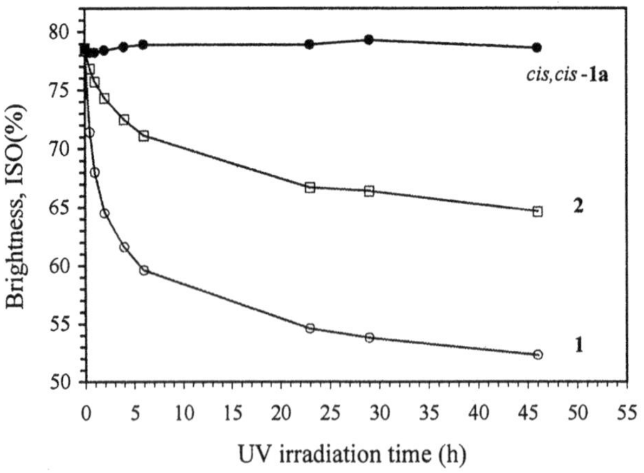

Figure 3. Brightness stability of the papers impregnated with the model phenol **1**, the methylated phenol **2** and the hydrogenated phenol, *cis,cis*-**1a**.

The slow reaction with lignin and the colloidal nature of the catalysis prevent the use of our current catalytic systems for the hydrogenation of lignin aromatics in pulps. Some of these systems, however, may be useful for introducing a higher photostability and a narrower molecular weight distribution to industrial lignin that can be further modified to give value-added lignin by-products/materials.

4. CATALYTIC HYDROGENOLYSIS OF LIGNIN PHENOLS

Lignin phenol is one of the key functional groups responsible for the light-induced yellowing of mechanical wood pulps and papers. Alkylation of lignin phenols with dimethylsulfate, Me_2SO_4, a powerful but very toxic alkylating agent has been shown to retard the yellowing in wood, newspapers and groundwood pulps,[13, 52] but a large excess of the reagent (in weight, greater than that of the pulp)[52, 53] is required because of the rapid hydrolysis of Me_2SO_4 in the aqueous alkaline medium needed for the reaction. In addition, the alkylated materials often have a darker color because of the alkali-induced formation of chromophores in lignin during the reaction. Other alkylating agents such as olefin oxides[11] and diazomethane[18] were found to be less effective than dimethylsulfate, an we have recently used 5-chloro-1-phenyl-*1H*-tetrazole as an alkali-stable agent for the alkylation of lignin model phenols.[54] Acetylation of lignin phenols with acetic anhydride can give a significant yellowing inhibition at a much lower chemical charge than alkylation, but the reaction is not selective. Acetylation of the hydroxyl groups of cellulose also occurs, and this reduces the physical strength of the acetylated pulps and papers unless the reaction is carried out on already-made papers.

We have considered the use of catalytic hydrogenolysis (Scheme 2, method B) as a possible method for the selective modification of lignin phenols (Scheme 3).[55] Under this consideration, triflic anhydride or acylating agents capable of polarizing the lignin Ar-OH bond would first be used to triflate or acylate both the lignin phenols and the cellulosic hydroxyl groups. A catalytic hydrogenolysis would then be applied to cleave selectively the lignin trifluorosulfonoxyl ($-OSO_2CF_3$) or acyloxyl group. Selective hydrogenolysis of the lignin acyloxyl bond is to be expected because the hydrogenolysis requires the coordination of the catalyst to the aromatic functionality.[56] Finally, a mild alkaline treatment could be used to hydrolyze the cellulose ester to liberate the cellulosic hydroxyl groups (Scheme 3). Cacchi *et al.* have shown that aryl triflates of simple phenols can be catalytically cleaved via hydrogenolysis in the presence of a palladium catalyst.[56]

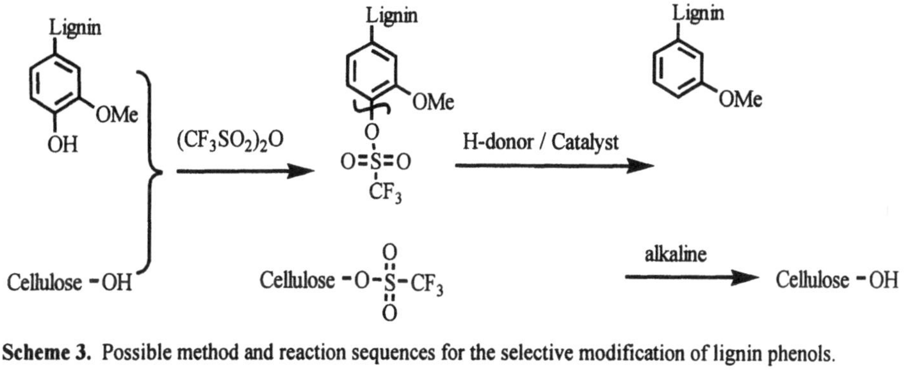

Scheme 3. Possible method and reaction sequences for the selective modification of lignin phenols.

Reaction of the lignin model phenols 2-methoxy-4-propylphenol (**1**), acetovanillone (**3**) and coniferyl aldehyde (**4**) with triflic anhydride gave the expected triflated derivatives **1'**, **3'** and **4'** in good isolated yields (> 60%).[55] Subsequent treatment of these derivatives with formic acid and triethylamine in the presence of a catalytic amount of palladium acetate and triphenylphosphine gave the hydrogenolyzed products **1"**, **3"** and **4"** in > 57% isolated yields (Scheme 4). The formic acid and triethylamine were used to form triethylammonium formate that served as the hydrogen donor in the hydrogenolysis. The Pd catalyst was homogeneous; during the reaction, no metal deposit was found.

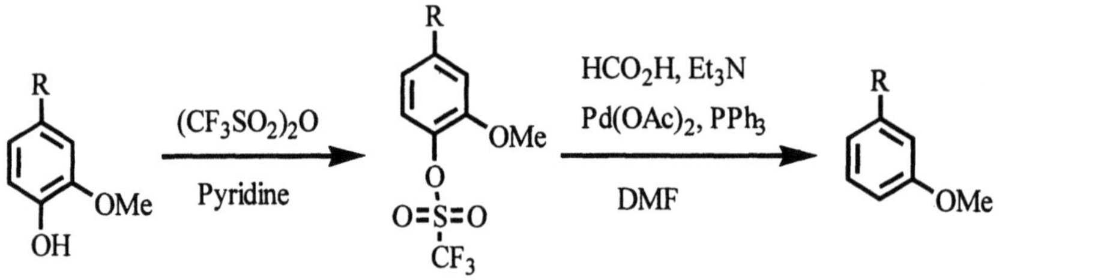

1: R = Pr
3: R = COMe
4: R = CH=CHCHO

1': R = Pr
3': R = COMe
4': R = CH=CHCHO

1": R = Pr
3": R = COMe
4": R = CH=CHCHO

Scheme 4. Catalytic hydrogenolysis of the phenolic hydroxyl groups in various lignin model compounds.

The hydrogenolysis/removal of the phenolic hydroxyl groups dramatically improved the photostability of the model compounds.[55] For example, when compound **1"** was impregnated onto a Whatman (No.1) filter paper and subjected to UV (~ 253 nm) irradiation, the paper showed much improved brightness stability compared to the paper impregnated with the phenol **1** (Figure 4). The sample also showed a slightly improved brightness stability/photostability than the paper impregnated with the methylated derivative **2** or the acetylated derivative **5** (Scheme 5 and Figure 4).[55]

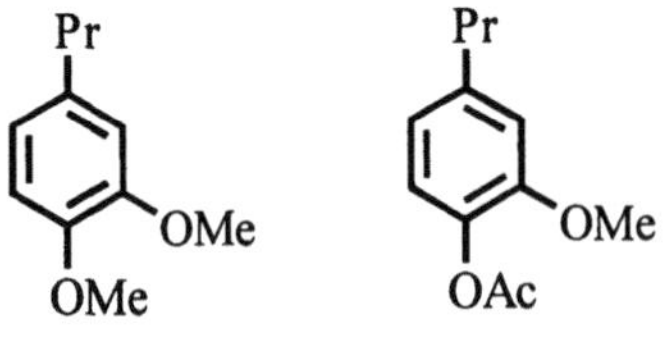

2 **5**

Scheme 5. Structures of the methylated and acetylated derivatives of **1**.

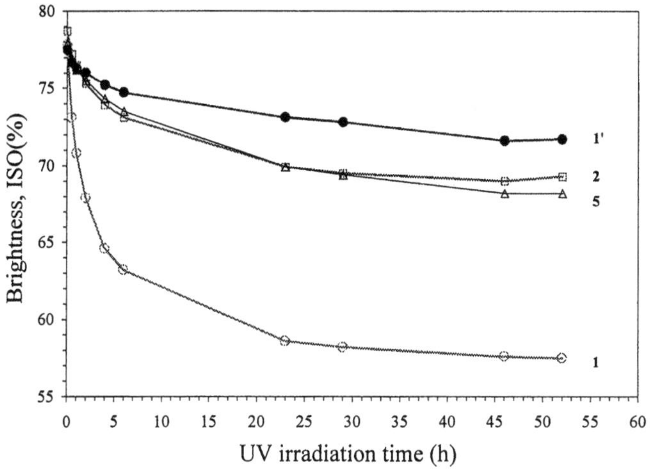

Figure 4. Brightness stability of the papers impregnated with compound **1"**, phenol **1**, the methylated derivative **2** and the acetylated derivative **5**.

When milled wood lignin (25 mg) dissolved in pyridine (1 mL) was treated with triflic anhydride (84 mg, 0.29 mmol) under similar conditions to those used for the reactions with the model compounds, darkening of the lignin was observed; the lignin solution turned from light yellow to deep red, indicating that reaction(s) other than triflation had also occurred.[55] Attempts to triflate a spruce, peroxide-bleached, thermomechancial pulp (BTMP) that had been solvent-exchanged to pyridine via an ethanol treatment also produced a significant darkening of the pulp. The discoloration encountered in the attempted reactions with milled wood lignin and mechanical pulps is likely due to acid-catalyzed chromophore formation in the lignin; triflic acid is expected to form upon the reaction of triflic anhydride with the hydroxyl groups in lignin and/or cellulose.

Next, we attempted to replace triflic anhydride with an acylating agent less capable of causing the darkening of lignin and pulps. The acetyl analogue of **1'**, 1-acetoxy-2-methoxy-4-propylbenzene (**5**), was prepared from the reaction of acetic anhydride with **1** in pyridine. When **5** was treated with HCO_2H and Et_3N in the presence of $Pd(OAc)_2$ and PPh_3, no reaction was observed, suggesting that the electron-withdrawing (polarizing) ability of the acetoxyl (-OAc) group was not strong enough for the subsequent hydrogenolysis to occur. Efforts to prepare the trifluoroacetyl analogue of **1'** from **1** and trifluoroacetic anhydride at various temperatures in the presence or absence of pyridine also failed (Scheme 6).[55]

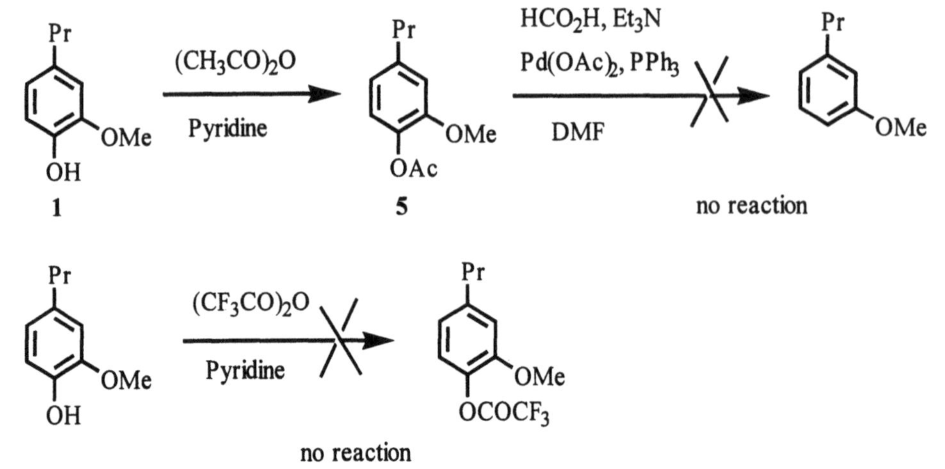

Scheme 6. Attempts to replace triflic anhydride with acetic anhydride or trifluoroacetic anhydride in the catalytic hydrogenolysis of **1**.

5. CATALYTIC H_2-REDUCTION OF α-CARBONYL/HYDROXYL GROUPS

The third method (method C, Scheme 2) we have developed is the catalytic H_2-reduction of lignin α-carbonyls or α-hydroxyls to α-methylene groups. This method is aimed at both the bleaching and the yellowing inhibition of mechanical wood pulps. Currently, bleaching of mechanical wood pulps in pulp mills is accomplished either reductively by sodium hydrosulfite (dithionite), $Na_2S_2O_4$, or oxidatively by alkaline hydrogen peroxide.[57, 58] During sodium hydrosulfite bleaching, the carbonyl groups (-C=O) in coniferaldehydes and quinones are reduced to the hydroxyl groups (-CH-OH), forming the coniferalcohols and hydroquinones, respectively. Reductive bleaching in this manner, however, is only temporary and reversible, because the reduced products/hydroxyl groups can undergo autooxidation readily to give the original chromophores/carbonyl groups. The minimization of air-contact on hydrosulfite-bleached mechanical pulps is often required to avoid a dramatic loss of brightness during storage.[57] Upon exposure to light, both hydrosulfite-bleached and peroxide-bleached mechanical pulps and papers lose their brightness (turn yellow) rapidly via the many pathways including those shown in Scheme 1. If the lignin carbonyls (C=O) can be reduced to methylene (CH_2) instead of just hydroxyl (CH-OH) groups, a bleaching effect similar to that of sodium hydrosulfite should be achievable, while producing a bleached pulp with improved photostability.

5.1. Catalytic Reduction of α-Carbonyl/Hydroxyl Group in Lignin Model Compounds under Heterogeneous Conditions

Catalytic conversion of the α-carbonyl group in acetovanillone (**3**) and 3,4-dimethoxyacetophenone (**6**) to the α-methylene group was achieved following a procedure reported for the reduction of simple aryl ketones,[59] which uses ammonium formate as the hydrogen source in acetic acid solution in the presence of a catalytic amount of Pd/C (Scheme 7). The yields for the methylene compounds **3m** and **6m** were

53.4 and 80.6%, respectively.[60] The presumed intermediates, the α-hydroxyl derivatives **3h** and **6h** were not detected, likely because of their facile conversion to **3m** and **6m**. Indeed, when subjected to the same reaction conditions, 1-(4'-hydroxy-3'-methoxyphenyl)ethanol (**3h**) was rapidly converted to **3m**.[60]

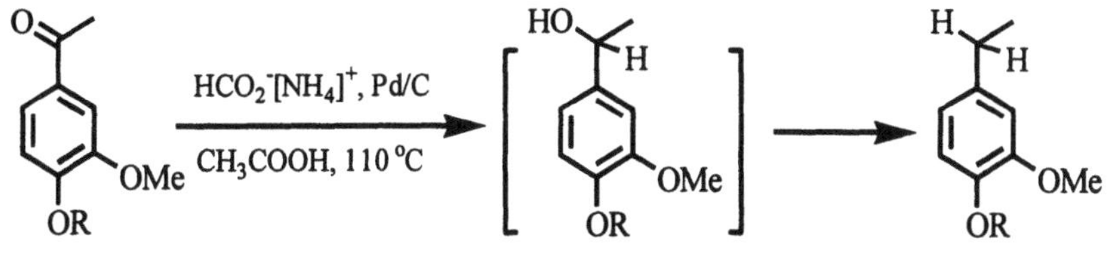

3: R = H
6: R = Me

3h: R = H
6h: R = Me

3m: R = H
6m: R = Me

Scheme 7. Catalytic conversion of the α-carbonyl to α-methylene group in lignin model compounds.

Although the above catalytic conversion represents an improvement over a previous non-catalytic method[61] in terms of the reductant cost (HCO_2NH_4 vs. Et_3SiH or $NaBH_4$) and the reaction pH/acidity (CH_3COOH vs. CF_3COOH), the catalyst is heterogeneous and is not suitable for reaction with pulp substrates. To see if such a chemical conversion could be catalyzed homogeneously, reaction of **6** was attempted in the presence of soluble $Pd(OAc)_2$ instead of the insoluble Pd/C; however, Pd metal precipitated, and no conversion of **6** was obtained.[60] Attempts to stabilize $Pd(OAc)_2$ with 2 equivalents of PPh_3 using either ammonium formate or H_2 (100 psi) as the reductant also failed to prevent the formation of Pd metal or affect any conversion of **6** to **6h** or **6m**.[60]

We have also evaluated a catalytic system based on the use of a rhodium diene complex and β-cyclodextrin that was reported to affect the conversion of C=O to CH_2 in simple aryl ketones.[62] Reduction of **6** at room temperature (~ 20 °C) under 1 atm of H_2 in the presence of di-μ-chloro-bis(η^4-1,5-hexadiene)dirhodium(I) and β-cyclodextrin gave **6m** as the major product and **6h** as the minor one (Scheme 8). For example, after ~16 h, **6** (at 60% conversion) was converted to **6h** (11%) and **6m** (49%). During the reaction, the solution turned from yellow (the initial color of the Rh catalyst solution) to black within 30 min, while the formation of metallic Rh was observed within 2 h. Further tests on this catalytic system suggested that the reaction was catalysed by colloidal Rh.[60]

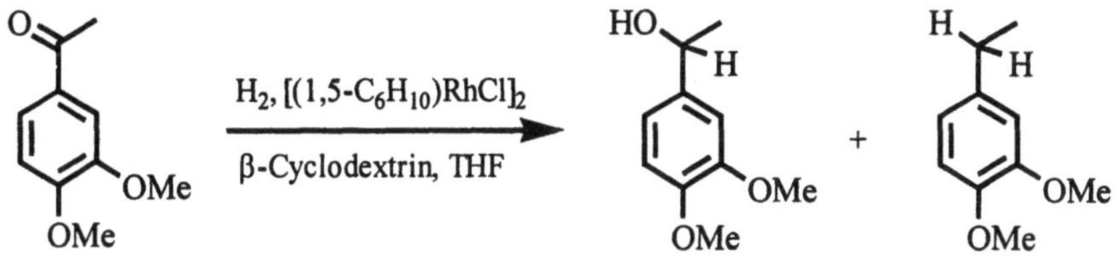

6 **6h** **6m**

Scheme 8. Catalytic H_2-reduction of the α-carbonyl group in 3,4-dimethoxyacetophenone (**6**).

5.2. Effect of Catalytic Reduction of α-Carbonyl/Hydroxyl Groups on Photostability

The effect of catalytic reduction of α-carbonyl and α-hydroxyl groups on the photostability of lignin model compounds was evaluated by impregnating the model compounds onto Whatman (No. 1) filter papers and then exposing the papers to UV (~253 nm) or fluorescent light (~ 570 nm).[60] Figure 5 shows the %ISO brightness values of the papers coated with **6**, **6h** and **6m**, respectively, versus the UV light exposure times. Reduction of the α-carbonyl to α-hydroxyl group (**6** to **6h**) provided only a small photostabilization to the paper, while a much higher photostabilization was obtained by reduction of the α-carbonyl to the α-methylene group (**6** to **6m**), particularly in the earlier stage of yellowing (loss of brightness). The advantage of fully reducing the α-carbonyl to α-methylene became much more evident when the papers coated with the model compounds were subjected to fluorescent light used to mimic paper-use in an office environment (Figure 6). Under such light exposure, the reduction of the α-carbonyl to the α-hydroxyl group was detrimental to the brightness stability of the paper after a short, initial stage of yellowing, but reduction to the α-methylene group provided complete brightness stability to the paper.

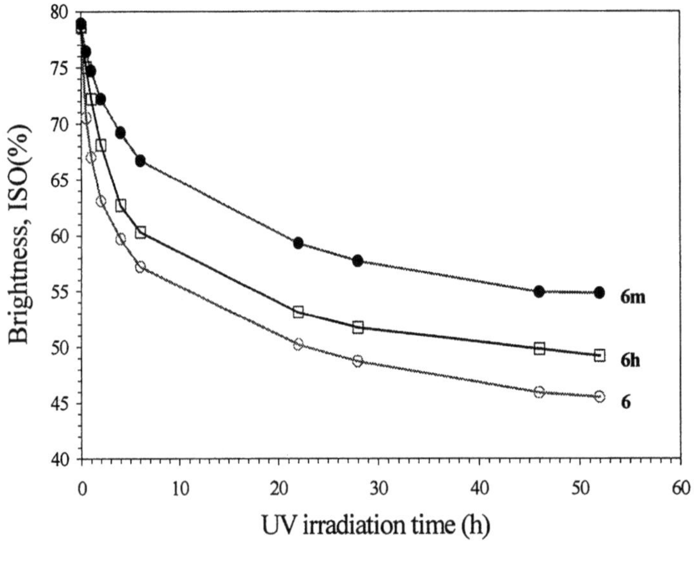

Figure 5. Brightness stability of the papers impregnated with **6**, **6h** and **6m** under UV irradiation.

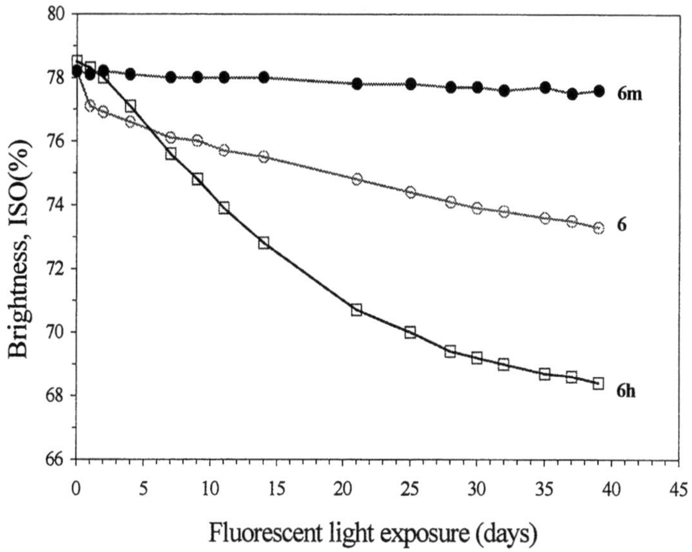

Figure 6. Brightness stability of the papers impregnated with **6**, **6h** and **6m** under fluorescent light.

5.3. Development of Water-soluble, Homogeneous Catalysts

In order to apply a method such as the catalytic, H_2-reduction of lignin α-carbonyl/hydroxyls to α-methylene groups within mechanical wood pulps, a water-soluble, homogeneous catalyst is required. Over the past few years, we have studied a large number of known catalysts as well as many new ones synthesized in our laboratories. Among them, the most promising catalyst is a new, water-soluble Ru complex.[63] This complex, 'Ru-Ligand$_n$' effectively catalyzes the H_2-reduction of the α-carbonyl to the α-methylene group in acetovanillone (**3**) under H_2 (500 psi) at 80 °C in a buffered aqueous solution to give predominantly (76%) or exclusively (> 99%) of the α-methylene product (**3m**), depending on the reaction conditions (Scheme 9). During the reaction, the solution remains homogeneous *without any precipitation of Ru metal.*

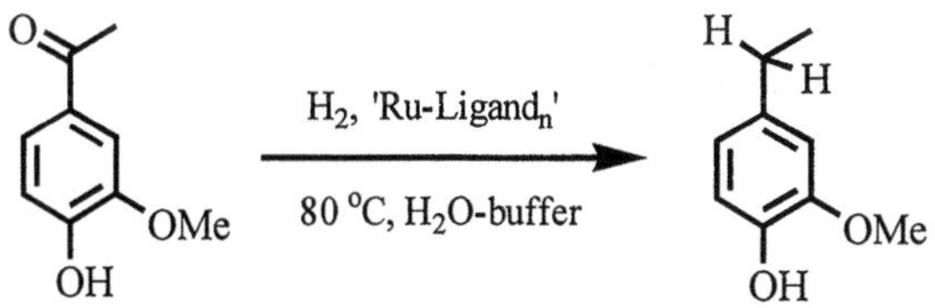

3 **3m**

Scheme 9. Aqueous, homogeneous H_2-reduction of acetovanillone (**3**).

More importantly, this Ru complex also catalyzes the H_2-reduction of lignin chromophores in mechanical wood pulps. Figure 7 shows the %ISO brightness values of an aspen chemithermomechanical pulp (CTMP) versus the amounts of the Ru complex applied to the H_2-treatment of the pulp. The unbleached aspen CTMP had an ISO brightness of 65.8%. When the pulp was subjected to just hydrogen (340 psi) at 1.5% 'consistency' (percentage of pulp in a mixture of pulp and water) in H_2O at 100 °C for 3 h, the ISO brightness dropped to 59.3% due to the heat-induced darkening of the pulp. In the presence of the Ru complex, for example, 0.25% Ru on pulp, the H_2-treatment gave a pulp with an ISO brightness of 72.5% (Figure 7). This is the first example of pulp bleaching with H_2 ('H_2-bleaching'), and we are still in the process of optimizing the reaction conditions and evaluating the potential of the system.[63]

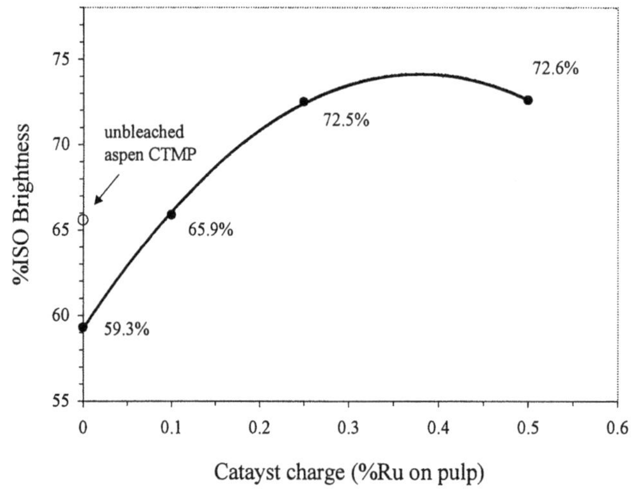

Figure 7. Catalytic, H_2-bleaching of aspen CTMP in the presence of a water-soluble Ru catalyst.

6. CONCLUSIONS

The three methods developed by us provide a complete or partial photostabilization to the various lignin model compounds studied. The hydrogenation of lignin aromatics was the most effective, but the reaction was hampered by the heterogeneous nature and/or the large size of the catalyst, as well as by the rather slow conversion. Although several catalysts were claimed in the literature to be homogeneous for the hydrogenation of simple aromatic compounds, our detailed studies revealed that none of them was truly so, particularly when applied to the hydrogenation of lignin aromatics, except perhaps for the sulfonated triphenylphosphine systems (e.g. **XI** in Table 2). Development of a water-soluble, homogeneous catalyst capable of hydrogenating the aromatic rings of lignin in isolated lignin samples and in mechanical wood pulps remains a significant challenge.

Catalytic hydrogenolysis of lignin phenols may provide a more selective photostabilization to lignin-containing mechanical wood pulps, but the requirement of derivatizing the lignin phenols with triflic anhydride prior to the hydrogenolysis makes it difficult to extend the method beyond model compounds because of the darkening of lignin and pulp by triflic anhydride.

Among the three methods developed by us, catalytic H_2-reduction of lignin α-carbonyl (C=O) and/or α-hydroxyl (CH-OH) to α-methylene (CH_2) groups using a water-soluble, homogeneous, pulp-compatible catalyst is the most promising. Future efforts should focus on the understanding and optimization of this catalytic transformation.

7. ACKNOWLEDGEMENTS

We thank: the Natural Science and Engineering Research Council of Canada for financial support via the Mechanical and Chemimechanical Pulps Network and a Research Grant; Paprican for financial support; Johnson Matthey Ltd. and Colonial Metals Inc. for loans of $RuCl_3$; and the postdoctoral fellows Drs. Terrance Y. H. Wong, Graham C. Cairns and Andy Z. Lu for their major contributions to this research.

8. REFERENCES

1. C. J. Biermann, *Handbook of Pulping and Papermaking* (Academic Press, San Diego, 1996).
2. *International Fact and Price Book* (Pulp Paper Intern., Miller Freeman, Brussels, 2001), p. 278.
3. I. A. Pearl, *The Chemistry of Lignin* (Marcel Dekker, New York, 1967), p. 292.
4. G. J. Leary, Recent progress in understanding and inhibiting the light-induced yellowing of mechanical pulps, *J. Pulp Paper Sci.* **26**(6), J154-J160 (1994), and references therein.
5. K. P. Kringstad and S. Y. Lin, Mechanism in the yellowing of high-yield pulps by light – structure and reactivity of free radical intermediates in the photodegradation of lignin, *Tappi J.* **53**(12), 2296-2301 (1970).
6. A. Castellan, N. Colombo, P. Fornier De Violet, A. Nourmamode, and H. Bouas-Laurent, Photodegradation of lignin: A photochemical study of bleached CTMP and lignin model molecules, *Proc. 5th Intern. Symp. Wood Pulping Chem. Vol. 1* (Raleigh, North Carolina, 1989), pp. 421-430.
7. J. Gierer and S. Y. Lin, Photodegradation of lignin. A contribution to the mechanism of chromophore formation, *Svensk Papperstidn.* **75**, 233-239 (1972).
8. C. Vanucci, P. Fornier De Violet, H. Bouas-Laurent, and A. Castellan, Photodegradation of lignin. A photophysical and photochemical study of a non-phenolic α-carbonyl β-O-4 lignin model dimer: 3,4-dimethyoxy-α-(2'-methoxyphenoxy)acetophenone, *J. Photochem. Photobiol. A: Chem.* **41**, 251-265 (1988).
9. J. C. Scaiano, J. C. Netto-Ferreira, and V. Wintgens, Fragmentation of ketyl radicals derived from α-phenoxyacetophenone: an important mode of decay for lignin-related radicals? *J. Photochem. Photobiol. A: Chem.* **59**, 265-268 (1991).
10. J. A. Schmidt and C. Heitner, Light-induced yellowing of mechanical and ultra-high yield pulps. Part 2. Radical-induced cleavage of etherified guaiacylglycerol-β-arylether groups is the main degradative pathway, *J. Wood Chem. Technol.* **13**(3), 309-325 (1993).
11. R. P. Singh, Bleaching of groundwood pulp: Investigation of methods based on chemical modification of the wood pulp, *Tappi J.* **49**(7), 281-286 (1966).
12. V. Loras, Bleaching and stabilization of flash-dried mechanical pulp, *Pulp Paper Mag. Can.* **69**(1), 57-63 (1968).
13. G. J. Leary, The yellowing of wood by light: part II, *Tappi J.* **51**(6), 257-260 (1968).
14. A. F. A. Wallis and R. H. Wearne, Hydroxyethylation of lignin in mechanical pulps, *Appita J.* **36**(3), 192-196 (1982).
15. A. Castellan, A. Nourmamode, C. Noutary, and D. Lachenal, Improvement of the photostability of peroxide bleached chemithermomechanical pulps using cationic phase transfer catalyst and hydroxyl blocking agents, *Cellulose Chem. Technol.* **26**, 451-459 (1992).

16. M. Paulsson, R. Simonson, and U. Westermark, Chemical modification of lignin-rich paper, Part 2. Photostabilization by acetylation of paper made from spruce TMP and aspen CTMP, *Nordic Pulp Paper Res. J.* **11**(2), 109-114 (1996).
17. S. Y. Lin and K. P. Kringstad, Stabilization of lignin and lignin model compounds to photodegradation, *Tappi J.* **53**(9), 1675-1677 (1970).
18. U. Tschirner and C.W. Dence, Attempts to photostabilize Norway spruce TMP by chemical modification, *Paper & Timber* **4**, 338-346 (1988).
19. A. Hatakka, A. Mettala, L. Rouhiainen, and L. Viikari, The effect of quinone-reducing and phenol-methylating enzymes on the yellowing of mechanical pulp, *Holzforschung* **48**, 82-88 (1994).
20. S. Omori, R. C. Francis, and C. W. Dence, Studies on the mechanism of the photoyellowing of bleached mechanical and chemimechanical pulps. II. The role of photosensitizers, *Holzforschung* **45**(suppl.), 93-100 (1991).
21. J. A. Moulijn, P. W. N. M. van Leeuwen, and R. A. van Santen, *Catalysis: An Integrated Approach to Homogeneous, Heterogeneous and Industrial Catalysis* (Elsevier, Amsterdam, 1993).
22. R. A. Sheldon, Consider the environmental quotient, *Chemtech.* **24**(3), 38-47 (1994).
23. W. Keim, Vor- und Nachtelle der homogenen Obergangsmetallkatalyse, dargestellt am SHOP-Prozeβ, *Chem. Ing. Tech.* **56**, 850-853 (1984).
24. M. Lougheed and J. R. Presley, The upgrading of a bleach plant: mill experience with a two-stage bleach plant retrofit, *Proc. Tappi Pulping Conf. Book 1* (Seattle, Washington, 1989), pp. 261-265.
25. J. N. Armor, The multiple roles for catalysis in the production of H_2, *Appl. Catal. A.* **176**, 159-176 (1999).
26. M. Jacoby, Fuel cells heading for sale, *Chem. Eng. News* (June 14), 31-37 (1999).
27. S. W. Eachus and C. W. Dence, Hydrogenation of lignin model compounds in the presence of a homogeneous catalyst, *Holzforschung* **29**(2), 41-48 (1975).
28. D. G. B. Boocock, D. Mackay, M. McPherson, S. Nadeau, and R. Thurier, Direct hydrogenation of hybrid poplar wood to liquid and gaseous fuels, *Can. J. Chem. Eng.* **57**(2), 98-101 (1979).
29. J. M. Pepper and M. D. Rahman, Lignin and related compounds. XI. Selective degradation of aspen poplar lignin by catalytic hydrogenolysis, *Cellulose Chem. Technol.* **21**, 233-239 (1987).
30. J. S. Shattai and W. W. Zmierczak, Process for conversion of lignin to reformulated hydrocarbon gasoline, U.S. Patent: 5,959,167 (1999).
31. E. E. Harris, J. D'Ianni, and H. Adkins, Reaction of hardwood lignin with hydrogen, *J. Am. Chem. Soc.* **60**, 1467-1470 (1938).
32. H. P. Godard, J. L. MaCarthy, and H. Hibbert, Hydrogenation of wood, *J. Am. Chem. Soc.* **62**, 988 (1940).
33. H. Greenfield, Studies in nuclear hydrogenation, *Ann. N.Y. Acad. Sci.* **214**, 233-242 (1973).
34. T. Q. Hu, B. R. James, and C.-L. Lee, Towards inhibition of yellowing of mechanical pulps. Part I: Catalytic hydrogenation of lignin model compounds under mild conditions, *J. Pulp Paper Sci.* **23**(4), J153-J156 (1997).
35. T. Q. Hu, B. R. James, and C.-L. Lee, Towards inhibition of yellowing of mechanical pulps. Part II: Water-soluble catalysts for the hydrogenation of lignin model compounds, *J. Pulp Paper Sci.* **23**(5), J200-J205 (1997).
36. B. R. James, Y. Wang, C. S. Alexander, and T. Q. Hu, Catalytic hydrogenation of aromatic rings in lignin, *Chem. Ind.* **75**, 233-242 (1998).
37. T. Q. Hu, B. R. James, and Y. Wang, Towards inhibition of yellowing of mechanical pulps. Part III: Hydrogenation of milled wood lignin, *J. Pulp Paper Sci.* **25**(9), 312-317 (1999).
38. K. R. Januszkiewicz and H. Alper, Exceedingly mild, selective and stereospecific phase-transfer-catalyzed hydrogenation of arenes, *Organometallics* **2**, 1055-1057 (1983).
39. B. R. James, Y. Wang, and T. Q. Hu, Hydrogenation of substituted aromatics, *Chem. Ind.* **68**, 423-428 (1996).
40. J. Schulz, A. Roucoux, and H. Patin, Unprecedented efficient hydrogenation of arenes in biphasic liquid-liquid catalysis by re-usable aqueous colloidal suspensions of rhodium, *Chem. Commun.* 535-536 (1999).
41. R. J. Bonilla, B. R. James, and P. G. Jessop, Colloid-catalysed arene hydrogenation in aqueous/supercritical fluid biphasic media, *Chem. Commun.* 941-942 (2000)
42. T. Q. Hu, B. R. James, S. J. Rettig, and C.-L. Lee, Stereoselective hydrogenation of lignin degradation model compounds, *Can. J. Chem.* **75**, 1234-1239 (1997).
43. M. A. Bennett, T.-N. Huang, A. K. Smith, and T. W. Turney, An arene hydrido-complex of ruthenium(II) as catalyst for homogeneous hydrogenation of benzene and olefins, *J. Chem. Soc. Dalton Trans.* 233-234 (1974).
44. M. A. Bennett, T. N. Huang, and T. W. Turney, Dinuclear arene hydrido-complexes of ruthenium(II): reactions with olefins and catalysis of homogeneous hydrogenation of arenes, *J. Chem. Soc. Chem. Commun.* 312-314 (1979).

45. M. A. Bennett, Reducing the aromatic ring, *Chemtech* **7**, 444-446 (1980).
46. G. Meister, G. Rheinwald, H. Stoeckli-Evans, and G. Suss-Fink, Hydrogen activation by arene ruthenium complexes in aqueous solution, Part 2. Build-up of cationic tri- and tetra-nuclear ruthenium clusters with hydrido ligands, *J. Chem. Soc. Dalton Trans.* 3215-3223 (1994).
47. L. Plasseraud and G. Suss-Fink, Catalytic hydrogenation of benzene derivatives under biphasic conditions, *J. Organomet. Chem.* **539**, 163-170 (1997).
48. T. Y. H. Wong, R. Pratt, C. G. Leong, B. R. James, and T. Q. Hu, Catalytic hydrogenation of lignin aromatics using Ru-arene complexes, *Chem. Ind.* **82**, 255-266 (2000).
49. A. Bjorkman, Studies on finely divided wood, Part I. Extraction of lignin with neutral solvents, *Svensk Papperstidn.* **59**, 477-485 (1956).
50. H. H. Brownell, Isolation of milled wood lignin and lignin carbohydrate complex. Part II of a study on the lignin-carbohydrate bond, *Tappi* **48**(9), 513-519 (1965).
51. T. Q. Hu and B. R. James, Towards inhibition of yellowing of mechanical pulps. Part IV: Photostability of hydrogenated lignin model compounds, *J. Pulp Paper Sci.* **26**(5), 173-175 (2000).
52. D. H. Andrews and R.P. Singh, Canadian Patent: 810,571 (1966).
53. J. A. Schmidt and C. Heitner, Light-induced yellowing of mechanical and ultra-high yield pulps. Part 1. Effect of methylation, $NaBH_4$ reduction and ascorbic acid on chromophore formation, *J. Wood Chem. Tech.* **11**(4), 397 (1991).
54. A. Z. Lu, T. Q. Hu, D. A. Osmond, B. O. Patrick, and B. R. James, Tetrazole ethers from lignin model phenols: synthesis, crystal structures, and photostability, *Can. J. Chem.* **79**, 1201-1206 (2001).
55. T. Q. Hu, G. R. Cairns, and B. R. James, Removal of phenolic hydroxyl groups in lignin model compounds and its effect on photostability, *Holzforschung* **54**, 127-132 (2000); and unpublished results.
56. S. Cacchi, P. G. Ciattini, E. Morera, and G. Ortar, Palladium-catalysed triethylammonium formate reduction of aryl triflates. A selective method for the deoxygenation of phenols, *Tet. Lett.* **27**(45), 5541-5544 (1986).
57. M. E. Ellis, The technology of mechanical pulp bleaching, chapter 2: hydrosulfite (dithionite) bleaching, in: *Pulp Bleaching – Principle and Practice*, edited by C. W. Dence and D. Reeve (Tappi Press, Atlanta, 1996), pp. 491-512.
58. J. R. Presley and R. T. Hill, The technology of mechanical pulp bleaching, chapter 1: peroxide bleaching of (chemi)mechanical pulps, in: *Pulp Bleaching – Principle and Practice*, edited by C. W. Dence and D. Reeve (Tappi Press, Atlanta, 1996), pp. 457-489.
59. S. Ram and L. D. Spicer, Reduction of aldehydes and ketones to methylene derivatives using ammonium formate as a catalytic hydrogen transfer agent, *Tet. Lett.* **29**(31), 3741-3744 (1988).
60. T. Q. Hu, G. Cairns, and B. R. James, Towards inhibition of yellowing of mechanical pulps – catalytic modification of lignin non-aromatic units, *Proc. 11th Intern. Symp. Wood Pulping Chem. - Vol. I* (Nice, France, 2001), pp. 219-222; and unpublished results.
61. T. Q. Hu, G. Leary, and D. Wong, A new approach towards the yellowing inhibition of mechanical pulps. Part I: Selective removal of α-hydroxyl and α-carbonyl groups in lignin model compounds, *Holzforschung* **53**(1), 43-48 (1999).
62. H. A. Zahaika and H. Alper, β-Cyclodextrin-promoted, rhodium(I)-catalyzed conversion of carbonyl compounds to hydrocarbons under remarkably mild conditions, *Organometallics* **5**, 1908-1911 (1986).
63. T. Q. Hu, B. R. James, T. H. Y. Wong, and A. Z. Lu, Bleaching of mechanical wood pulps with hydrogen in the presence of a ruthenium catalyst, *Patent Disclosure, C198*, Pulp and Paper Research Institute of Canada, February 2000; and results to be published.

CHARACTERISTICS, INDUSTRIAL SOURCES, AND UTILIZATION OF LIGNINS FROM NON-WOOD PLANTS

Jairo H. Lora*

1. INTRODUCTION

Non-wood fibers were the first material ever used by mankind for paper production and had played a dominant role until about one hundred years ago. Today, non-wood fibers such as bagasse and straw remain an important pulping raw material in certain developing countries with limited forest resources. In addition, non-wood fibers such as abaca, sisal, flax, jute, and hemp are used to produce specialty papers in several countries.

When compared with wood, non-wood fibres are at a logistic disadvantage due to their high bulkiness, and the fact that they are produced seasonally, which gives rise to transportation and storage problems. As a result, non-wood fibre mills are normally small in capacity, and can not benefit from the economies of scale enjoyed by wood-based operations.

The vast majority of non-wood fiber mills do not have a recovery system for the pulping chemicals because of the capital-intensive nature of such processes. As a result, the spent pulping liquors are an environmental liability. Although for many years mills have been allowed to release the liquors into the environment with little or no treatment, the situation is changing. China and India, the main producers of non-wood pulps are enacting legislation to decrease the pollution from small mills. The lignin present in the liquor constitutes a potential co-product that, if recovered, could not only decrease the environmental impact of non-wood pulping operations but also contribute to their financial well being. In this chapter we will review the characteristics of non-wood lignins, their sources and uses.

* Jairo H. Lora, Granit S. A., 7 Camby Chase, Media, PA 19063-4545, USA.

Chemical Modification, Properties, and Usage of Lignin
Edited by Thomas Q. Hu, Kluwer Academic/Plenum Publishers, 2002

2. CHARACTERISTICS

Although non-wood lignins have been the subject of considerable less study than lignins from wood, significant progress has been made in the understanding of their native structure. From the standpoint of chemical composition, non-wood lignins are more complex than their wood counterparts. Softwoods contain predominantly guaiacyl-type lignins and hardwoods possess a mixture of guaiacyl and syringyl-type lignins, while most non-wood lignins have significant amounts of p-hydroxyphenylpropane nuclei in addition to syringyl and guaiacyl units.

Degradation techniques are often used to infer the type and quantity of aromatic nuclei in lignins. Thioacidolysis and nitrobenzene oxidation are analytical methods that give estimates of the type and abundance of the aromatic nuclei involved in units linked by β-O-4 bonds and in non-condensed units, respectively. Tables 1 and 2 illustrate the results obtained when various authors applied these techniques to a range of non-wood lignins, either *in situ* or after isolation. As can be seen, the proportion of the three types of nuclei varies for different non-wood species. While kenaf and jute are high in non-condensed syringyl units, and have low, or lack, *p*-hydroxyphenylpropane, wheat straw and bagasse are high in p-hydroxyphenylpropane and have syringyl and guaiacyl non-condensed units in roughly equal quantities. The data also illustrates the fact that for a given species there is variability depending on cultivar type and on the specific part of the plant.

Table 1. Molar ratios of syringyl (S), guaiacyl (G), and p-hydroxyphenylpropane (H) units upon thiacidolysis of non-wood lignins

Species	S	G	H	Ref
Wheat straw cv. Champlain - *in situ*	46	49	5	1
Wheat straw cv. Capitole - *in situ*	53	43	4	1
Wheat straw – alkali soluble	55	40	5	2
Wheat straw - *in situ*	43	49	8	3
Rice straw cv. Balilla norm - *in situ*	40	45	15	1
Rice straw cv. Balilla brittle - *in situ*	47	45	8	1
Kenaf –bast fibers – cv Everglades	78	22	n. d.	4
Kenaf – inner bast – cv. Everglades	60	40	n. d.	4
Kenaf –woody core- cv. Everglades	58	42	n. d.	4
Kenaf –bast fibers – cv Aokawa-3	79	21	n. d.	4
Kenaf – inner bast – cv. Aokawa-3	60	40	n. d.	4
Kenaf –woody core- cv. Aokawa-3	58	42	n. d.	4
Kenaf –bast fibers – cv Metsa	71	29	n. d.	4
Kenaf – inner bast – cv. Metsa	70	30	n. d.	4
Kenaf –woody core- cv. Metsa	66	34	n. d.	4
Kenaf – cell wall preparation	84	16	n. d.	5

n. d. = not detected.

Table 2. Molar ratios of syringyl (S), guaiacyl (G), and p-hydroxyphenylpropane (H) nitrobenzene oxidation products

Species	S	G	H	Ref
Wheat straw internodes - *in situ*	45	46	9	6
Wheat straw nodes - *in situ*,	52	36	13	6
Wheat straw - alkali soluble	46	43	11	2
Wheat straw - cell wall residue	37	41	22	2
Wheat straw – milled straw lignin	37	48	15	7
Wheat straw – *in situ*	43	50	7	3
Wheat straw – milled straw lignin	34	51	15	3
Wheat straw – *in situ*	33	52	15	8
Wheat straw – milled straw lignin	43	49	8	9
Wheat straw - alkali soluble	48	48	4	9
Wheat straw – acetic acid lignin	45	44	11	9
Bagasse – Kraft lignin	35	38	27	10
Kenaf - bast fibers	79	20	1	11
Kenaf – woody core	66	31	3	11
Kenaf –bast fibers – cv Everglades	79	21	n. d.	4
Kenaf – inner bast – cv. Everglades	58	42	n. d.	4
Kenaf –woody core- cv. Everglades	57	43	n. d.	4
Kenaf –bast fibers – cv Aokawa-3	80	20	n. d.	4
Kenaf – inner bast – cv. Aokawa-3	60	40	n. d.	4
Kenaf –woody core- cv. Aokawa-3	58	42	n. d.	4
Kenaf –bast fibers – cv Metsa	68	32	n. d.	4
Kenaf – inner bast – cv. Metsa	72	28	n. d.	4
Kenaf –woody core- cv. Metsa	66	34	n. d.	4
Hemp – bast fibers	47	53	n. d.	12
Hemp – woody core	44	56	n. d.	12
Jute - woody core	58	42	n. d.	13
Jute - bast fibers	67	33	n. d.	13

n. d. = not detected.

There is a general agreement in the literature that in bamboo, and in grasses, such as wheat straw, sugar cane bagasse, and reed canary grass, among others, the majority of the p-hydroxyphenylpropane units appear as p-coumaric ester groups linked to the lignin core structure.[9, 14-17] Recent work has established that the ester linkage is exclusively connected to the γ-position of the lignin aliphatic chain.[18] A smaller proportion of p-coumaric acid groups is connected to the lignin via ether groups or carbon-to-carbon bonds. Other phenolic acids such as ferulic acid esters and ethers are also incorporated in certain non-wood lignin.[16, 19-23]

Phenolic acids etherified to the lignin are a crosslinking bridge for cell wall polymers, since such acids can also be attached to the carbohydrates via an ester link.[24, 25] Such lignin carbohydrate complexes are very abundant in certain non-woods. For instance, most of the lignin in rice straw is chemically bonded to the hemicelluloses, according to Yu et al.[26, 27] The occurrence of lignin-carbohydrate complexes presents a challenge for the industrial production of high purity lignins.

Another structural feature that may occur in native non-wood lignins is acylation. For instance, according to Ralph, kenaf lignin is about 50% acetylated.[18]

Based on degradation studies, inferences can be made about the degree of condensation of *in situ* and isolated lignins. Thus, wheat straw lignins have fewer β-O-4 linkages and are more condensed (i.e., have more carbon-to-carbon bonds) than those from wood.[2, 21]

NMR and FTIR have been used to elucidate the structure of non-wood lignins. Most of the ^{13}C NMR work found in the literature has been of a qualitative nature, including studies on wheat straw[21], jute[13] and bagasse.[16, 28] Quantitative ^{13}C NMR techniques have been used to estimate condensed and non-condensed aromatic nuclei in rice straw lignin. The results show S:G:H ratios of 24:34:42 for the non-condensed units, and 20:44:36 for the total (condensed + non-condensed) nuclei.[29] Wheat straw lignins before and after biodegradation have been characterized by ^{31}P NMR, a technique that gives a good quantitative picture of the nature of the different hydroxyl groups present.[30]

Compared with the work on native non-wood lignins, a lot less has been published on the properties of non-wood lignins recovered from industrial black liquors. Properties such as chemical functionality, purity, and molecular weight play a significant role in determining performance in (and suitability for) specific applications. These parameters are determined not only by the structure and characteristics of the native lignin, but also by the pulping and lignin recovery processes. Table 3 shows the data on aliphatic (R) and aromatic (Ar) hydroxyl, carboxyl, and methoxyl group for non-wood lignins obtained from pulping liquors from various commercial and semi-commercial processes. The hydroxyl and carboxyl groups are very instrumental for the chemical modification of lignin for various industrial uses.

Table 3. Functional groups in selected non-wood lignins (moles per lignin C_9 unit)

Species	Pulping Process	R-OH	Ar-OH	COOH	OCH_3	Ref
Wheat straw	Soda	0.26	0.38	0.28	0.71	31
Wheat straw	Soda	0.68	0.56	0.35	0.89	32
Wheat straw	Soda-AQ	0.87	0.75	n. m.	1.02	9
Wheat straw	Alkaline Sulfite with AQ and formaldehyde	n. m.	n. m.	0.26	n. m.	33
Wheat straw	Solvent (Acetic Acid)	0.65	0.59	n. m	1.02	9
Wheat straw	Solvent (Alcohol)	0.62-0.82	0.39-0.65	0.09-0.12	0.75-0.79	34
Hemp	Soda	0.57	0.49	0.14	0.86	34
Bagasse	Kraft	n. m.	0.51	0.21	0.85	10
Bagasse	Autohydrolysis Extraction	0.34	0.44	n. m.	0.56	35
Bagasse	Solvent (Acetic Acid)	n. m.	0.77	n. m.	0.80	36
Reed	Solvent (Alcohol)	0.52-0.56	0.50-0.62	0.08-0.09	0.95-1.01	34
Kenaf (whole)	Solvent (Alcohol)	0.76	0.47	0.04	1	37
Kenaf (core)	Solvent (Alcohol)	0.58	0.43	0.09	0.95	37

n. m.= not measured.

Wheat straw lignosulfonates from alkaline sulfite pulping in the presence of AQ and formaldehyde have a high sulfonic acid content (about 0.83 mol per C_9 unit) with nearly 80% on the side chain and the balance on the aromatic ring.[33]

The non-substituted C-3 and C-5 aromatic positions of the lignin molecule are important for lignin derivatization since those are the sites where alkali-catalyzed aromatic substitution reactions can take place, as needed when trying to increase the reactivity of the lignin by methylolation (for instance, for phenolic resins) or methylsulfonation (for dispersants). Data available on this key parameter suggests that some non-wood lignins may have equal or greater number of non-substituted C-3 and C-5 aromatic carbons than softwood kraft, which is considered a benchmark for industrial production of lignin derivatives (See Table 4).

Table 4. Number of non-substituted C-3 and C-5 positions in non-wood lignins and a softwood kraft lignin

Species	Pulping Process	Non-substituted C-3 & C-5 per C_9	Ref
Bagasse	Soda (mild pulping)	0.7	38
Wheat straw	Soda	0.45	32
Wheat straw	Solvent (Alcohol)	0.64 – 0.83	34
Reed	Solvent (Alcohol)	1.0 – 1.1	34
Pine	Kraft	0.7	32

Another important factor in the potential utilization of lignins is their purity. Non-wood lignins may have three significant extraneous components: sugars, silica and proteins.[21, 40] As discussed earlier, native non-wood lignins may have co-valent bonds with the hemicelluloses. Therefore, it is not surprising to find lignin-carbohydrate complexes in soda pulping liquors,[39] which may lead to the presence of sugars in non-wood lignin preparations. Gosselink et al. reported that soda lignins recovered at the pilot level from industrial black liquors from flax, hemp and wheat straw have 2.1 to 3.2% sugars, 1.9 to 4.9% ash (due in part to silica) and 0.76 to 1.61% nitrogen (equivalent to 4.8 to 10.1% protein).[41] Such levels of contaminants are considered an improvement over soda lignins obtained in the past, which had purity less suitable for value-added utilization.[42] Even higher purity can be attained for non-wood organosolv lignins, which can usually be produced with less than 1.0% sugars and ash.[37]

Other properties of relevance to lignin utilization include molecular weight and thermal characteristics. Although the molecular weight data presented in Table 5 is somewhat scattered, it suggests that non-wood lignins have a relatively low molecular weight, with organosolv lignins at the lower end of the range. A recent round robin among several European labs has identified significant variability in results of molecular weight determinations by size exclusion chromatography for non-wood lignins, which points out to the need for standard procedures in this area.[41]

Lignosulfonates obtained by alkaline sulfite pulping of wheat straw in the presence of formaldehyde and AQ are also reported to have lower molecular weight than wood lignosulfonates and methylsulfonated kraft lignins.[33]

Little information has been published on the thermal properties of non-wood lignins. The data in Table 6 suggests that soda lignins have a higher glass transition temperature than organosolv lignins, which is probably related to the lower molecular weight of the latter. Abächerli has recently reported that alkali flax, sisal, abaca, and jute lignins exhibit thermal flow at 120 to 180 °C.[46]

Table 5. Weight (M_w) and number (M_n) average molecular weight of selected non-wood lignins

Species	Pulping Process	M_w	M_n	Ref
Wheat straw	Soda	4400	1200	32
Wheat straw	Soda	Not reported	1400	41
Wheat straw	Soda-AQ	3270	1770	9
Wheat straw	Solvent (Acetic Acid)	4430	2020	9
Wheat straw	Solvent (Alcohol)	Not reported	800	43
Flax	Soda	Not reported	600	41
Hemp	Soda	Not reported	600	41
Bagasse	Solvent (Acetone)	2740	840	44
Bagasse	Solvent (Acetic Acid)	Not reported	830	36
Bagasse	Autohydrolysis	4360	420	45
Reed	Solvent (Alcohol)	1410-1480	650-680	34

Table 6. Glass transition temperature of selected non-wood lignins

Species	Pulping Process	Glass Transition Temperature, °C	Ref
Wheat straw	Soda-AQ	160 (crude), 185 (purified)	9
Wheat straw	Solvent (Acetic acid)	142 (crude), 172 (purified)	9
Bagasse	Solvent (Acetic acid)	170	36
Hemp	Soda	158 (1 bar), 103 (20 bar)	34
Reed	Solvent (alcohol)	96 – 97 (1 bar), 63 - 72 (20 bar)	34
Wheat straw	Solvent (alcohol)	106 – 122 (1 bar), 61 – 89 (20 bar)	34
Kenaf	Solvent (alcohol)	66 – 70 (1 bar), 104 – 108 (20 bar)	37

3. INDUSTRIAL SOURCES

Most industrial non-wood pulping is done with alkaline processes, including the soda, soda-AQ, alkaline sulfite, and kraft processes. Soda and soda-AQ are the predominant processes, since in general non-wood fibers are easy to delignify because of their relatively low lignin content and their accessible morphology. Soda pulping is almost always employed for the most widely used feedstocks such as sugarcane bagasse and wheat straw. The alkaline sulfite process is often used to pulp specialty fibers such as abaca, flax, and hemp. In some countries some non-wood mills switch pulping processes depending on the relative price of sodium sulfite and caustic soda. The kraft process is used commercially for bamboo and kenaf.

The liquors potentially available for lignin recovery are primarily from small soda mills that cannot afford a recovery system and that may be under pressure to either treat their effluent or close down. A second source is larger mills that may have a recovery process but want to incrementally expand their pulping capacity and need an economic way to handle the excess liquor generated.

Several researchers have proposed lignin recovery from soda black liquors as a strategy to diminish the Chemical Oxygen Demand (COD) of the effluents. The proposed processes typically involve lignin precipitation by acidification followed by liquid solids separation.[47-50] One of the difficulties of this procedure when dealing with non-wood fibers is that the spent liquors from alkaline pulping often contain silica, which may co-precipitate with the lignin, rendering it of lower quality. According to Gratzl and Chen,[42] precipitation processes used in Asia have not been successful to separate lignin from silica. The LPS® precipitation process recently introduced commercially in Europe is claimed to have solved this problem and to have obtained soda lignins with relatively low ash and silica content.[51]

Although lignin recovery eliminates a substantial portion of the effluent load, it is by no means a complete solution to the environmental problems of non-wood mills, and additional treatment may be required to achieve acceptable discharge limits. Fortunately, the remaining effluent appears to be amenable to biological treatment. Indian and Chinese researchers have reported on the successful biodegradation of such an effluent with anaerobic as well as photosynthetic organisms.[52-54]

In the case of alkaline sulfite pulping liquors, Chinese researchers have also proposed lignin removal by acidification with sulfur dioxide[55] or with sulfuric acid[56] as an effective way to reduce the pollution load. Surprisingly good results are claimed, in spite of the more hydrophilic nature of the lignin present in these liquors.

One further variant on alkaline sulfite pulping involves the use of formaldehyde and anthraquinone as pulping additives, which leads to the formation of highly sulfonated lignins in the spent liquors.[33, 57] It is doubtful that a pulping method involving formaldehyde could have widespread industrial use, but the liquors obtained have properties suitable for a variety of applications and, therefore, whole liquor utilization has been proposed as a preferred way for disposal.

More recently, novel pulping processes have been proposed for processing non-wood fibres. For instance, autocatalyzed organosolv pulping is reported to be particularly suitable for non-wood fibers, since this technique is suitable for small scale operation, given the simplified recovery process, immunity to silica accumulation problems, and the generation of high purity co-products, such as lignin. One example of autocatalyzed solvent pulping is the Alcell® process, in which aqueous ethanol is used as the pulping medium and lignin is recovered by precipitation by adjustment of the solvent concentration, the pH, and temperature of the spent cooking liquor. This process has been proposed for wheat straw, reed, kenaf, and sugar cane bagasse among others.[37, 58-60] Other organosolv processes that have been proposed for non-wood fibres and offer the potential of lignin recovery include pulping with organic acids.[61-63] In spite of the promise offered by organosolv pulping processes, their commercial implementation has not yet occurred.

Biomass fractionation technologies have also been tried on non-wood fibers and can potentially generate lignins of interesting properties. Autohydrolysis of biomass (i.e., treatment with water or steam at high temperature without added catalysts) can be used to solubilize hemicelluloses (which can be further converted to furfural) and generate a residue from which lignin can be extracted with solvents or dilute alkali.[64] A South African company developed a lignin called Sucrolin from residues from industrial furfural production from sugar cane bagasse in the 1980's.[65] A similar process, based on steam explosion can also be used to obtain an alkali extractable lignin from biomass. This is under investigation on wheat straw at the pilot level in Italy.[66]

Regarding the commercial availability of non-wood lignins, there are several suggestions in the literature on the industrial implementation of the precipitation of soda lignin in Asia for environmental management reasons.[67-69] It is not known if lignin recovered in such operations is currently being marketed. Concerning Europe, in the Fall of 2000 a lignin recovery plant using the LPS® process started operating at a small flax pulping mill in France. This unit has started generating lignins in solution and powder form for a range of industrial markets.[70]

4. UTILIZATION

Perhaps because of the difficulties encountered in efficiently recovering high quality lignins from non-wood fibers,[42] some of the attempts for lignin utilization have included the direct use of the liquors without much further processing. Chinese researchers have identified the potential use of black liquor from soda cooking of wheat straw, rice straw, and Chinese alpine rush in tertiary oil recovery. Black liquor flooding was found to be similar to alkali flooding, but followed a much more complicated mechanism due to its more complex composition.[31, 71, 72] Concentrated and spray-dried black liquors have been used to manufacture gels to reduce the permeability of underground strata and thus aid in the recovery of oil by steam flooding. The gel manufacture process involves a reaction with phenol, formaldehyde and a crosslinking agent. Partially hydrolyzed polyacrylamide and activated clay can be added to improve viscoelastic properties. The gels obtained had good strength properties and were stable to heat (250 ° C for at least 168 h), water, salt, caustic and acid. Lignin-based gels were tested in 10 oil wells in China in 1991-1992 and found to be capable of increasing oil output with a useful life of six months.[73, 74]

One of the approaches that have been attempted for lignin utilization involves the direct use of spent liquors in agriculture. Thus, liquors from ammonia sulfite pulping have been tried for irrigation and fertilization, but efforts in this direction have proven to be impractical due to seasonal demand considerations.[42]

Although the use of whole liquors may be feasible and justified under specific circumstances, the use of isolated lignins is what offers the potential for most widespread and versatile utilization. One current effort on the use of lignins recovered from existing pulp mills is the Valoir project, which is a consortium of nine industrial companies and research institutions, partially funded by the European Community. This consortium is pursuing the use of non-wood lignins as emulsifiers,[75] as components of thermoplastics,[76] and in the concrete and wood panels industries.

Non-wood lignins have been evaluated as additives for the manufacture of biodegradable, starch-based plastic materials for packaging and other applications. For instance, the use of Alcell® wheat straw lignin in starch composite films results in an improvement in water resistance properties of the films and in certain plasticizing effects.[43] An extension of this work to alkaline lignins has shown that wheat straw and hemp lignins can reduce the glass transition temperature of starch/glycerol films by 60-70%.[77]

Lignin has several reactive groups that can be used for modification reactions to obtain useful derivatives. Since lignin from wood is almost exclusively marketed as sulfonated derivatives (either lignosulfonates or methylsulfonated kraft) the conversion of isolated non-wood lignins to sulfonated derivatives has been the focus of research and development efforts.

Straw soda black liquor and lignin isolated therefrom were sulfonated to produce water reducers for concrete (i.e., dispersants that allow the use of less water when making concrete, which results in a stronger product). The product obtained from the sulfonated liquor was similar in performance to wood-derived calcium lignosulfonate, while the sulfonated lignin product performed similar to naphthalene-sulfonate, which is considered a high-quality water-reducer.[78]

The reactivity of soda straw lignin and pine kraft lignin in methylsulfonation and sulfonation were compared by He *et al.*[32] Significant differences between the two types of lignins were noticed in methylsulfonation, with pine kraft lignin being much more reactive than straw lignin. This is attributed to the higher number of non-substituted ortho positions to the phenolic OH (i.e., C-3 and C-5) in the former. In the case of high temperature sulfonation, straw and pine lignins behaved similarly. Regardless of its lower reactivity, it was possible to generate from straw soda lignin products with excellent water solubility characteristics and superior dispersing properties. For instance, at the same sulfonation level, straw lignin had lower surface tension than kraft lignin, probably because of the higher carboxylic acid content of the non-wood lignin. A methylsulfonation pilot plant for bamboo with a capacity of 2 tons per day was built in China in 1992.[42]

Reed lignosulfonates have been grafted with acrylic acid using a redox co-polymerization initiated with peroxide in a one-step as well as in a continuous process. The products obtained have potential use as water treatment chemicals, with the higher molecular weight products exhibiting better dispersing properties.[79]

Modified lignosulfonates obtained by pulping straw with sodium sulfite in the presence of formaldehyde are characterized by their high degree of sulfonation (4.15 mmol of sulfonic acid groups per gram of lignin) and low molecular weight. Their application in drilling muds and as dispersants (for instance, for titanium dioxide) has been studied by Chinese researchers.[33, 80]

Given its aromatic structure and renewable nature, lignin has been extensively evaluated as a partial replacement for phenol in the manufacture of phenolic resins. Studies on the utilization of bagasse soda and autohydrolysis lignins in phenolic resins have shown that a number of strategies can be employed to activate the lignin molecule and enhance its reactivity. These include methylolation, phenolation, methylolation followed by reaction with phenol or resorcinol, and reacting with pitch residues from coal gasification. Depending on the modification method employed, the resulting phenolic polymers can be used as cold set adhesives, thermosets for wood panels, and saturants for paper laminates, among others.[81-87] Bagasse lignins obtained by pulping in acetic acid were successfully used in phenolic molded products up to a 40% phenol replacement level with improvement in mechanical properties. The lignins were shown to chemically react with the phenol polymer chain, acting as chain extender rather than as filler.[36]

A few authors have reported the preparation and characterization of miscellaneous non-wood lignin derivatives. Thus, bagasse kraft lignin has been graft co-polymerized with maleic anhydride.[10] Brazilian researchers have prepared polyamides by polymerization of organosolv sugarcane bagasse lignin that had been aminated with formaldehyde and ammonia (Mannich reaction).[46] The carboxymethylation of sisal soda lignin was reported by researchers at the Swiss Federal Institute of Technology in Lausanne, who demonstrated good reproducibility and correlation between reaction conditions and product characteristics.[88]

Glasser and Jain acylated sugarcane bagasse autohydrolysis lignin and demonstrated that the derivative had a lower thermal softening temperature relative to the parent lignin. Acylation, therefore, can be used to prepare derivatives capable of thermal flow with polyethylene and with potential use in polymer blends.[89]

In a different approach to the use of lignin in plastics, de Jong et al. have proposed the use of straw soda lignins to dye polyolefins, such as polypropylene, with reactive dyes. The mechanical properties of the polymer were not negatively impacted, with the mechanical stability to UV radiation actually showing an improvement.[90]

Egyptian researchers prepared several rice straw lignin derivatives, including carboxylated (using acetic and propionic acid), cyanoethyl, and amino lignins and evaluated them in a variety of applications. They found that the propionated lignin was very good for paper sizing, and the amino-lignin derivative performed well as a flocculating aid and as an additive to enhance water retention in drilling muds.[91]

Straw lignin has also been tried as a polyol in polyurethane foams manufacture. Straw lignin from alkali oxygen pulping has been reported to have higher aliphatic hydroxyl content than lignin from conventional soda cooks, and therefore has greater reactivity with isocyanates.[92]

Straw soda-AQ lignin was modified to form chelating resins for binding metals.[93] In general the best results were obtained by a modification sequence involving ultrafiltration, acid-catalyzed phenolation, Mannich reaction with various amino acids, and carboxymethylation. The products obtained were found to bind copper preferentially. Another proposed approach to the production of chelants involves the oxidation of sugar cane bagasse lignin obtained by organosolv pulping in acetic acid.[94]

The use of isolated lignins and their derivatives (rather than just crude black liquors) in agriculture has been proposed as a potential large lignin market. For instance, Brazilian and British researchers have shown the advantages of using lignin as a controlled release matrix for herbicides, such as 2,4-D and Ametryn, in several studies using sugar cane bagasse lignin.[45, 95, 96]

Cuban workers have proposed sugar cane bagasse lignin as a soil improving agent.[97] Oxidative ammonolysis (i.e., reaction of lignin with oxygen and ammonia) has been proposed as a procedure to generate nitrogen-rich, partially oxidized lignins. The products are claimed to be effective soil conditioners and slow nitrogen-releasing fertilizers.[42]

Popa et al. found that alkaline flax lignin after nitration increases germination of bean seeds and biomass yield by about 30%.[98] Bagasse soda lignin is used in Cuba as an antidiarrheic for cattle, and apparently there are other medicinal and veterinary products under development, such as an anti-mycotic cream and medicaments for pigs.[97] The area of biological activity of lignins and lignin derivatives is certainly a promising field, and one in which much work remains to be done.

5. CONCLUSIONS

Recovery of non-wood lignins from existing pulp mills could alleviate environmental problems, and could contribute to the viability of many small operations. Non-wood lignins are polymers with great potential as chemicals and as materials. Their versatile chemical functionality and wide range of physical properties offer multiple avenues for the development of specialty products. To realize the full potential of non-wood lignins, it is necessary to implement processes that deliver them with high and consistent quality and to increase the understanding of the relationship between their chemical structure and performance.

6. REFERENCES

1. C. Lapierre and B. Monties, Structural informations gained from the thioacidolysis of grass lignins and their relation with alkali solubility, in: *Tappi Proceedings, International Symposium on Wood and Pulping Chemistry* (Tappi Press, Atlanta, 1989), pp. 615-621.
2. C. Lapierre, D. Jouin, and B. Monties, Lignin characterization of wheat straw samples as determined by chemical degradation procedures, in: *Physicochemical Characterisation of Plant Residues for Industrial and Feed Use*, edited by A. Chesson and E. R. Orskov (Elsevier Applied Science, London and New York, 1989), pp. 118-130.
3. H. Zhai and Y. Z. Lai, The characteristics of wheat straw lignin in situ, in: *Proceedings 8th International Symposium on Wood and Pulping Chemistry, Volume II Poster Presentations* (Helsinki, 1995), pp. 119–124.
4. N. Nishimura, A. Izumi, and K. Kuroda, Structural characterization of kenaf lignin: differences among kenaf varieties, *Industrial Crops and Products*, in press (2001).
5. J. Ralph, R. D. Hatfield, F. Lu, J. H. Grabber, H. G. Jung, and J. S. Han, Kenaf's amazing lignin, in: *Proceedings 8th International Symposium on Wood and Pulping Chemistry, Volume II Poster Presentations* (Helsinki, 1995), pp. 125–128.
6. E. Billa and B. Monties, Structural variability of lignin and associated phenolic acids in wheat straw, *Cellul. Chem. Technol.* **29**, 305-314 (1995).
7. Z. Z. Lee and X.Q. Pan, Characterization of wheat straw lignin, in: *International Symposium on Wood and Pulping Chemistry Proceedings, Technical Papers* (Vancouver, 1985), pp. 153-154.
8. W. He, J. Xue, Q. Lee, D. Tai, and Z. Lee, A study on the structure of wheat straw lignin by nuclear exchange method, in: *Proceedings 3rd International Non-Wood Fiber Pulping and Papermaking Conference, Volume 1* (International Academic Publishers, Beijing, 1996) pp. 65-71.
9. X.-J. Pan and Y. Sano, Comparison of acetic acid lignin with milled wood and alkali lignins from wheat straw, *Holzforschung* **54**, 61-65 (2000).
10. L. Huawu and C. Jiaxiang, Study on graft co-polymerization of bagasse kraft lignin with maleic anhydride, *Cellul. Chem. Technol.* **25**, 79-84 (1991).
11. A. M. L. Seca, J. A. S. Cavaleiro, F. M. J. Dominguez, A. J. D. Silvestre, D. V. Evtuguin, and C. P. Neto, Structural characteristics of the bark and core kenaf lignin (variety Salvador), in: *Proceedings 9th International Symposium on Wood and Pulping Chemistry, Poster Presentations* (Montreal, 1997), pp. 132-1 to 132-4.
12. S. J. Song, X. Z. Nie, and M. Y. Zhang, Studies on the lignin chemical characteristics of hemp, in: *Proceedings 7th International Symposium on Wood and Pulping Chemistry, Volume III* (Beijing, 1993), pp. 329-333.
13. A. Islam and K. V. Sarkanen, The isolation and characterization of the lignins of jute (*Corchorus capsularis*), *Holzforschung* **47**, 123-132 (1993).
14. M. Shimada, T. Fukuzuka, and T. Higuchi, Ester linkages of p-coumaric acid in bamboo and grass lignins, *Tappi* **54**(1), 72-78 (1971).
15. H. H. Nimz, D. Robert, O. Faix, and M. Nemr, Carbon-13 NMR spectra of lignins. 8. Structural differences between lignins of hardwoods, softwoods, grasses and compression wood, *Holzforschung* **35**, 16-26 (1981).
16. N. Fernandez, R. Mörck, S.C. Johnsrud, and K. P. Kringstad, Carbon-13 NMR study on lignin from bagasse, *Holzforschung* **44**, 35-38 (1990).

17. S. Galkin, E. Ämmälahti, I. Kilpeläinen, G. Brunow, and A. Hatakka, Characterisation of milled wood lignin from reed canary grass *(Phalaris arundinacea)*, *Holzforschung* **51**, 130-134 (1997).
18. J. Ralph, Recent advances in characterizing non-traditional lignins, in: *Proceedings 9th International Symposium on Wood and Pulping Chemistry, Oral Presentations* (Montreal, 1997), pp. PL2-1 to PL2-7.
19. A. Scalbert, B. Monties, J. Y. Lallemand, E. Guittet, and C. Rolando, Ether linkage between phenolic acids and lignin fractions from wheat straw, *Phytochemistry* **24**, 1359-1362 (1985).
20. A. Scalbert, B. Monties, C. Rolando, and A. Sierra-Escudero, Formation of ether linkage between phenolic acids and graminaeae lignin: a possible mechanism involving quinone methides, *Holzforschung* **40**, 191-195 (1986).
21. A. Scalbert, B. Monties, E. Guittet, and J. Y. Lallemand, Comparison of wheat straw lignin preparations. I. Chemical and spectroscopic characterization, *Holzforschung* **40**, 119-127 (1986).
22. T. A. Gorshkova, V. V. Salnikov, N. M. Pogodina, S. B. Chemikosova, E. V. Yablokova, A. V. Ulanov, M. V. Ageeva, J. E. G. Van Dam, and V. V. Lozovaya, Composition and distribution of cell wall phenolic compounds in flax (*Linum usitatissimum L.*) stem tissues, *Annals of Botany* **85**, 477-486 (2000).
23. J. M. Lawther, R. C. Sun, and W. B. Banks, Extraction and comparative characterization of ball-milled lignin (LM), enzyme lignin (LE) and alkali lignin (LA) from wheat straw, *Cellul. Chem. Technol.* **30**, 395-410 (1996).
24. J. Jacquet, B. Pollet, C. Lapierre, F. Mhamdi, and C. Rolando, New ether-linked ferulic acid-coniferyl alcohol dimers identified in grass straws, *J. Agric. Food Chem.* **43**, 2746-2751 (1995).
25. T. B. T. Lam, K. Iiyama, and B. A. Stone, Quantitative determination of lignin polysaccharide associations in wheat internode cell walls: ferulic acid bridges, in: *Appita Proceedings, 6th International Symposium on Wood and Pulping Chemistry, Volume 1* (Melbourne, 1991), pp. 29-33.
26. H. S. Yu, S. Y. Fu, X. H. Wen, B. Shen, X. Y. Huang, and Y. L. Shi, Characterization of lignin-carbohydrate complexes in rice straw, in: *Proceedings 8th International Symposium on Wood and Pulping Chemistry, Volume I Oral Presentations* (Helsinki, 1995), pp. 139 – 146.
27. X. H. Wen, H. S. Yu, S. Y. Fu, X. Y. Huang, and Z. Y. Cai, A two dimensional ^{1}H-^{13}C NMR study of rice straw lignin, in: *Proceedings 7th International Symposium on Wood and Pulping Chemistry, Volume III* (Beijing, 1993), pp. 495 – 502.
28. Y. I. Rumyantseva, R. G. Zhbankov, R. Teeiaere, and R Krus, Comparative analysis of bagasse alkaline and Björkman lignins structure by spectroscopic methods, *Cellul. Chem. Technol.* **28**, 425-434 (1994).
29. S. Y. Fu, X. H. Wen, H. S. Yu, X. Y. Huang, and Z. Y. Cai, Application of quantitative ^{13}C-NMR technique to rice straw lignin characterization, in: *Proceedings 7th International Symposium on Wood and Pulping Chemistry, Volume II* (Beijing, 1993), pp. 788-794.
30. C. Crestini, G. G. Sermanni, and D. S. Argyropoulos, Biodegradation of wheat lignin by Lentinus edodes, in: *Proceedings 9th International Symposium on Wood and Pulping Chemistry, Poster Presentations* (Montreal, 1997), pp. 16-1 to 16-4.
31. J. Wei, F. Chen, and M. Baoqi, Heavy crude oil sweepment with alkaline black liquor, in: *Proceedings 7th International Symposium on Wood and Pulping Chemistry, Volume II* (Beijing, 1993), pp. 863-868.
32. W. He, D. Tai, and S. Lin, A comparative study of the reactivities of straw and wood alkali lignin in sulfonation reactions, in: *Appita Proceedings, 6th International Symposium on Wood and Pulping Chemistry, Volume* 1 (Melbourne, 1991), pp. 509-515.
33. W. He, Y. Jin, Z. Z. Lee, and D. Tai, Chemical structure and properties of SF lignosulfonate, in: *Proceedings 3rd International Non-Wood Fiber Pulping and Papermaking Conference, Volume 2* (International Academic Publishers, Beijing, 1996), pp. 693-702.
34. J. Lora, Unpublished data.
35. O. Faix, D.S. Argyropoulos, D. Robert, and V. Neirinck, Determination of hydroxyl groups in lignins – Evaluation of ^{1}H-, ^{13}C-, ^{31}P-NMR, FTIR and wet chemical methods, *Holzforschung* **48**, 387-394 (1994).
36. R. S. G. Piccolo, F. Santos, and E. Frollini, Sugar cane bagasse lignin in resol-type resin: alternative application for lignin-phenol-formaldehyde resins, *J. M. S. Pure Appl. Chem.* **A34**(1), 153-164 (1997).
37. J. H. Lora, S. R. Winner, and E. K. Pye, Organosolv pulping of kenaf by the Alcell® process, in: *Kenaf Properties, Processing and Products*, edited by T. Sellers and N. A. Reichert (Mississippi State University Publications Department, Mississippi State, 1999), pp. 285-291.
38. G. H. Van der Klashorst and H. F. Strauss, Properties and potential utilization of industrial soda bagasse lignin, *Holzforschung* **40**(6), 375-382 (1986).
39. R. K. Kulkarni, R. K. Jain, A. Gupta, A. K. Dixit, and R. Pant, Studies on lignin-carbohydrate complexes in bagasse and their influence on black liquor properties, in: *Proceedings 7th International Symposium on Wood and Pulping Chemistry, Volume I* (Beijing, 1993), pp. 523-531.
40. F. Chen, Chemical characteristics of soda lignin from Chinese alpine rush, in: *Proceedings 7th International Symposium on Wood and Pulping Chemistry, Volume III* (Beijing, 1993), pp. 555 - 559.

41. R. J. A. Gosselink, A. Abächerli, H. Semke, R. Malherbe, A. Nadif, and J. E. G. van Dam, Characterization of sulfur-free lignins from alkaline pulping of annual fibre crops, in: *Proceedings, Fifth International Forum* (The International Lignin Institute, Bordeaux, France, 2000).
42. J. S. Gratzl and C. L. Chen, Technical lignins from non-wood sources, in: *Proceedings 3rd International Non-Wood Fiber Pulping and Papermaking Conference, Supplementary Volume* (International Academic Publishers, Beijing 1996), pp. 15-21.
43. S. Baumberger, C. Lapierre, J. Lora, and B. Monties, Novel utilization of wheat straw organosolv lignin in starch composites, in: *Proceedings 10th International Symposium Wood and Pulping Chemistry, Volume I Oral Presentations* (Yokohama, 1999), pp. 680-683.
44. W. R. Trevelin and A. A. S. Curvelo, Poliamidas a partir de bagaço de cana de açucar, in: *Anais do 2° Congresso Brasileiro de Polímeros, Volume 1* (Sao Paulo, 1993), pp. 126-129.
45. N. Duran, J. A. de Souza, and A. Ferraz, Sugar cane bagasse lignin: characterization and use as a controlled release device for 2,4-D and its biodegradation, in: *Proceedings 3rd International Non-Wood Fiber Pulping and Papermaking Conference, Supplementary Volume* (International Academic Publishers, Beijing, 1996), pp. 10-14.
46. A. Abächerli, New lignins from agricultural plants, *Proceedings Fifth International Forum* (The International Lignin Institute, Bordeaux, France, 2000).
47. J.-Y. Lu, Y.-M. Hu, and P.-K. Du, Lignin separator and its application: A technique of separating lignin from black liquor, *China Pulp Pap.* **14**(5), 19-23 (1995).
48. J. Lan, Ligno-sulfonates from straw-based alkaline pulping black liquors, in: *Proceedings Second International Non-Wood Fiber Pulping and Papermaking Conference, Volume II* (CTAPI, Shanghai, 1992), pp. 986-992.
49. S. Liancan Sun et al., *Chinese Patent* 1050064 (1989).
50. L. Jinyi et al., *Chinese Patent* 1050241 (1990).
51. A. Abächerli and F. Doppenberg, Method for preparing alkaline solutions containing aromatic polymers, *PCT WO 98/42912* (1998).
52. R. K. Jain, A. Gupta, R. M. Mathur, and A. G. Kulkarni, Lignin removal process – an emerging eco-friendly technology for non-wood based pulp and paper mills, in: *Proceedings 4th International Non-Wood Fiber Pulping and Papermaking Conference* (CTAPI, Jinan, 2000), pp. 208-216.
53. S, Wu, Y. Hu, and W. Liang, A preliminary study on the bio-treatment of spent liquor from the preparation of wheat straw alkali lignin, in: *Proceedings 3rd International Non-Wood Fiber Pulping and Papermaking Conference, Volume 2* (International Academic Publishers, Beijing, 1996), pp. 631-636.
54. L. Xie, Y. Chen, and S. Wu, Studies on the treatment of black liquor from wheat straw soda-AQ pulping process by photosynthetic bacterium, in: *Proceedings 4th International Non-Wood Fiber Pulping and Papermaking Conference* (CTAPI, Jinan, 2000), pp. 331-337.
55. B.-Y. Liu, R.-N. Yang, and Y.-Z. Zhang, Method for reducing pollution load in wheat straw pulping, in: *Proceedings 3rd International Non-Wood Fiber Pulping and Papermaking Conference, Volume 1* (International Academic Publishers, Beijing, 1996), pp. 270-275.
56. D. Tan and L. Deng, A study on the mechanism of the wet cracking of wheat straw alkaline sulfite pulping spent liquor, in: *Proceedings 3rd International Non-Wood Fiber Pulping and Papermaking Conference, Volume 2* (International Academic Publishers, Beijing, 1996), pp. 662-665.
57. W. He, D. Tai and Z. Lee, Effect of formaldehyde during alkaline sulfite pulping of wheat straw, in: *Proceedings 7th International Symposium on Wood and Pulping Chemistry, Volume 1* (Beijing, 1993), pp. 523-531.
58. S. R. Winner, L. Minogue, and J. H. Lora, Alcell pulping of annual fibers, in: *Proceedings 9th International Symposium on Wood and Pulping Chemistry, Poster Presentations* (Montreal, 1997), pp. 120-1 to 120-4.
59. J. H. Lora and E. K. Pye, The Alcell® process: an environmentally sound approach to annual fibres pulping, in: *Proceedings Non-wood Fibres for Industry Conference, Volume II* (Pira International, Surrey, 1994), Paper 15.
60. S. R. Winner, G.C. Goyal, E. K. Pye, and J. H. Lora, Pulping of agricultural residues by the ALCELL process, in: *Proceedings Tappi Pulping Conference* (Orlando, 1991), pp. 435-439.
61. R. A. M. C. De Groote, A. A. S. Curvelo, P. C. Frangiosa, M. F. T. Pinto, and M. D. Zambon, Molecular weight distribution of sugar cane bagasse lignin in the Acetosolv process, in: *Appita Proceedings, 6th International Symposium on Wood and Pulping Chemistry, Volume 2* (Melbourne, 1991), pp. 385-391.
62. A. Seisto and K. Poppius-Levlin, Milox pulping of agricultural plants, in: *Proceedings 8th International Symposium on Wood and Pulping Chemistry, Volume I Oral Presentations* (Helsinki, 1995), pp. 407-414.
63. P. Rousu, P. and K. Teemu, Formic acid based pulping technology – Efficient, economical and environmentally friendly, in: *Proceedings 4th International Non-Wood Fiber Pulping and Papermaking Conference* (CTAPI, Jinan, 2000), pp. 33-42.

64. J. H. Lora and M. Wayman, Delignification of hardwoods by autohydrolysis and extraction, *Tappi* **61**(6) 47-50 (1978).
65. R. Reiman, Sucrolin – a new lignin from sugar cane bagasse, in: *Proceedings EWLP'90, Utilization and Analysis of Lignins* (Hamburg, 1990), pp. 29-37.
66. C. Bonini and M. D'Auria, Low cost steam exploded lignin from straw: degradation and use, in: *Proceedings Fifth International Forum* (The International Lignin Institute, Bordeaux, France, 2000).
67. F. Chen, J. Miao, and Z. Lu, Liquefaction of wheat straw soda lignin in organic solvent, in: *Proceedings 4th International Non-Wood Fiber Pulping and Papermaking Conference* (CTAPI, Jinan, 2000), pp. 253-259.
68. L. Ma and X. Li, Treatment of mill effluents from reed BKP process, in: *Proceedings 4th International Non-Wood Fiber Pulping and Papermaking Conference* (CTAPI, Jinan, 2000), pp. 616-621.
69. C. Wu, G. Yang, Z. Li, and J. Chen, Recycling of effluents from biological and flocculation treatments of wheat straw pulp wastewater, in: *Proceedings 4th International Non-Wood Fiber Pulping and Papermaking Conference* (CTAPI, Jinan, 2000), pp. 613-615.
70. J. H. Lora, A. Abächerli, and F. Doppenberg, Debottlenecking the recovery system of soda pulp mills by lignin recovery and wet oxidation – Application to non-wood fibers black liquors, in: *Tappi 2000 Pulping Process & Product Quality Conference – Abstract Book* (Boston, 2000), p. 27.
71. F. Chen and B. Ma, The mechanisms of sweeping oil from underground strata with non-wood black liquor, in: *Proceedings 3rd International Non-Wood Fiber Pulping and Papermaking Conference, Volume 2* (International Academic Publishers, Beijing, 1996), pp. 683-689.
72. W. Ma and F. Chen, Technical methods for increasing oil recovery with non-wood black liquor, in: *Proceedings 3rd International Non-Wood Fiber Pulping and Papermaking Conference, Supplementary Volume* (International Academic Publishers, Beijing, 1996), pp. 38-43.
73. X. Wang and B. Ma, Preparation of high temperature gel in alkali black liquor, in: *Appita Proceedings, 6th International Symposium on Wood and Pulping Chemistry, Volume 1* (Melbourne, 1991), pp. 473-475.
74. B. Ma, X. Wang, and X. Zhang, Studies on the characteristics of gel from alkali lignin", *Proceedings 7th International Symposium on Wood and Pulping Chemistry, Volume II* (Beijing, 1993), pp. 534-538.
75. P. Käuper, A. Nadif, A., and D. Hunkeler, Sulfur-free lignin from alkaline pulping as emulsifiers, in: *Proceedings Fifth International Forum* (The International Lignin Institute, Bordeaux, France, 2000).
76. H. Nitz and R. Mülhaupt, Lignin based polymer compounds and liquid wood, in: *Proceedings Fifth International Forum* (The International Lignin Institute, Bordeaux, France, 2000).
77. S. Baumberger, Lignin utilization in starch thermoplastics: molecular origin of compatibility, in: *Proceedings Fifth International Forum* (The International Lignin Institute, Bordeaux, France, 2000).
78. D. Y. Chang and S. Y. N. Chan, Straw pulp waste liquor as a water-reducing admixture, *Magazine of Concrete Research* **47**(171), 113-118 (1995).
79. J. Xue and D. Tai, Studies on grafting copolymerization of lignosulfonate, in: *Proceedings 7th International Symposium on Wood and Pulping Chemistry, Volume III* (Beijing, 1993), pp. 539-544.
80. J. Zhang and Z. Lee, The effectiveness of SEP lignin in drilling fluids, in: *Proceedings 3rd International Non-Wood Fiber Pulping and Papermaking Conference, Volume 2* (International Academic Publishers, Beijing, 1996), pp. 690-692.
81. G. H. van der Klashorst, Lignin formaldehyde wood adhesives, in: *Wood Adhesives – Chemistry and Technology, Vol. 2*, edited by A. Pizzi, Chapter 6 (Marcel Dekker, New York, 1989).
82. R. E. Ysbrandy, R. D. Sanderson, and G. F. R. Gerischer, DSC thermal analysis of phenol and phenol-lignin extended resols and their physical behavior in paper laminates, *Das Papier* **45**, Jahrgang, Heft 2, 62-67 (1991).
83. R. E. Ysbrandy, R. D. Sanderson, and G. F. R. Gerischer, Adhesives from autohydrolysis bagasse lignin, a renewable resource, *Holzforschung* **46**(3), 249-252 (1992).
84. R. E. Ysbrandy, R. D. Sanderson, R. L. Wright, and G. F. R. Gerischer, Thermal reactivity and molecular mass analyses of phenosolvan pitch resoles and phenolated pitch/lignin mixtures, *Holzforschung* **48**(3), 244-253 (1994).
85. R. E. Ysbrandy, G. F. R. Gerischer, and R. D. Sanderson, Reactive filler/binder from inexpensive by-products for use in high-pressure phenolic paper laminates, in: *Proceedings Pacific Rim Bio-based Composites Symposium, FRI Bulletin No. 177* (Rotorua, New Zealand, 1992), pp. 364-375.
86. J.-S. M. Kazayawoko, B. Riedl, J. Poliquin, A. O. Barry, and L. M. Matuana, A lignin-phenol-formaldehyde binder for particleboard. Part 1. Thermal characteristics, *Holzforschung* **46**(3), 257-262 (1992).
87. J.-S. M. Kazayawoko, B. Riedl, and J. Poliquin, A lignin-phenol-formaldehyde binder for particleboard. Part 2. Physicomechanical properties, *Holzforschung* **46**(4), 349-354 (1992).
88. P. Käuper, A. Nadif, and D. Hunkeler, Carboxymethylation of a soda lignin from sisal crop, Unpublished data (2001).

89. W. G. Glasser and R. K. Jain, Lignin derivatives. I. Alkanoates, *Holzforschung* **47**, 225-233 (1993).
90. E. De Jong, E. Scott, R. Gosselink, and J. E. G. van Dam, Simultaneous colouring and U. V. stabilisation of materials using dyed lignin, in: *Proceedings Fifth International Forum* (The International Lignin Institute, Bordeaux, France, 2000).
91. M. A. El-Taraboulsi and M. M. Nassar, Lignin derivatives from desilicated rice straw soda black liquor, *Indian Pulp and Paper* **33**(5), 3-12 (1979).
92. K.-L. Chen, L. Zhang, S.-H. Yang, and S.-C. Li, Preparation of polyurethane based on lignin isolated from alkali-oxygen pulping wheat straw waste liquor, in: *Proceedings 4th International Non-Wood Fiber Pulping and Papermaking Conference* (CTAPI, Jinan, 2000), pp. 344-352.
93. X. Pan, D. Xie, L. Xie, and Y. Long, Chelate resins based on straw soda-AQ lignin, in: *Proceedings 7th International Symposium on Wood and Pulping Chemistry, Volume II* (Beijing, 1993), pp. 858-862.
94. A. R. Goncalves, M. P. Urbano, A. R. Cotrim, and F. Texeira da Silva, Oxidation of lignin –containing liquor from acetosolv pulping of sugar-cane bagasse, in: *Proceedings 9th International Symposium on Wood and Pulping Chemistry, Oral Presentations* (Montreal, 1997), pp. I4-1 to I4-3.
95. T. F. Da Silva and R. M. Wilkins, Use of lignins as support to herbicides in controlled release formulations, in: *Proceedings Second Brazilian Symposium on Chemistry of Lignin and Other Wood Components, Volume 3* (Campinas, 1992), pp. 165-172.
96. A. R. Cotrim, J. A. Souza, A. Ferraz, F. T. Da Silva, and R. Wilkins, The influence of different lignins on the controlled-release of 2,4-D and Ametryn herbicides, *Proceedings Third Brazilian Symposium on Chemistry of Lignin and other Wood Components, Volume 4* (Belo Horizonte, 1993), pp. 278-281.
97. N. Fernandez, Pulp and paper development from sugar cane bagasse, in: *Proceedings 3rd International Non-Wood Fiber Pulping and Papermaking Conference, Volume 1* (International Academic Publishers, Beijing, 1996), pp. 231-240.
98. V. I. Popa, I. Tudose, and N. Anghel, Influence of polyphenols and lignin products on the biosynthesis processes, in: *Proceedings Fifth International Forum* (The International Lignin Institute, Bordeaux France, 2000).

INDEX

MIX
Papier aus verantwortungsvollen Quellen
Paper from responsible sources
FSC® C105338

If you have any concerns about our products,
you can contact us on
ProductSafety@springernature.com

In case Publisher is established outside the EU,
the EU authorized representative is:
Springer Nature Customer Service Center GmbH
Europaplatz 3, 69115 Heidelberg, Germany

Printed by Libri Plureos GmbH
in Hamburg, Germany